WATER QUALITY MODELING FOR WASTELOAD ALLOCATIONS AND TMDLs

WATER QUALITY MODELING FOR WASTELOAD ALLOCATIONS AND TMDLs

Wu-Seng Lung, PhD, PE

University of Virginia

John Wiley & Sons, Inc.

New York / Chichester / Weinheim / Brisbane / Singapore / Toronto

Library of Congress Cataloging-in-Publication Data

Lung, Wu-Seng.
    Water quality modeling for wasteload allocations / by Wu-Seng Lung.
        p. cm.
    Includes bibliogrphical references.
    ISBN-13: 978-0-471-15883-7

    1. Water quality—Mathematical models.   I. Title: Wasteload allocations.   II. Title.
TD370.L86   2001
628.1′682—dc21   2001024024

10 9 8 7 6 5 4 3 2 1

To my wife, Kathy

CONTENTS

Preface / xi

Acknowledgments / xv

1 INTRODUCTION / 1

1.1 Using a More Robust Model / 2

1.2 Developing the Stream Reaeration Coefficient / 4

1.3 Post Auditing the Water Quality Response—A Bigger Picture / 4

1.4 New Horizon in Water Quality Modeling / 6

1.5 It Is the Modeling Skills and Data, Not the Model, That Matters / 7

2 TOTAL MAXIMUM DAILY LOADS (TMDLs) / 8

2.1 Evolution of Water Quality Modeling in Water Pollution Control / 8

2.2 What Is a TMDL? / 9

2.3 Water Quality Endpoints for TMDLs / 11

2.4 Water Quality Modeling for TMDL / 13

2.5 A TMDL Modeling Study / 15

3 DERIVATION OF MASS TRANSPORT COEFFICIENTS / 21

3.1 One-Dimensional Advective Transport in Streams, Rivers, and Estuaries / 22

3.2 One-Dimensional Longitudinal Dispersion Coefficient / 30

3.3 Lateral Dispersion Coefficient in Rivers and Estuaries / 34

3.4 Vertical Diffusion Coefficient / 36

3.5   Two-Dimensional Vertically Integrated Mass Transport / 43

3.6   Two-Dimensional Longitudinal/Vertical Mass Transport / 44

3.7   Need a Hydrodynamic Model? / 61

3.8   Linking a Hydrodynamic Model with a Water Quality Model / 68

4   DERIVATION OF KINETIC COEFFICIENTS / 71

4.1   Biochemical Oxygen Demand and $CBOD_u$ to $CBOD_5$ Ratio / 72

4.2   Nitrification in Wastewater and Receiving Water / 84

4.3   Reaeration Coefficient / 89

4.4   Saturation Dissolved Oxygen Level / 98

4.5   Sediment Oxygen Demand / 99

4.6   Phytoplankton and Dissolved Oxygen / 103

4.7   Dissolved Oxygen Impact / 109

4.8   Phytoplankton/Nutrient Kinetics / 110

4.9   Modeling Sediment Diagenesis Processes / 119

4.10   Calibrating the Nutrient/Eutrophication Kinetics / 124

4.11   Post Audit of BOD and Eutrophication Models / 125

5   COMPUTATIONAL TOOLS AND ACCESSORIES / 130

5.1   Computer Hardware and Capability / 130

5.2   Modeling Frameworks and FORTRAN Compiler / 132

5.3   Geographical Information System / 133

5.4   Post Processing and Visualization / 136

6   USING THE STREAM AND QUAL2E MODELS / 143

6.1   Equations Behind the STREAM Model / 144

6.2   Implementing the Governing Equation in STREAM Code / 145

6.3   STREAM Model Input Structure / 146

6.4   Application of STREAM to Rock Creek / 149

6.5   Application of STREAM to the Roanoke River / 158

6.6   Application of the QUAL2E Model to the Blackstone River / 172

7   USING THE WASP AND EUTRO MODELS / 188

7.1   The WASP Model / 188

7.2   Key Attributes / 189

7.3   Mass Transport Modeling Applications / 192

7.4   Water Quality Modeling Application / 221

7.5   Numerical Tagging / 241

7.6   Modeling Sediment-Water Interactions / 252

7.7   Incorporating Nonpoint Source Loads / 257

7.8   Linking a Hydrodynamic Model with the WASP/EUTRO Model / 261

7.9   Summary Remarks / 268

8 USING THE HAR03 MODEL / 270

8.1   Model Capabilities / 270

8.2   Input Data Structure and Model Configuration / 271

8.3   Near-Field Mixing Zone Modeling / 275

8.4   Modeling Mixing Zone in the James River Using HAR3 / 281

8.5   Far-Field Modeling / 291

8.6   Summary / 313

References and Further Readings / 314

About the Author / 325

Index / 327

PREFACE

The current federal water pollution control law was developed in 1972 and has been expanded, updated, modified, and augmented at regular intervals. Water quality modeling, which establishes how pollutants will interact with and alter the receiving waters, has been an important pillar for implementing and enforcing the regulations since 1972 and before. In the 1980s the U.S. Environmental Protection Agency (EPA) developed a series of national guidance manuals for performing wasteload allocations. The series consists of multiple volumes designed to assist modelers in modeling streams, lakes, and estuaries for a variety of water quality problems. EPA's National Center for Exposure Assessment Modeling (CEAM) in Athens, Georgia supports a number of water quality models by conducting workshops and publishing user manuals on water quality models such as QUAL2E and WASP. EPA also assists modelers with manuals on screening methods for water quality modeling and kinetic rates, constants, and formulations.

Despite these efforts, there are not enough skilled modelers working in state agencies and consulting firms to meet the demand for performing wasteload allocations. It is in these agencies and the consulting firms they hire that the practice of modeling has its more important impact, since most permit processes now require wasteload allocation analyses to varying degrees. This unfortunate situation is further exacerbated by tight schedules, as over 40,000 total maximum daily loads (TMDLs) should be calculated over the next 8 to 15 years across the country. Furthermore, that figure does not include the water quality modeling work associated with routine NPDES permitting.

Fortunately, two excellent textbooks are available to the modeling profession: *Principles of Surface Water Quality Modeling and Control* by Bob Thomann and John Mueller, published in 1987, and *Surface Water-Quality Modeling* by Steven Chapra, published in 1997. These books have helped to shape the profession. However, in this as in all engineering fields, there is always the need to explore beyond the textbook materials and classroom notes; this need is met via engineering practice. Attending a modeling workshop for a week does not necessarily provide an engineer with suffi-cient experience to satisfactorily operate a model. Engineering practice is essential in water quality modeling. The rationale behind this book is to present something extra, tricks of the trade if you will, that can help the modeling practitioners perform water quality modeling correctly and more efficiently.

One unique aspect of water quality modeling is the importance of its data. To il-lustrate the significance of data, one could compare water quality modeling with hy-drodynamic modeling. A hydrodynamic model has at least five fundamental equations: three momentum equations, one continuity equation, and the equation of state. In a discipline founded centuries ago, these equations constitute the fundamen-tal elements of a hydrodynamic model. On the other hand, a water quality model is based on one equation: mass balance. Beyond that, there are varying degrees of em-piricism in the water quality model, that is, various ways of using data to formulate the water column kinetics. Due to this empirical nature, adequate, correct data are the key to model development, configuration, calibration, and validation.

A modeler can learn the cause-and-effect relationship between the pollutant loads and the receiving water response simply by examining the data, looking for spatial and temporal trends, and formulating a conceptual model. It is this data reading process that leads to physical insights and eventually evolves into a water quality model. Numerical computation is simply a tool to assist the modeler in quantifying the physical, chemical, and biological processes. My training at Hydroscience, Inc., a respected consulting firm, under the late Dr. Donald J. O'Connor, taught me how important data are to water quality modeling. I should also point out that while data are important, they are not the absolute measure in obtaining a full understanding of the receiving water system. We should not believe the data completely by fitting the model results through the data points. And we must keep in mind that it is equally im-portant to comprehend any differences between the model results and data. With that understanding, this book strongly emphasizes using data to independently derive model coefficient values (hydrodynamics and water column kinetics).

Unlike other textbooks that serve the profession with modeling concepts and the-ories, this book focuses on how to use models correctly. A number of commonly used water quality models in wasteload allocations and TMDL development are thor-oughly discussed in this book. A significant wealth of data is displayed to demon-strate the use of models as well. The materials presented are derived from my research and practice over the past 25 years in this profession. I have also used them since 1983 in two graduate water quality modeling courses that I teach at the Uni-versity of Virginia. It is my hope that this book can become a companion volume to the other textbooks for the modeling profession, serving graduate students, practi-tioners, and regulatory staff.

The materials are organized in the following order. Following Chapter 1, which describes the need for water quality modeling in water quality management, Chapter 2 focuses on water pollution control regulations and their relationships to water qual-ity modeling and wasteload allocations for the determination of TMDLs. The material emphasizes how to derive the model coefficients to quantify physical, chemical, and biological processes of a variety of water quality problems.

Chapter 3 covers the derivation of mass transport coefficients of different water bodies, including streams, rivers, lakes, impoundments, estuaries, and coastal waters. Data requirements for the derivation are also covered. Techniques to develop mass transport coefficients from field data or from hydrodynamic models are presented for rivers, lakes, and estuaries.

Derivation of kinetic coefficients in the water column is presented in Chapter 4. Case studies are used to demonstrate how to independently develop the kinetic coefficients in the water column. In addition, a discussion on sediment-water interactions and their processes is presented.

Microcomputers are being used in water quality modeling; hence Chapter 5, on computer requirements, is included. One of the beneficial uses of microcomputers in water quality modeling is post processing of model results. Hardware and software advances have made graphics interfaces and visualization easy to use, thereby significantly reducing the modeler's effort in effectively communicating model results.

The rest of the book details the use of a number of commonly applied water quality modeling frameworks, using case studies to clarify the principles. In Chapter 6, the simple STREAM model for biochemical oxygen demand/dissolved oxygen (BOD/DO) analysis, which is recommended by the EPA (1995) guidance manual for small streams, is the first one presented. Although the modeling framework is simple, it has proved successful in many wasteload allocation studies. Select case studies for the STREAM model are included. Also presented in Chapter 6 is the QUAL2E model—a popular program for use in stream and river wasteload allocations under steady-state conditions.

Chapter 7 covers another modeling framework, WASP, one of the most widely used models in TMDL work to date. Its eutrophication module, EUTRO, has been applied to numerous water quality modeling studies by consulting firms as well as regulatory agencies across the country. Its versatility and flexibility make this modeling framework a very useful tool. Also included in the presentation is some advanced use of the WASP/EUTRO modeling framework to offer additional insights into the water quality problems. This chapter concludes with material on using other hydrodynamic models to drive the WASP/EUTRO model.

The last chapter is on the HAR03 model. While HAR03 is primarily used for the steady-state, multidimensional analysis of mass transport and water quality constituents characterized with first-order kinetics, it is also used for wasteload allocations in water quality–based toxics control analysis. HAR03 is particularly useful in modeling toxic chemicals for mixing zone analyses of small wastewater flows with negligible momentum. For many wastewater dischargers, mixing zone modeling is the key (vs. far-field modeling) to the NPDES permit of toxic discharges.

ACKNOWLEDGMENTS

Dr. Ray Canale (Michigan) introduced me to water quality modeling when I took his graduate course at the University of Michigan. He also encouraged me to work at Hydroscience following my PhD work. My mentors at Hydroscience were Drs. Bob Thomann and Donald O'Connor (Manhattan College). Nauth Panday of the Maryland Department of the Environment and Bob Ambrose of EPA at Athens, Georgia, helped me throughout the writing of this book with their thoughtful reviews of the manuscript and insightful discussions on different models. I am also grateful to Bob Sobeck for his reviews and comments on the first two chapters. Monica Wedo (now at the University of Texas) performed many rounds of editorial reviews of the manuscript. This book is dedicated in memory of the late Dr. Donald O'Connor, my teacher, friend, and colleague.

Finally, I would like to thank my lovely wife, Kathy, and our three children, Connie, Tina, and Vicki, for their continuing love and support. It is their constant encouragement throughout my professional career that has made this book a reality.

CHAPTER 1

INTRODUCTION

The message of this chapter is: water quality modeling applied correctly for water quality management. Decision makers rely on modeling results these days to formulate their management strategy. Further, model results for management use are being scrutinized closely, not only by technical reviewers, but also by regulatory agencies, stakeholders, and the general public. Perhaps more importantly, the Environmental Protection Agency's (EPA's) final total maximum daily load (TMDL) rule, promulgated on July 13, 2000 (40 CFR Part 9 et al.), clearly calls for a monitoring and modeling plan to be an integral part of the TMDL's implementation plan. Model results are the backbone of a TMDL.

Water quality modeling has been around for several decades. It is important to apply this technology correctly to water quality management. The cost and effort of a water quality modeling study is usually a small fraction of the implementation cost for a water quality management strategy. It is this aspect that has made water quality modeling an attractive approach for developing a management strategy. It is also because of this aspect that a large sum of money is always at stake, demanding accurate model results to support the costly implementation. For example, it is not difficult to understand how significant amounts of time and effort have been spent over the years on the Chesapeake Bay modeling project to restore the Bay. Continuing reviews and scrutinies have been conducted to ensure that correct results are used to advise the Bay decision-makers.

A number of successful water quality modeling case studies are presented below to demonstrate this point.

1.1 USING A MORE ROBUST MODEL

While the water quality modeling technology has been developed for and applied to wasteload allocations and TMDLs, there is a shortage of the skilled practitioners who can perform the task. As a result, the available modeling technology may not be used properly in practice. A good example of not utilizing available modeling technology for current regulatory needs can be found in some states' River Basin Water Quality Management Plans (WQMPs). For example, these Plans in the Commonwealth of Virginia, developed in mid 1970s, determined allowable point source carbonaceous biochemical oxygen demand (CBOD) loading rates using the Tennessee Valley Authority's (TVA's) flat water equation. The equation, an empirical formula originally developed from a statistical multiple-regression analysis of 15 assimilative capacity studies in the TVA region, provides an estimate of allowable 5-day biochemical oxygen demand (BOD_5) loading rates using a minimum amount of field data:

$$Y = 10138 \frac{(DO_{mix})^{1.094} Q^{0.864} S^{0.06}}{T^{1.423} (DO_{min})^{1.474}} \qquad (1\text{-}1)$$

where
Y = assimilative capacity of the stream (lb BOD_5/day)
DO_{mix} = dissolved oxygen concentration (mg/L) in the stream following complete mixing
Q = sum of stream flow and waste flow (cfs)
S = stream bed slope (ft/ft)
T = stream water temperature (°C)
DO_{min} = minimum allowable DO of the stream (mg/L)

There are a number of technical issues concerned with using Eq. 1-1 in wasteload allocations. First, waste characteristics are not considered in the equation. What is the ratio of ultimate CBOD ($CBOD_u$) to 5-day CBOD ($CBOD_5$) in the wastewater? This ratio is a good indicator of the wastewater characteristics and strongly affects the in-stream deoxygenation rate of CBOD (K_d). Without this factor, Eq. 1-1 does not address how rapidly the waste is being stabilized in the stream, which in turn plays an important role in determining the DO sag. Consider two waste discharges, one with a high K_d and the other with a low K_d rate. With everything else being equal, Eq. 1-1 would yield the same wasteloads for the discharger. Yet they should not be the same in reality. Thus the TVA equation is not robust enough for wasteload allocations. It does not have the fundamental attributes to characterize the BOD/DO relationship!

Another important factor in stream BOD/DO modeling is time of travel. Again, nothing in Eq. 1-1 reflects the stream velocity. The stream bed slope in Eq. 1-1 is only an indirect indicator of time of travel. Admittedly, limited field data was a primary factor of using the regression equation. However, the River Basin Management Plans based on this calculation soon became laws in Virginia. Today, the permit writers still adhere to this equation in their NPDES permit process because this equation is part of the laws! A recent modeling study (including field data collection) submitted to the state regulatory agency finally convinced the regulators to consider revising the dated river basin plan by incorporating the model results (see the case study in Chapter 6).

The TVA equation was also applied in 1975 to the Kerr Reservoir near Clarksville, Virginia, to quantify the assimilative capacity in the Roanoke River Basin WQMP, resulting in another misuse of available modeling technology in practice. The most obvious difference between conditions at Clarksville and those represented by the TVA equation is the variable slope. The slope of the original riverbed has little significance to low flow water quality conditions in the reservoir. The dissolved oxygen sag (DO sag) variable is another important parameter that is inappropriately represented by the TVA equation applied to Kerr. DO sag in the TVA equation is associated with the downstream direction, whereas the important direction for consideration of dissolved oxygen in Kerr is the vertical direction. Hence, there is a significant disparity between what the TVA equation represents and conditions at Clarksville.

The Kerr Reservoir at Clarksville transitions from riverine conditions to lentic conditions and as such exhibits water quality characteristics associated with limnological waters. Specifically, during late summer, the upper waters of Kerr become supersaturated with dissolved oxygen while the lower waters have dissolved oxygen levels that, at certain locations, deplete to anoxic conditions at bottom. During winter months, the entire water column becomes oxygen rich as deep as the bottom. These seasonal variations describe the natural oxygen cycles of a lake, or lentic, body of water, where algae, feeding on nutrients, mediate oxygen supplies along with other biological, chemical, and physical processes. Comparatively, the TVA equation has no appropriate mechanism to account for the significant processes in the Kerr Reservoir.

To correct this problem, the EPA's WASP/EUTRO5 model was used to configure the water column into a two-layer system, with additional longitudinal segmentation along the reservoir. Most importantly, field data were collected to support this effort. The resulting water quality model is able to reproduce the dissolved oxygen stratification in the water column and the calibrated model eventually served as a tool for water quality management.

Perhaps the most significant outcome of these two modeling studies is that the river basin plans have been modified by the Virginia Department of Environmental Quality (VDEQ) with more robust modeling results. Looking back at the original WQMPs, the Department's technical support document (SWCB, 1976) states, "Without detailed field data, this [TVA equation] predictive methodology was judged to be the best for use in this Plan in lieu of a detailed water quality model utilizing questionable assumptions." Further, the WQMP states:

> Before expending large sums of money for expanding and upgrading treatment plants or constructing new treatment plants, it would be wise to expend the time and costs necessary to collect the field data (above and below the existing or proposed discharge) and to review the existing water quality data. In most instances, the collection of new data

and a review of the existing data (including water quality modeling) will expend only a fraction of a percent of the total initial capital outlay for the facility.

With the available modeling technology and field data collection, there is no reason to rely on the outdated WQMPs.

1.2 DEVELOPING THE STREAM REAERATION COEFFICIENT

In the modeling study of the Shirtee Creek in Alabama, a modeler selected a wrong empirical equation to calculate the reaeration coefficient for wasteload allocation. The decision to use the Langbien and Durum (1967) equation for that small creek was based primarily on the fact that the equation provided a better fit of observed dissolved oxygen data than did the Tsivoglou (Tsivoglou and Neal, 1976) equation. This is curve fitting, not water quality modeling. The modeler ignored physical insights and forced the model calculations to match the data. The modeling outcome yielded an allowable CBOD loading rate for the point source discharge much lower than it should have been. This is a more subtle issue than using the TVA equation and is one of the many deficiencies encountered these days. This mistake was finally corrected when another party submitted a modeling study using the Tsivoglou equation for reaeration (see the case study in Chapter 4).

The above cases demonstrate the need for properly trained modelers, in practical applications, a need that will increase in light of the significant number of skilled modeling analysts being required in the TMDL work for the next two decades.

1.3 POST AUDITING THE WATER QUALITY RESPONSE—A BIGGER PICTURE

Point source nutrient controls were developed in mid 1980s for the James Estuary in Virginia. Prior to that time, most municipal wastewater treatment plants in the James River Basin had secondary treatment to reduce CBOD loads. A series of debates went on through the mid 1980s regarding a phosphate detergent ban for the basin. A water quality modeling study by Lung (1986) and Lung and Testerman (1989) concluded that a phosphate detergent ban would not provide any water quality benefit to the estuary, that is, it would not significantly reduce the algal biomass in the Upper James Estuary. Why? The system is light limited, not phosphorus limited. Therefore it would take additional effort such as phosphorus removal at the treatment plants to achieve an appreciable reduction in algal biomass in terms of chlorophyll *a*.

The General Assembly of the Commonwealth of Virginia passed a phosphate detergent ban, effective January 1, 1988. Subsequently, Virginia Water Control Board required that those plants with flows greater than 1 mgd implement phosphorus removal. By 1996, the algal biomass in the James Estuary started to show signs of a small reduction in algal biomass (Figure 1-1). The model used by Lung (1986) was run again with the nutrient load reductions following the phosphate detergent ban and

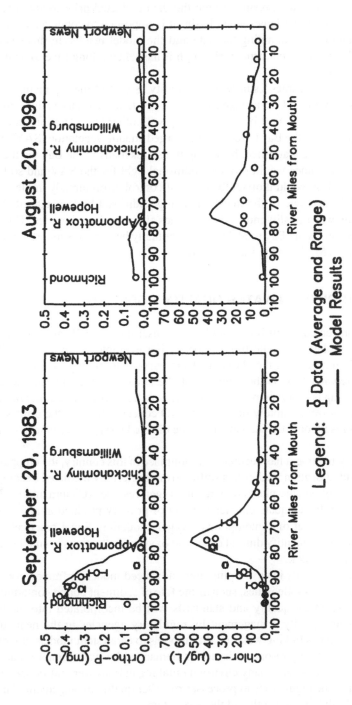

Figure 1-1 Model post audit results of the James Estuary: 1983 (pre-control) versus 1996 (post-control) conditions for orthophosphate and chlorophyll *a*.

phosphorus removal, and its results mimic the data collected prior to and after the phosphorus controls in the James River basin. It is important to conduct a post audit of the water quality by examining the data and the model results. In this case, the water quality benefits are confirmed, although it has taken a long time to achieve a small improvement.

One must view the James Estuary water quality as part of the big picture as the James is a tributary to the Chesapeake Bay. Since the James Estuary is light limited, the nutrient control effort must be viewed along with other considerations currently being contemplated for the Bay to restore submerged aquatic vegetation (SAV) by removing solids from the watershed. Reducing the solids loads to the Bay via the tributaries such as the James would provide additional light for the SAV and so to the suspended algae. But would it offset the nutrient control effort already implemented in the James? That is one of the questions being addressed by the state-of-the-art Chesapeake Bay water quality model that includes all the major tributaries. This is an important question, as the Bay model would eventually be used to support the development of many TMDLs in the Bay watershed.

1.4 NEW HORIZON IN WATER QUALITY MODELING

The material presented in this book should prove useful for TMDL modeling at present, but it also addresses some future aspects. For example, many well stabilized wastewaters do not produce the classic DO sag curve, particularly in small streams when the wastewater flow is the majority of the stream flow during dry weather. Rather, the DO sag is observed near the discharge outfall. Efficient stream reaeration in shallow waters would rapidly increase the DO concentrations in the downstream direction. As a result, only the second half of the classic DO sag curve is seen in many streams today.

As secondary treatment has become commonplace in this country, the emphasis of many wasteload allocation studies has shifted from bacterially mediated decomposition to the effect of aquatic plants on stream oxygen reserves (Chapra and Di Toro, 1991). Consequently, adequate characterization of primary production and respiration is an essential facet of contemporary oxygen modeling. Some practical, inexpensive techniques to quantify diurnal dissolved oxygen fluctuations in small streams are presented in this book.

Recent research efforts in developing more advanced modeling frameworks continue to address more subtle issues, such as the fate of sediment bound contaminants. It is clear that sediment quality and standards will be incorporated into modeling analyses for water quality management by regulatory agencies in the near future. New frontiers, such as linking water quality models to ecosystem models, are on the horizon to address living resources issues as demanded by water quality managers and decision makers. In fact, many environmental regulations were introduced in recent years to link the regulations to processes of water quality management planning and to protect both human health and the ecosystem.

1.5   IT IS THE MODELING SKILLS AND DATA, NOT THE MODEL CODE, THAT MATTER

One may draw a comparison between water quality modeling and hydraulic calculations. While water quality modeling requires a significant amount of model parameter tuning and calibration, using the HEC models is a more straightforward task. Many software packages for hydraulic computations are for sale in the classified section of *Civil Engineering* magazine, but there have been no water quality models for sale in the advertisements. Many subtle, technical issues are encountered in water quality modeling, making its application less straightforward than application of the hydraulic calculation models. Providing technical support to water quality model applications is a key element in distributing programs. In fact, technical assistance is part of the modeling skills that are at the heart of this book.

In recent years, several international institutions have begun marketing their hydrodynamic and water quality models, and their software packages are offered at substantial costs, yet they do not provide the source code. As demonstrated in the many case studies presented in this book, it is always necessary to reconfigure the model in water quality modeling applications. Without the source code, users are more or less working in the dark, without the benefit of a full understanding of the model. On the other hand, many U.S. government agencies offer their code on the Web for free distribution. For example, the QUAL2E and WASP modeling frameworks are available free from the U.S. EPA. The CE-QUAL-W2 model can be downloaded from the Army Corps of Engineers Waterway Experiment Station Web site for anyone to use. Of course, the source code is there. (Note that all the case studies used in this book are from these freely available models.)

As stated in the Preface, field data are essential to water quality modeling. An experienced modeler would feel frustrated with his or her work without data support. How do you know your model results are right? Modeling does not generate data. It only interprets data to produce the big picture. Over the course of the development of the Chesapeake Bay model, the modelers have continued working with the Bay scientists to learn, for example, how to model zooplankton or how to get reasonable primary production rates for suspended algae, to name just two instances. In establishing many TMDLs, the first order of business is to collect sufficient data to support the modeling effort. When is the data sufficient? Never. That is why the skills of the modeler are needed to make the judgement. In most cases, the modeler must navigate a narrow passage between the extent of available data and data needs if the modeling effort is to produce a successful modeling study.

Dr. Robert Thomann of Manhattan College once said:

> The key to the success of a water quality modeling study is how the modeler derives the model coefficient and parameter values and how the model results are interpreted to make sense of the modeling outcome. The model code is secondary. One could even hire a computer programmer to write the code.

CHAPTER 2

TOTAL MAXIMUM DAILY LOADS (TMDLs)

2.1 EVOLUTION OF WATER QUALITY MODELING IN WATER POLLUTION CONTROL

Before 1972, water pollution control efforts were based on achievement of ambient water quality standards. Arguably, this was economically efficient since each source would control its discharge only to the degree necessary to meet local water quality standards (Schroeder, 1981). However, the approach proved virtually impossible to administer, because of the difficulties in translating ambient standards into end-of-pipe effluent limits for individual dischargers. The result was regulatory frustration and very little cleanup. One of the factors that contributed to the frustration was that the "translating" technology, water quality modeling, was not fully developed.

In 1972, the Federal Water Pollution Control Act was amended to require a minimum level of control based on available treatment technology. Thus the need to determine impact on the receiving water, often the most difficult determination to make, was largely eliminated. The result was substantial reduction in pollution, even if this was achieved in areas in which cleanup was not needed to meet water quality goals. Over the next half dozen years, secondary treatment was promulgated as the minimum level for all publicly owned treatment works (POTWs) on the assumption that the expected water quality responses were worth the expenditure. Effluent limitation guidelines were established for industrial wastewaters. Wasteload allocation was used specifically for those instances in which there was some doubt that the water quality standards could be achieved by secondary treatment alone.

The pendulum continued to swing between wasteload allocation and effluent requirements. In 1977, the Clean Water Act was passed and water quality standards for toxic substances were prepared. It was clear by 1982 that the Environmental Protection Agency (EPA) strongly favored water quality–based effluent limitations, rather

8

than technology-based limits, as basic water quality pollution control strategy, thereby reversing the regulatory trend. By that time, water quality modeling technology had advanced significantly and was ready to address a variety of water quality problems.

There is little doubt that the Clean Water Act of 1977 and its across-the-board abatement approach of secondary treatment required for municipal wastewaters has made substantial improvement in the water quality of many natural water systems. However, there remain a significant number of waterbodies that are water quality–limited, thereby requiring additional work, leading to the development of so-called total maximum daily load (TMDL), as outlined in Section 303 of the Clean Water Act, to safeguard the quality of these waterbodies. Figure 2-1 shows the time lines for the water pollution control regulations and the evolution of water quality modeling technology.

2.2   WHAT IS A TMDL?

While the initial provisions of Section 303 amplified the process of establishing state water quality standards, Section 303(d) added a prescription for using these standards to upgrade waters that remained polluted after the application of technology-based requirements. In short, Section 303(d) requires the following steps (Houck, 1999). The states will:

1. Identify waters that are and will remain polluted after the application of technology standards, i.e., effluent limits;
2. Prioritize these waters, taking into account the severity of their pollution; and
3. Establish TMDLs for these waters at levels necessary to meet applicable water quality standards, accounting for seasonal variations and with a margin of safety to reflect lack of certainty about discharges and water quality.

States are to submit their inventories and TMDLs to the EPA for approval. If the EPA does not approve, the agency is to promulgate them itself for incorporation into state planning. Under Section 303(e) of the Clean Water Act, states are to develop plans for all waters that include, inter alia, (1) discharge limitations at least as stringent as the requirements of its water quality standards and (2) TMDLs. In the event a state fails to develop the list or to develop TMDLs, the EPA is obligated to do so. In general, water quality standards, as set out in the TMDL program, lay largely dormant— at least until the 1990s, when they were activated by citizen suits demanding implementation in order to control water pollution from nonpoint sources that cumulatively impaired waters nationwide (Houck, 1999).

A TMDL is essentially a "pollution budget" designed to restore the health of the polluted body of water. A TMDL is an estimate of the maximum pollutant loading from point and nonpoint sources that receiving waters can accept without violating water quality standards. Determining a TMDL is difficult for a combination of point

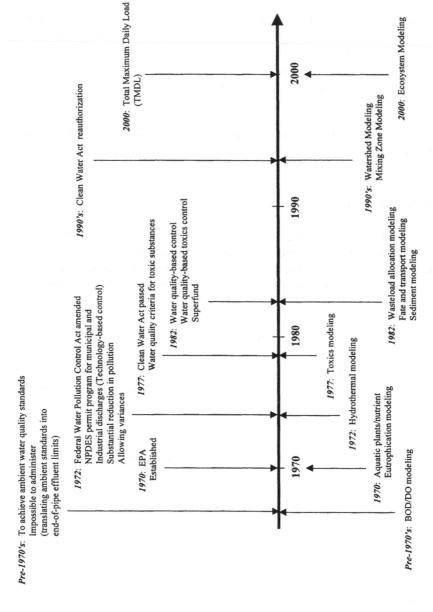

Figure 2-1 History of water pollution control and water quality modeling applications.

and nonpoint pollutant sources because of the fundamentally different nature of the two sources. Although the TMDL requirement has been in existence for 20 years, most implementation has focused on point source requirements rather than on nonpoint loadings.

Environmental groups have become impatient with the TMDL process. For example, in the lawsuit, *Sierra Club et al. v. Hankinson*, the court found that EPA's failure to disapprove Georgia's inadequate TMDL submissions was in violation of the Administrative Procedure Act and the Clean Water Act. On September 3, 1996, the court ordered EPA to ensure that TMDLs were established for all 303(d) listed waters within five years and to ensure that they are implemented through the NPDES permitting program of the state. In July 1997, the parties signed a consent decree that would supersede the district court's order of September 3, 1996. The decree establishes a schedule for developing TMDLs in each of Georgia's watershed basins from 1998 to 2005. The decree provides that the EPA move to develop the TMDLs if Georgia does not do so. Similar situations have been reported in many other states across the country.

TMDLs are established to achieve and maintain water quality standards. A water quality standard is the combination of a designated use for a particular body of water and the water quality criteria designed to protect that use. Designated uses include activities such as swimming, protecting a drinking water supply, and shellfish propagation and harvest. Water quality criteria consist of narrative statements and numeric values designed to protect such designated uses. Criteria may differ among waters with different designated uses.

Development of TMDLs typically requires the use of one or more water quality models. Often these models are sophisticated and new users may require assistance and/or training. Nearly all of these TMDLs are for contaminants for which major contributions come from nonpoint sources (e.g., nitrogen, phosphorus, and pathogens).

2.3 WATER QUALITY ENDPOINTS FOR TMDLs

One of the key elements in a nutrient TMDL is specific application of nutrient enrichment endpoints, which are influenced by a broad array of factors and processes, including physical factors, biological factors, and human impacts. Given these diverse and wide-ranging effects, measurement endpoints have been classified based on their relationship to assessment endpoints and to system function. An endpoint:

1. Quantifies an assessment endpoint,
2. Quantifies a physical, chemical, or biological factor or process critical to defining the ecosystem,
3. Is an early indicator of incipient system change, or
4. Is an indicator of historic trends.

From the TMDL standpoint, items that quantify an assessment endpoint are most suitable. They include color, Secchi depth, light attenuation coefficient, total suspended solids, dissolved oxygen concentration, biochemical oxygen demand, sediment oxygen demand, algal biomass (i.e., chlorophyll *a*), algal species composition, aquatic vegetation biomass, and fish population. In practice, dissolved oxygen concentration and algal biomass are the most commonly applied as assessment endpoints in TMDLs.

Dissolved oxygen concentration in natural waters is a function of physical and chemical factors that determine solubility and the transport of oxygen across the air-water interface as well as the mixing of surface and deep waters. In addition, biological processes mediated by suspended phytoplankton in the water column, benthic algae, and aquatic plants produce oxygen via photosynthesis. Plant, animal, and bacterial respiration consume oxygen, as do autotrophic bacterially mediated processes such as nitrification. Other bacterial processes leading to the reduction of sulfate and carbon dioxide produce sulfide and methane, which react with and provide additional demands on dissolved oxygen. Considering especially these biological processes, nutrient enrichment is often signaled by excessive oxygen production in surface waters, leading to supersaturation in some cases, and by hypoxia or anoxia in deep waters when excessive plant production is consumed. Thus dissolved oxygen concentration is one of the major endpoints for assessment of eutrophication in natural waters. Since algal and plant photosynthesis are closely related to dissolved oxygen in eutrophication assessment, it is clear that algal biomass is another important endpoint for eutrophication assessment. Thus dissolved oxygen and chlorophyll *a* have been used as the major endpoints in nutrient TMDLs.

Nutrient concentrations have rarely been used as endpoints in TMDLs. One exception is nitrate as the national drinking water standards set a 10 mg/L concentration as the limit. Some states are contemplating using low dissolved inorganic nitrogen (DIN) and dissolved inorganic phosphorus (DIP) concentrations as receiving water endpoints. Use of low DIN and DIP levels in TMDL development is rare and unusual and should be viewed as specific to the particular body of water. In receiving waters where light and/or transport (flow) are the limiting factor(s) for algal growth, meeting low DIN and DIP standards would require a substantial reduction of point and nonpoint nutrient loads but may not necessarily improve the water quality in terms of the chlorophyll *a* and dissolved oxygen concentrations. Careful water quality modeling is therefore required to develop TMDLs for these waters.

In the Chesapeake Bay region, historical water quality data are being analyzed to determine the status of total nitrogen, DIN, total phosphorus, DIP, chlorophyll *a*, and total suspended solids concentrations and Secchi depth. Unfortunately, there are no scientifically established goals for "good" and "poor" concentrations for any of these parameters to use as assessment endpoints. Instead, to determine current water quality status of each parameter, a Bay-wide scale was devised for each salinity zone (tidal fresh, oligohaline, and mesohaline). This relative scale uses the distribution of all of the data available Bay-wide from 1985 to 1996. For each salinity regime, the 5th and 95th nutrient percentile concentrations are developed for each parameter. As values below and above these percentile concentrations are considered to be outliers, the baseline is developed using values between the 5th and 95th percentile concentrations. Since higher nutrient, chlorophyll *a*, and total suspended solids levels are

TABLE 2-1 Percentiles from the Chesapeake Bay Data (1985–1996)

Water Quality Parameter	Salinity Zone	5th Percentile	95th Percentile
Chlorophyll *a* (μg/L)	Tidal fresh	1.0	47.0
	Oligohaline	1.49	70.7
	Mesohaline	2.59	31.6
Secchi depth (meter)	Tidal fresh	0.25	1.3
	Oligohaline	0.2	1.0
	Mesohaline	0.4	2.6
Total nitrogen (mg/L)	Tidal fresh	0.566	3.57
	Oligohaline	0.625	3.76
	Mesohaline	0.425	1.55
Total phosphorus (mg/L)	Tidal fresh	0.038	0.245
	Oligohaline	0.040	0.254
	Mesohaline	0.016	0.119
DIN (mg/L)	Tidal fresh	0.098	2.86
	Oligohaline	0.025	2.43
	Mesohaline	0.011	0.887
DIP (mg/L)	Tidal fresh	0.005	0.095
	Oligohaline	0.002	0.054
	Mesohaline	0.002	0.030
Total suspended solids (mg/L)	Tidal fresh	3.75	47
	Oligohaline	7.0	76
	Mesohaline	3.2	37

It should be pointed out that using the 5th percentile concentration as the poor endpoint would be extremely conservative in TMDL development. Reasonable endpoint concentrations should be between the 5th percentile and 95th percentile values.

considered to indicate worse water quality conditions than lower levels, the 95th percentile concentration is considered to be the "poor" endpoint and the 5th percentile concentration is considered to be the "good" endpoint (Table 2-1). For Secchi depth, the reverse is true, and so the 5th percentile level is considered to be the "poor" endpoint and the 95th percentile level is considered to be the "good" endpoint.

2.4 WATER QUALITY MODELING FOR TMDL

Development of TMDLs typically requires the use of one or more environmental fate and transport models, including watershed, hydrodynamic, and receiving water models. In many situations, simplified models can be used to address the water quality problems. In other cases, sophisticated models are required and assistance in using these models and/or training is needed. There is a number of modeling frameworks currently available from the EPA, and they are being widely used in many TMDL studies nationwide:

1. Watershed models—BASINS and HSPF
2. Hydrodynamic models—CE-QUAL-W2 and EFDC
3. Mass transport models—HAR03 and WASP
4. Receiving water models—STREAM, QUAL2E, and WASP/EUTRO

In addition, the EPA is conducting water quality research to support TMDL development (U.S. EPA, 2000). The Ecosystems Research Division—Athens (ERD—Athens) of the EPA Office of Research and Development's National Exposure Research Laboratory (NERL) is developing a TMDL program called Technology Required for Alternative Analyses for a Changing Environment (TRACE) to achieve the following goal (U.S. EPA 2000):

> Develop and maintain a comprehensive technical support capability within NERL that directly links environmental TMDL exposure research activities and products for the EPA Office of Water and Regional Offices to be used for implementation of policy, regulatory development, remediation, and enforcement needs and activities.

Questions that are closely related to TMDL and are going to be addressed in the TRACE program include:

- How do we extrapolate the stream concentration-discharge dynamic relationship from hillslope to watershed to river basin scales? How do we optimally utilize existing and remote-sensed data to simulate multiscale, hydrochemical ecosystems?
- What are the sensitive hydrologic, geochemical, landscape modifications and habitat changes that mediate exposure/response of watershed/regional ecosystems? How do we quantify the interplay between these changes to the equations of motion for flow and solute transport, including nonlinear controls and responses?
- How do we characterize a stressor, such as land use changes or climate/meteorological changes, for exposure assessment?
- What are the multicomponent concentration-storage or concentration-discharge relationships at the watershed or river-reach scale for interactions of hydrochemical systems and how do we separate or distinguish among essential spatial and temporal components influencing these relationships?
- What are the geochemical kinetic processes that influence these exchanges between soil organic material and relatively important TMDL gases (carbon, nitrogen)? How are these processes modified by natural and human-induced changes in earth/ecological systems? How do geochemical processes mediate the exposure and response of earth/ecological systems to climate change and other stressors?

- What geochemical processes control nitrogen and phosphorus cycling? How do land uses or management regimes modify exposures to nonpoint stressor?
- How do biogeochemical processes mediate the exposure and response of regional ecosystems to natural and anthropogenic stressors, including land use and climate change, multimedia pollutant inputs, and so on? Processes include carbon and nitrogen cycling, storage, and release, organic pollutant transformation/fate, metal speciation/fate, and plant/soil–pollutant interactions.

Their research plan outlines the following tasks:

1. Couple models to provide a multimedia modeling system, including landscape characterization and water quality modeling
2. Advance the use of hydrologic analysis to support watershed assessment, ecological evaluation, and TMDLs
3. Develop TMDL modeling approaches for nutrients and toxicants
4. Develop TMDL modeling approaches for sediments
5. Develop TMDL modeling approaches for pathogens
6. Develop watershed scale field data on sediments, nutrients, fecal coliforms, and toxics in the South Fort Broad River, Georgia
7. Develop and evaluate riparian buffer models to evaluate best management practices and estimate buffer effects for TMDLs and basin planning

Note that the first five tasks are generic in nature, having a broad impact on the TMDL program. Tasks 6 and 7 are more or less site-specific studies, serving as a protocol for the modeling practice.

2.5   A TMDL MODELING STUDY

The Transquaking River is located in Dorchester County, Maryland (Figure 2-2). It originates south of East New Market area and finally drains to the Chesapeake Bay, through the Fishing Bay, roughly seven miles due south of Bestpitch. The river is approximately 23.2 miles in length, from its confluence with the Fishing Bay to the head of the tide. The Transquaking River watershed is about 110.8 square miles. The land uses in the watershed consist of forest and other herbaceous (62.6%), mixed agriculture (33.1%), water (2.6%), and urban (1.7%). The Transquaking River is tidal throughout its navigable reach, which extends from the highly depositional delta area at its mouth for approximately 20.5 miles upstream to an area known as Higgins Mill Pond.

The total nitrogen load coming from nonpoint sources is 545,113 lb/yr and the total nonpoint phosphorus load is 41,987 lb/yr. There is only one point source in the watershed, Darling International, Inc., contributing 354,050 lb/yr of nitrogen and 1,825 lb/yr of phosphorus to the basin.

The river is impaired by the nutrients (nitrogen and phosphorus), which cause excessive algal blooms and violation of the dissolved oxygen standard. Figure 2-3 pre-

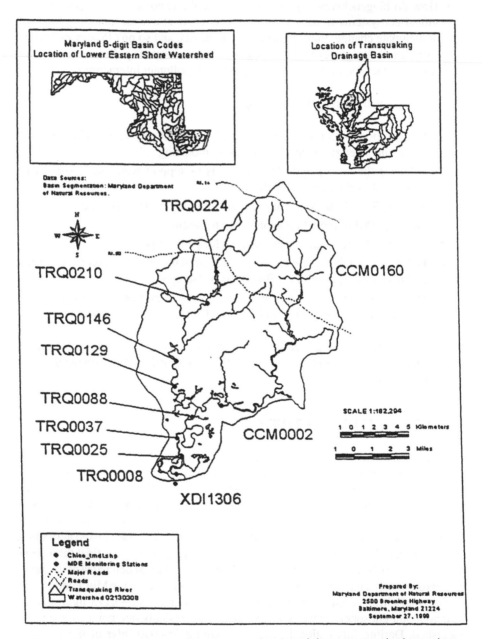

Figure 2-2 The Transquaking River watershed and receiving water monitoring stations.

sents a longitudinal profile of chlorophyll *a* data sampled during the 1998 field surveys to support the TMDL effort. As the data indicate, chlorophyll *a* concentrations in the lower 10 miles are below 50 µg/L. Much higher chlorophyll *a* levels, reaching a peak over 140 µg/L, occur in the upstream reach. Dissolved oxygen concentrations along the river are also shown in Figure 2-3, indicating a DO sag (below 4 mg/L) at about 5 miles from the mouth. The Transquaking River was first identified on the

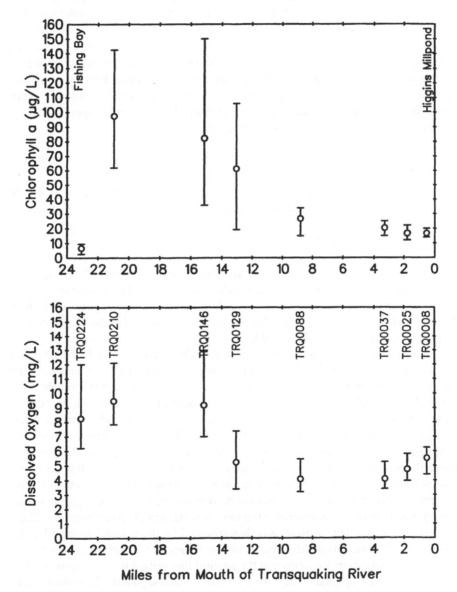

Figure 2-3   Longitudinal dissolved oxygen and chlorophyll *a* concentration profiles in the Transquaking River, MD.

303(d) list submitted to EPA by the Maryland Department of the Environment (MDE) in 1996.

The water quality goal of these TMDLs is to reduce high chlorophyll *a* concentrations and maintain dissolved oxygen concentrations at levels that support the designated uses of the Transquaking River. The TMDLs were determined using the WASP5 modeling framework.

Data to support the modeling effort was collected during six water quality surveys from March 12, 1998 to September 15, 1998. The WASP/EUTRO5 model was configured for the Transquaking River to simulate seasonal steady-state water quality conditions using the data from the six surveys. The river is divided into 28 segments, receiving nonpoint flows and nutrient loads from 18 watershed segments (Figure 2-4).

The objective of the nutrient TMDLs for the Transquaking River is to reduce nutrient inputs to a level that will ensure the maintenance of the dissolved oxygen standards and reduce the frequency and magnitude of algal blooms. Specifically, the nutrient TMDLs for the Transquaking River are intended to:

1. Assure that a minimum dissolved oxygen level of 5 mg/L is maintained throughout the Transquaking River system, and
2. Reduce peak chlorophyll *a* levels (a surrogate for algal biomass) to below 50 µg/L. (Note that the MDE established permit limits based on maintaining chlorophyll *a* concentrations below a maximum level of 100 µg/L, with an ideal goal of less than 50 µg/L.)

The dissolved oxygen level is based on specific numeric criteria for Use I & II waters set forth in the Code of Maryland Regulations 28.08.02. The chlorophyll *a* level is based on the designated use of Transquaking River, and guidelines set forth by Thomann and Mueller (1987) and by the EPA's *Technical Guidance Manual for Developing TMDLs* (U.S. EPA, 1995).

For the model runs where the nutrient loads to the system were reduced, a method was developed to quantify the reductions in sediment nutrient fluxes and sediment oxygen demand (SOD). First, initial estimates were made of the total organic nitrogen and organic phosphorus settling to the sediments, from particulate nutrient organics, living algae, and phaeophytin, in each WASP/EUTRO5 segment. These estimates were derived by running the expected condition scenario once with correct settling of organics and chlorophyll *a*, then again with no settling. The difference between the two runs was what was assumed to settle to the sediments. All phaeophytin was assumed to settle to the bottom. The amount of phaeophytin was estimated from in-stream water quality data. To calculate the organic loads from the algae, it was assumed that the nitrogen to chlorophyll *a* ratio was 10, and the phosphorus to chlorophyll *a* ratio was 1. This analysis was then repeated for the reduced nutrient loading conditions. The percentage difference between the amount of nutrients that settled in the expected condition scenarios and the amount that settled in the reduced loading scenarios was then applied to the nutrient fluxes in each segment. The reduced nutrient scenarios were then run again with the updated fluxes. A new amount of settled organics was calculated, and new fluxes were calculated. The process was repeated

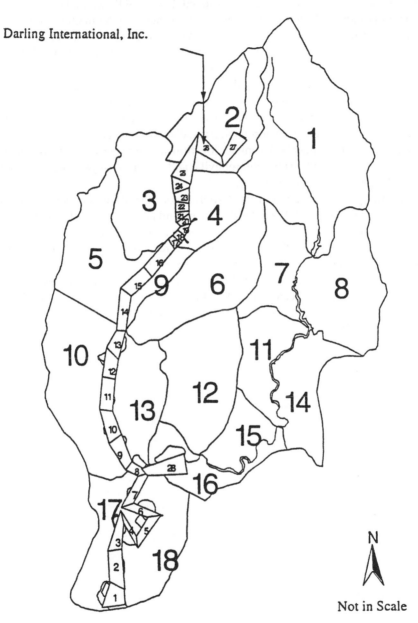

Figure 2-4   The watershed segments and WASP/EUTRO5 model segmentation for the Trans-quaking River TMDL study.

several times, until the reduced fluxes remained constant. Along with reductions in nutrient fluxes from the sediment, when the nutrient loads to the system are reduced, the SOD will also be reduced. It was assumed that the SOD would be reduced in the same proportion as the nitrogen fluxes, to a minimum of 0.5 gm O_2 m^{-2} day^{-1}.

In the scenario representing improved conditions in the stream flow during low flow, the total nonpoint source loads were based on the data from the 1998 survey. A margin of safety of 5% was included in the load calculation. The nitrogen and phosphorus loads were reduced to meet the chlorophyll a and dissolved oxygen standards in the receiving water. In another scenario with improved conditions under the average annual flow, nonpoint nutrient loads were decreased by 35% and a 3% margin of safety was used to meet the chlorophyll a and dissolved oxygen standards. The model results indicate that under both summer low flow and average annual flow conditions, the water quality standards for chlorophyll a and dissolved oxygen are met at all stations along the Transquaking River.

Following the completion of the TMDL report, the MDE conducted a public review of the proposed TMDLs to limit nitrogen and phosphorus loadings to the Transquaking River. The public comment period was open from November 12, 1999 through December 13, 1999. MDE received and responded to one set of written comments. The TMDLs have since been approved by the U.S. EPA and at this writing, they are being incorporated into the State's continuing Planning Process, pursuant to Section 303(e) of the Clean Water Act. In the future, the established TMDLs will support point and nonpoint source measures needed to restore water quality in the Transquaking River.

CHAPTER 3

DERIVATION OF MASS TRANSPORT COEFFICIENTS

Mass transport is a key component of the water quality model. Many mechanisms are responsible for mass transport in natural water systems, but a presentation of the theories and concepts behind these mechanisms is beyond the scope of this book. Rather, the emphasis of this chapter is on the approach and procedures commonly used in the water quality profession to quantify the mass transport coefficients. This chapter discusses one- , two- , and three-dimensional configuration of mass transport and derivation of mass transport coefficients. A number of methods, ranging from simplified mass transport analysis to sophisticated three-dimensional hydrodynamic modeling, are presented, along with case studies that illustrate the "how-to" procedure. These methods are applied to a variety of water bodies including streams, rivers, lakes, impoundments, estuaries, and coastal waters. This chapter also discusses linking a hydrodynamic model to a water quality model.

The observation by Orlob (1972) on estuarine modeling provides a perspective to the role of mass transport in water quality modeling:

> Success in verifying water quality models has generally been less than satisfactory, largely because of the paucity of good data, but also because of the greater dependence on empiricism in structuring the models. As noted previously, the relative dependence on the advection and diffusion terms in the mass transport equation is a function of the scale of the model. As the scale increases, i.e., segments become larger or phenomena are averaged over longer time steps, the dependence on the empirical effective diffusion coefficient increases. Verification of coarse scale mathematical models is usually accomplished by subjective adjustment of the coefficient until the model prediction agrees with prototype performance. Because the coefficients are derived from historic experience with the prototype, they are usually not considered reliable for prediction of prototype performance under conditions that differ markedly from the historic.

Following Orlob's observation, this modeling approach remained unchanged for 15 years. However, advancements in computational hardware and software in the last decade have significantly changed the approach—hydrodynamic models are now commonly used to drive the water quality modeling analysis (Thomann, 1998).

3.1   ONE-DIMENSIONAL ADVECTIVE TRANSPORT IN STREAMS, RIVERS, AND ESTUARIES

The key ingredient in one-dimensional mass transport in rivers and streams is the advective velocity, which depends on the hydraulic geometry of the system. In free-flowing streams and rivers, it is useful to develop the relationships between the average velocity and flow rate and between the average depth and flow rate. In a total maximum daily load (TMDL) modeling study of the South Fork South Branch Potomac River, West Virginia (Figure 3-1), hydraulic geometry was developed using available data from 4 U.S. Geological Survey (USGS) gaging stations along the river. Figure 3-2 shows the regression resulting in empirical equations for velocity, V, versus flow rate, Q, and depth, D, versus Q at four locations in the South Fork South Branch Potomac River in West Virginia. These regressions, represented by straight lines when plotted on a log-log scale, are taken as the one-dimensional mass transport model for advective flows. However, these hydraulic geometry data are not always readily available at all gaging stations, particularly for small rivers and streams. Extra effort may be required in contacting the USGS field offices to get the raw data when they are not computerized. While the data were readily available from USGS for the Moorefield station, data from the Sugar Grove station were only available in the form of a raw data sheet (Figure 3-3).

Computer models, such as the Army Corps of Engineers one-dimensional HEC-2 model, are commonly used to calculate the stream water profile and hydraulic geometry when field data are not obtained. HEC-2 is a robust modeling framework and has been successfully applied to numerous studies with good results. Figure 3-4 shows the HEC-2 model results from the Quinebaug River in Massachusetts for the portion River Mile 54 to River Mile 46. Note the pools created by the three dams: Cutlery Dam, American Optical Dam, and West Dudley Dam. An important result from the HEC-2 model calculation is a calculated time of travel. Figure 3-5 shows the time of travel for the Kalamazoo River from River Street to Main Street with three different flow conditions ranging from 7.19 m^3/s to 30.37 m^3/s. As expected the lowest flow rate results in the longest time of travel in the river.

A commonly used approach in evaluating one-dimensional mass transport in streams and rivers is modeling the concentration profile of a conservative substance to backcalculate the longitudinal advection and dispersion. This procedure is essentially a mass balance of the conservative substance that accounts for inflows and outflows as well as input and output loads of such a substance along the river. Figure 3-6 shows the results of a steady-state modeling analysis of the Blackstone River in Massachusetts and Rhode Island under three different flow conditions using chloride data measured in the summer and fall of 1991. The Environmental Protection

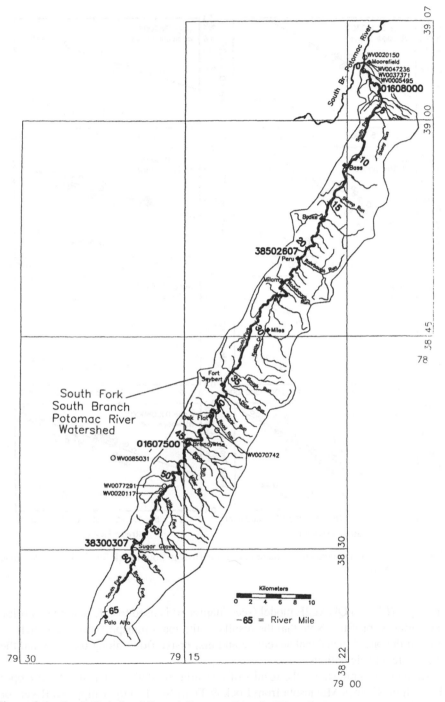

Figure 3-1   South Fork South Branch Potomac River in West Virginia, showing USGS gaging stations.

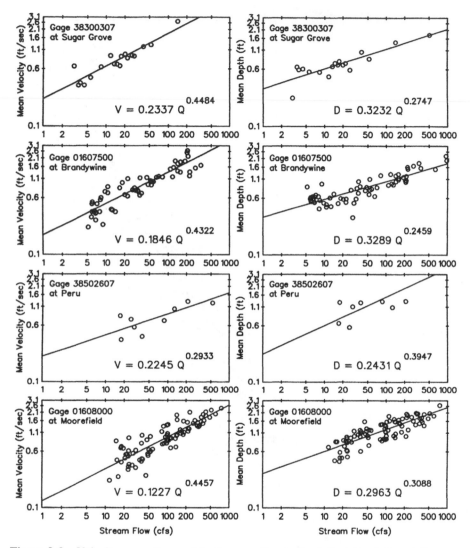

Figure 3-2   Velocity versus flow and depth versus flow in South Fork South Branch Potomac River, West Virginia.

Agency's (EPA's) QUAL2E model (see Chapter 6) is used to calculate the chloride concentration profiles. Note that the results of the mass transport modeling analysis confirm the one-dimensional advective and dispersive flows along the river and the time-of-travel calculations.

Another example shows the results of a similar modeling analysis of the Upper Mississippi River in Minnesota from Lock & Dam No. 1 to the Chippewa River for three summer flow conditions in 1988, 1990, and 1991 using specific conductivity data. Results from the steady-state modeling of conductivity using the EPA's WASP

9-275-F
(Apr. 93)

U.S. DEPARTMENT OF THE INTERIOR
U.S. GEOLOGICAL SURVEY

WATER RESOURCES DIVISION

Meas. No. _12_

Comp. by _____

Sta. No. _____ DISCHARGE MEASUREMENT NOTES   Checked by _____

South Fork South Branch @ Sugar Grove

Date _2-14._ 19 _95_ Party _LK Rogers (M) MM.Hes (Rxx)_
Width _35_ Area _94.5_ Vel. _0.67_ G.H. _____ Disch. _16.5_ _0.700_
Method _6_ No. secs. _27_ G.H. change _0_ in _.6_ hrs. Susp. _Rsd_
Method coef. _1.0_ Hor. angle coef. _1.0_ Susp. coef. _1.0_ Meter No. _F-1041_
Type of meter _Pygmy_ Date rated _____ Tag checked _____
Meter _____ ft. above bottom of wt. Spin before meas. _free_ after _✓_
Meas. Plots _____ % diff. from _____ rating. Levels obtained _____

GAGE READINGS						WATER QUALITY MEASUREMENTS
Time	Inside				Outside	No _____ Yes _____ Time _____
						Samples Collected
5^{10}					15.5 /	No _____ Yes _____ Time _____
1455					- .10 / 15.41	Method Used
						EDI _____ EWI _____ Other _____
5^{10}						SEDIMENT SAMPLES
1530					15.41	No _____ Yes _____ Time _____
						Method Used
						EDI _____ EWI _____ Other _____
Weighted M.G.H.						BIOLOGICAL SAMPLES
G.H. correction						Yes _____ Time _____
Correct M.G.H.						No _____ Type _____

Check bar. chain found _____ changed to _____ at _____
Wading, cable, ice, boat, upstr., downstr., side bridge _____ feet, mile, above below gage _RP_
Measurement rated excellent(2%), good (5%), fair (8%), poor (over 8%); based on the following cond:
Flow _____ fairly uniform
Cross section _____ fairly rough
Control _____ shore ice 50 % across stream
Gage operating _____ Weather _Cloudy 35°_
Intake/Orifice cleaned _____ Air _____ °C@ _____ Water _____ °C@
Record removed _____ Extreme Indicator: Max. _____ Min. _____
Manometer N₂ Pressure Tank _____ Feed _____ Bbl rate _____ per min. _____
CSG checked _____ Stick reading _____
Observer _None_
HWM _____ outside, in well
Remarks _____

G. H. of zero flow _____ ft.   Sheet No. _____ of _____ sheets

Figure 3-3 Raw data to develop hydraulic geometry in South Fork South Branch Potomac River, West Virginia.   *(continues)*

9-275-11
(Rev. 6-71)

**UNITED STATES
DEPARTMENT OF THE INTERIOR
GEOLOGICAL SURVEY**

WATER RESOURCES DIVISION

Meas. No. _11_

Comp. by _MDK_

Sta. No. .... **DISCHARGE MEASUREMENT NOTES** Checked by ....

South Fork Mill Br. (@ Soy? Grove

Date _12-14_ 19_98_ Party _MDK ___ _ _0846_

Width _48.0_ Area _40.6_ Vel. _1.30_ G.H. _-14.92_ Disch. _52.6_

Method _..._ No. secs. _30_ G.H. change ___ in _½_ hrs. Susp. _Rod_

Method coef. _1_ Hor. angle coef. _Not (.)_ Susp. coef. _1_ Meter No. __

Type of meter _____/_ Date rated _51____ or rod, other.

Meter ___ ft. above bottom of wt. Spin before meas. _1.3 5/_ after _OK_

Meas. plots ___ % diff. from ___ rating. (Wading) cable, ice, boat, (ups?t) downstr., side

bridge ___ feet, mile, above, below gage. Levels obtained _of bridge_

BASE GAGE READINGS				
Time		Recorder	Inside	Outside
1455	RP2	23.24	+7.00	= 16.24
	RP1		=	1492
Weighted M.G.H.				
G.H. correction				
Correct M.G.H.				-16.24

AUX. GAGE READINGS				
Time		Recorder	Inside	Outside
		N/A		
Weighted M.G.H.				
G.H. correction				
Correct M.G.H.				

Check-bar, chain found ___   Check-bar, chain found ___

changed to ___ at ___   changed to ___ at ___

Measurement rated excellent (2%), (good (5%)), fair (8%), poor (over 8%), based on following

conditions: Cross section _Even, Rough, Cobbles_

Flow _Subcritical, Steady_ Weather _Cloudy, Cool_

Other _Turbulent, Nonuniform_ Air ___ °F. @ ___

Gage _None_ Water ___ °F. @ ___

___ Record removed _None_ Intake flushed t ___

Observer _None_

Control _Clear, Much of Debris below control has washed out; probably large shift._

Remarks _None_

G.H. of zero flow _No_ ft. Sheet No. ___ of ___ sheets.

Figure 3-3   (continued)

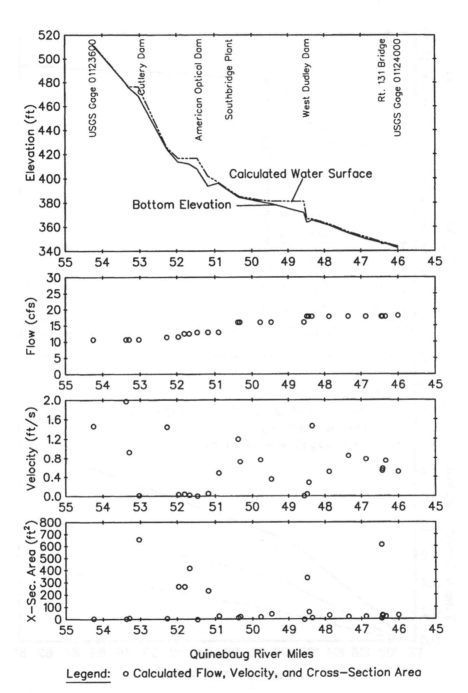

Figure 3-4   HEC-2 model input to determine one-dimensional hydraulic geometry and mass transport in Quinebaug River, Massachusetts.

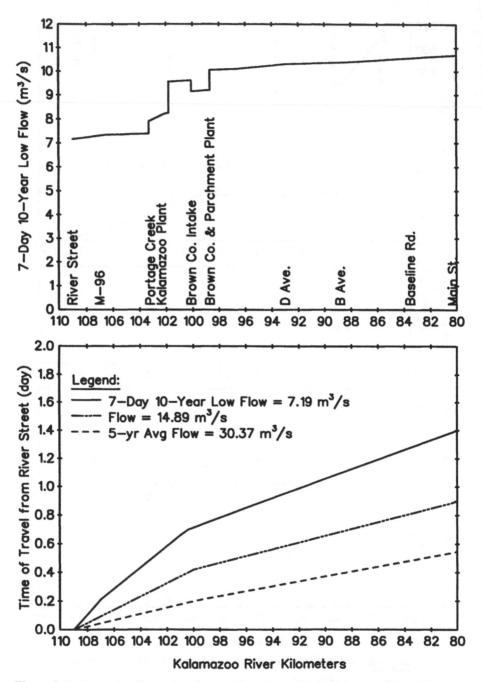

Figure 3-5   Seven-day 10-year low flow and time of travel in the Kalamazoo River, Michigan.

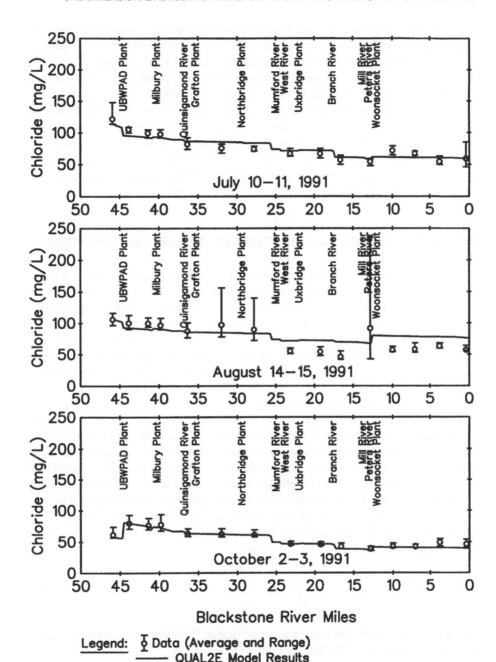

Figure 3-6   QUAL2E model calibration of one-dimensional mass transport in the Blackstone River, Massachusetts and Rhode Island, under three steady-state flow conditions.

modeling framework validate the time of travel and one-dimensional longitudinal dispersion coefficients for the river (Figure 3-7). The river flow during the summer months of 1988 was extremely low, close to the 7-day 10-year low flow value, while the river flows during the summer months of 1990 were much higher than normal. Further, flows in the summer of 1991 were higher than 1990. Note the significant impact of the Minnesota River on the concentration profile of conductivity during the summer 1988 flows, increasing the ambient conductivity concentration immediately below the junction of the Minnesota River. This impact on the ambient concentrations is minimized under high river flow conditions in 1990 and 1991. The impact of the Metro Plant, a significant point source, on the receiving water concentration gradually becomes less pronounced in 1990 and 1991 as well.

3.2 ONE-DIMENSIONAL LONGITUDINAL DISPERSION COEFFICIENT

Two major mechanisms cause longitudinal dispersion in rivers: vertical and transverse velocity shear. The rate of longitudinal dispersion reflects the balance between velocity shear (which acts to spread tracer along the channel) and transverse mixing (which promotes uniform concentrations and hence counteracts the effects of velocity shear). Thus if the rate of transverse mixing is very high (e.g., in a narrow sinuous channel) or if the velocity is uniform, then the rate of longitudinal dispersion is low (Rutherford, 1994). Conversely, if the rate of transverse mixing is low or if the velocity is highly nonuniform (e.g., in a wide channel with extensive shallows), then the rate of longitudinal dispersion is high.

   One-dimensional longitudinal dispersion in the water column due to velocity gradients in the vertical and lateral directions is also important in many riverine systems. While a number of empirical equations can be used to quantify this parameter, field studies remain the most accurate method for independently deriving this coefficient. A conservative tracer, such as the rhodamine WT dye or lithium, is most commonly used in field studies. Figure 3-8 shows the field-measured lithium concentration data reproduced by mass transport equations for the Grand Calumet River in Illinois. The mass balance equation for constant cross-sectional area, river flow, and dispersion with no other inputs except at the outfall is:

$$\frac{\partial C}{\partial t} = -U\frac{\partial C}{\partial x} + E_x\frac{\partial^2 C}{\partial x^2} - K \qquad (3\text{-}1)$$

where E_x is the longitudinal dispersion coefficient. If an instantaneous discharge of material occurs of mass M, then the solution to Eq. 3-1 is:

$$C(x,\ t) = \frac{M}{2A\sqrt{\pi E_x t}}\exp\left[\frac{-(x - Ut)^2}{4E_x t} - Kt\right] \qquad (3\text{-}2)$$

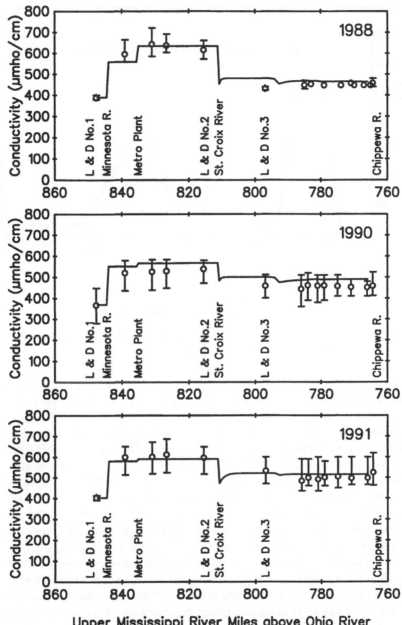

Figure 3-7   WASP model calibration of one-dimensional mass transport in the Upper Mississippi River under three steady-state flow conditions.

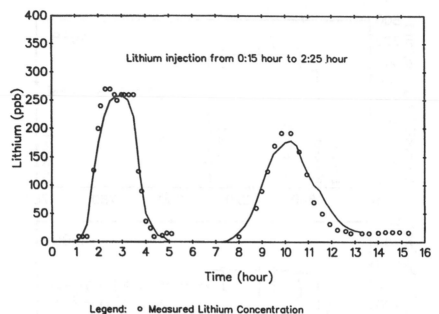

Figure 3-8 Computed and observed lithium concentration profiles to derive longitudinal dispersion coefficient in the Grand Calumet River, Illinois.

The longitudinal dispersion coefficient, E_x, which yielded the comparison between observed lithium and calculated results presented in Figure 3-8, was about 350 ft²/s (33.3 m²/s). This dispersion coefficient is typical of rivers (Rutherford, 1994).

Literature values are used for a preliminary estimate of longitudinal dispersion coefficients when field data are not available. Martin and McCutcheon (1999) provides an up-to-date values of one-dimentional longitudinal coefficients for estuaries.

The values listed in Martin and McCutcheon (1999) are useful when developing the longitudinal dispersion in a constant hydraulic geometry setting. In many physical settings, however, spatial variations of hydraulic geometry are significant. For example, many estuaries have been observed with an exponentially increasing cross-sectional area in the downstream direction. Lung (1993) presented a methodology to quantify a spatially constant, one-dimensional longitudinal dispersion coefficient in the Sacramento–San Joaquin Delta in California. Tidal effects also contribute to longitudinal dispersion of water quality constituents in one-dimensional estuaries. Since longitudinal gradients of salinity in an estuary reflect the dispersion, it is common to use salinity as a tracer to backcalculate the one-dimensional (particularly spatial variable) longitudinal dispersion coefficient. In general, a one-dimensional finite-segment (i.e., box) mass transport model is configured for estuarine systems. One of these modeling frameworks is the HAR03 model (see Chapter 8). Repeated adjustments of the one-dimensional longitudinal dispersion coefficients along an estuary would eventually calibrate a set of longitudinal dispersion coefficients for the estuary. Figure 3-9 shows

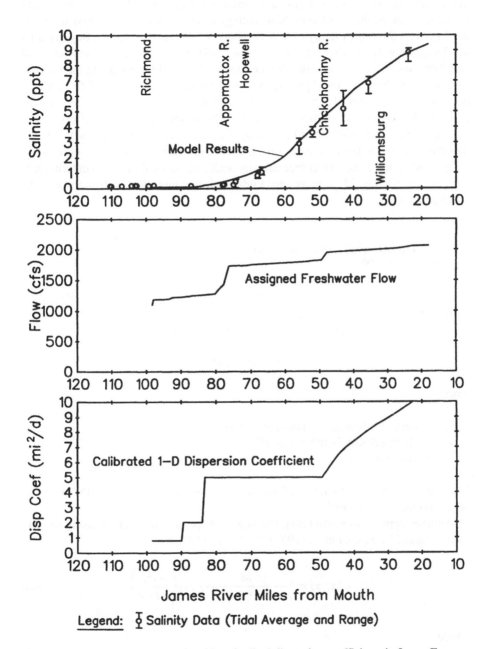

Figure 3-9   Derived one-dimensional longitudinal dispersion coefficients in James Estuary, Virginia.

such an analysis for the main channel of the James Estuary in Virginia under a tidally averaged steady-state condition (Lung, 1986). First, the model calculated salinity concentrations match the observed tidally averaged values closely. The mass transport model covers the entire estuary from Richmond, VA, to the Chesapeake Bay in a total of 63 segments. The freshwater flow is about 1,000 cfs near the fall line in Richmond and increases in the downstream direction with additional inflow along the estuary. Two major inflows are the Appomattox River (at River Mile 77) and the Chickahominy River (at River Mile 50). The freshwater flows are input to the model to calibrate the longitudinal dispersion coefficient. The calibrated longitudinal dispersion coefficient ranges from less than 1 mile²/day to about 10 mile²/day, consistent with the literature values. Note that the dispersion coefficient increases sharply below River Mile 50 when the James River increases its width substantially prior to entering into the Chesapeake Bay. More pronounced tidal actions cause the significant rise in the longitudinal dispersion.

3.3   LATERAL DISPERSION COEFFICIENT IN RIVERS AND ESTUARIES

Following the mixing in the vertical direction, the next stage of mixing in rivers and streams occurs in the lateral direction. One of the most commonly used equations to quantify lateral dispersion in rivers (Fischer et al., 1979) is:

$$D_y = 0.25du* \tag{3-3}$$

where

d = depth (or hydraulic radius) in meters
$u*$ = shear velocity in m/s = $(gds)^{0.5}$
s = channel slope

This equation has a ±50% range of accuracy in general, depending on the estimates of the parameters involved.

Another approach to estimating the lateral dispersion in rivers is using the following model by Fischer et al. (1979) and Neely (1982):

$$C(x, y) = \frac{M}{du(4\pi D_y x \, / \, u)^{1/2}} \exp\left(\frac{-y^2 u}{4D_y x}\right) \tag{3-4}$$

where

$C(x, y)$ = concentration of a conservative substance at any given location
M = mass of dye released / unit time
u = average velocity in the river

D_y = lateral dispersion coefficient across the river
x = distance downstream from the release of dye
y = distance in lateral direction
d = average depth in the river

Essentially, Rhodamine WT red dye is released in the open water of the river at a steady rate, defined as M. Field measurements of the dye concentrations (with a fluorometer) in the receiving water downstream from the release point provide a two-dimensional distribution of the dye, which is matched by Eq. 3-4. Subsequent adjustments of the lateral dispersion coefficient value in the equation eventually reproduce the two-dimensional distribution of dye concentrations.

In practice, specific conductivity may be used as a substitute for dye, particularly if a point source discharging into the river has a higher specific conductivity concentration than the ambient levels. This substitution for dye was used in a study of Burlington Industries, which discharges wastewater into the Banister River near Halifax, Virginia. To apply Eq. 3-4 to the Banister River for the determination of the lateral dispersion coefficient, the following data were used:

Total wastewater flow = 0.044 mgd

Effluent specific conductivity concentration = 730 μmho/cm

Ambient river conductivity concentration = 80 μmho/cm

River velocity near the bank = 0.30 ft/s (associated with a river flow of 189 cfs)

River depth = 2 ft

Following a series of model runs with Eq. 3-4, the model results shown in Figure 3-10 were achieved; four isopleth conductivity contours (labeled as 85, 90, 95, and

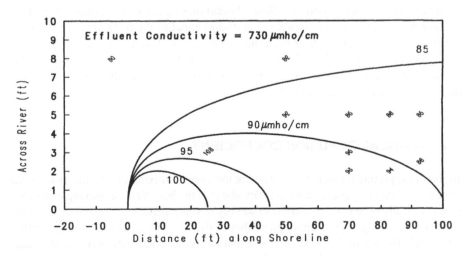

Figure 3-10 Modeling specific conductivity to calibrate lateral dispersion in the Banister River, Virginia.

100 μmho/cm) are displayed. Also shown for comparison are the measured conductivity levels (in smaller letters) in the vicinity of the outfall. The model calculated contours match the two-dimensional distribution of measured conductivity concentrations near the outfall with a lateral dispersion coefficient of 0.065 ft²/s. The conductivity plume, which attaches itself to the shore for over 100 ft, is reproduced by the model results. Approximately 10 ft from the shore in the lateral direction, the conductivity attenuates to approach the ambient level of 80 μmho/cm. Along the shore, the conductivity level approaches the ambient level at about 100 ft downstream from the outfall. Note that Eq. 3-4 is modified to accommodate the shore discharge instead of the center discharge in the river.

Equation 3-3 also provides an independent derivation of the lateral dispersion coefficient as follows. Based on a slope of 0.0003 as measured in the field, Eq. 3-3 yields a dispersion coefficient value equal to 0.069 ft²/s, thereby substantiating the value determined by Eq. 3-4.

For estuarine and tidal fresh systems, the following equation by Hamrick and Neilson (1989) is used to account for longitudinal dispersion:

$$C(x, y) = \frac{M}{\pi d \sqrt{D_x D_y}} \exp\left(\frac{ux}{2D_x}\right) K_0\left[\frac{u}{2\sqrt{D_x}}\left(\frac{x^2}{D_x} + \frac{y^2}{D_y}\right)^{0.5}\right] \tag{3-5}$$

where D_x is the longitudinal dispersion coefficient and K_0 is the modified Bessel function of the second kind of order zero. Again, Eq. 3-5 can be used to calibrate the lateral dispersion coefficient in an estuary or coastal system with a given longitudinal dispersion coefficient. Lung (1995) used Eq. 3-5 to determine a lateral dispersion coefficient value of 1 ft²/s in the James River Estuary below Richmond, Virginia.

Figure 3-11 shows the measured Rhodamine WT dye concentrations in the Delaware River Estuary used to calibrate lateral mixing across the estuary. Note that the dye was continuously released in the mid-channel of the river, thereby showing the highest concentrations near the center of the river. The data produced the measured dye profiles from July 23 to July 25. Such a data set is used with the WASP model (see the use of the WASP model in Chapter 7) to model the two-dimensional dye distributions in a time-variable mode.

3.4 VERTICAL DIFFUSION COEFFICIENT

In rivers the principal mechanisms causing the vertical mixing of a neutrally buoyant tracer is turbulence generated by velocity shear at the bed. Vertical mixing can be increased locally by secondary currents (notably at sharp bends) and by obstacles in the flow (such as bridge piers), although neither of these mechanisms has been quantified adequately (Rutherford, 1994). Where density stratification exists in rivers or estuaries, less dense water flows on top of the more dense water. Vertical mixing or dispersion is a function of the magnitude of this density difference or density gradient,

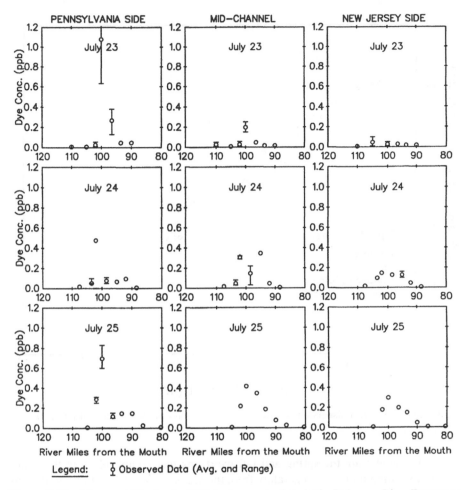

Figure 3-11 Measured Rhodamine WT dye concentrations in the Delaware River Estuary to determine lateral dispersion coefficient.

as shown in Figure 3-12. These empirical data show that where there is a small density gradient over depth (10^{-7}/m) there is significant vertical mixing, and where there is a larger density gradient (10^{-2}/m) vertical dispersion coefficients are reduced. Figure 3-12 can be used to provide a first estimate of the vertical dispersion coefficient in a stratified water system.

In many one-dimensional lake models, it is necessary to evaluate the vertical diffusion coefficient in the water column. In moderate climates, water columns with appreciable depths tend to stratify during the summer months with a distinctive thermocline inhibiting vertical mixing. Vertical diffusion coefficients are therefore commonly derived using temperature data. Figure 3-13 shows the temperature distribution in White Lake, Michigan, from May 1973 to June 1974. Strong thermal stratification in the summer months is clearly shown in the temperature distribution

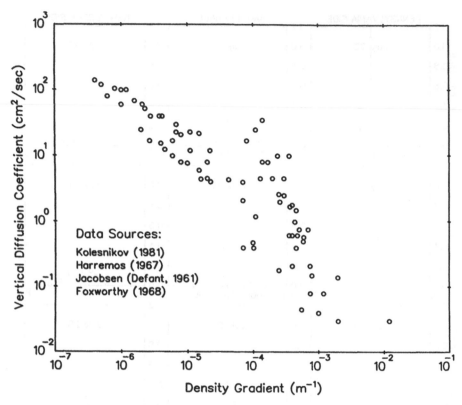

Figure 3-12 Vertical dispersion coefficient as a function of density gradient (from Hydro-Qual, 1982).

plot. Also shown are the spring and fall overturns that completely mix the water column, yielding uniform temperature from the water surface to the bottom. It is interesting that reverse stratification forms during winter months due to the fact the density of water is highest at 4 °C.

One widely used method to quantify vertical diffusion coefficients using the temperature data is the flux-gradient method first proposed by Jassby and Powell (1975). It is generally recognized that the thermal structure in lakes results from the interaction of solar heating and wind stress at the surface (Lerman and Stiller, 1969). Under the assumption of horizontal homogeneity, the equation describing the vertical transport of heat in a water column is:

$$\frac{\partial T}{\partial t} = \frac{\partial}{\partial z}\left(K\frac{\partial T}{\partial z}\right) - S_T \tag{3-6}$$

where T is temperature and S_T is a heat source or sink in the water column. One example of a heat source is solar radiation. The first term on the right-hand side of Eq. 3-6 represents the diffusional transport of heat in the vertical direction. The eddy dif-

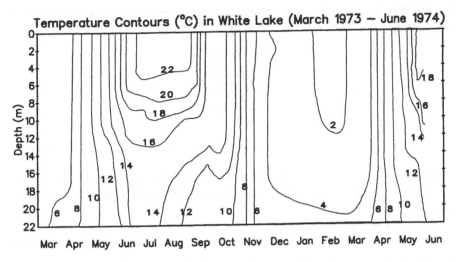

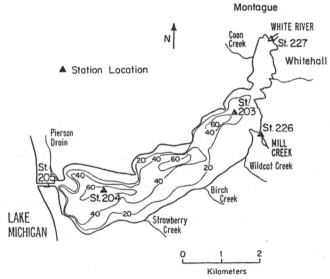

Figure 3-13 Temperature measurements in White Lake, Michigan (March 1973–June 1974).

fusion coefficient, K, is therefore a function of the thermal as well as the current structure in the lake. The heat source term, S_T, can be omitted in the water column except at the surface because light extinction usually limits penetration of solar radiation into deep water. Integration of Eq. 3-6 with respect to z from the bottom of the lake to depth z yields:

$$K \frac{\partial T}{\partial z} = \int_{-D}^{z} \frac{\partial T}{\partial t} \, dz$$

The heat flux into the sediments and the radiation absorbed by the sediments at $z = -D$ is not included. Thus K is calculated by:

$$K = \frac{\int_{-D}^{z} \left(\frac{\partial T}{\partial t} \right) dz}{\left(\frac{\partial T}{\partial z} \right)} \qquad (3\text{-}7)$$

where $\int_{-D}^{z} \left(\frac{\partial T}{\partial t} \right) dz$ represents the accumulated rate of change of stored heat between

z and the bottom of the lake and $\partial T/\partial z$ is the temperature gradient at depth z. Note that Eq. 3-7 does not apply at the surface because the solar radiation at $z = 0$ is not included.

The use of the data from White Lake, Michigan, is one example of applying this method to evaluate the vertical eddy diffusion coefficient. In the numerical calculation, the temperature gradient is calculated from the temperature profiles of White Lake using central-difference approximation at each meter. At the same time, the accumulated rate of change of stored heat from the bottom to each meter of depth is calculated. The eddy diffusion coefficient is then obtained by dividing the values of

$\sum_{z=-D}^{z} (\Delta T / \Delta t)$ by the corresponding temperature gradients. Table 3-1 shows a tab-

ulation of the results explaining the computation procedure with data collected in White Lake, Michigan. The vertical temperature profiles used to generate the temperature distribution in Figure 3-13 are shown in Figure 3-14 and they are used for Table 3-1 calculations. Results in Table 3-1 indicate that accurate temperature readings are essential to the successful application of the flux-gradient method. Sometimes negative values of K, which do not have physical meaning, in this case, can occur in the computation due to errors in the temperature gradient. Lerman and Stiller (1969) have estimated the errors in K due to random errors in temperature readings. Results from the flux-gradient method for evaluating K are shown in Figure 3-15. Temperature profiles are plotted together to give a complete picture of the seasonal variation of the thermal structure in White Lake. During the spring months, the lake remains nearly homothermal because wind action is able to mix the surface heat into great depths, yielding high eddy diffusion coefficient values throughout the water column. As heating continues, a thermocline forms and the surface temperature begins to increase rapidly because the upper layers are being heated at a more rapid rate than the deeper layers (whose temperature increases only slightly). The eddy diffusion coefficient reduces drastically in the deeper water because motion is limited there. Although it is not shown in Figure 3-14, the lake begins to lose heat after the surface temperature attains its maximum value. The initiation of free convection (with the gradual increase in eddy diffusivity) can be expected in the upper layers. In late fall, a large part of the lake is dominated by free convection. The lake is again homothermal and the eddy diffusivity varies little with depth.

The flux-gradient method yields a diffusion coefficient value of infinity, ∞, when $\partial T/\partial z$ approaches zero (see Eq. 3-7). For this situation, another method developed by

TABLE 3.1 Flux-Gradient Method to Determine Vertical Diffusion Coefficient

Depth (m)	T_1 (°C)	T_2 (°C)	$(T_2 - T_1)/\Delta t$ (10^{-6} °C/s)	$\Sigma[(T_2 - T_1)/\Delta t]\Delta z$ (10^{-6} °C m/s)	$\Delta T_1/\Delta z$ (°C/m)	$\Delta T_s/\Delta z$ (°C/m)	K_1 (cm²/s)	K_2 (cm²/s)
15	11.0	12.1	0.796	0.796	2.290	0.225	0.0035	0.0354
14	12.1	12.3	0.145	0.940	0.275	0.142	0.0034	0.0664
13	12.1	12.4	0.217	1.16	-0.0917	0.167	—	0.0694
12	12.1	12.7	0.434	1.59	-0.0083	0.483	—	0.0329
11	12.1	13.3	0.868	2.46	0.042	0.475	0.590	0.0518
10	12.2	13.7	1.09	3.54	0.158	0.625	0.224	0.0567
9	12.4	14.7	1.66	5.21	0.208	1.510	0.250	0.0345
8	12.6	16.4	2.75	7.96	0.200	1.040	0.398	0.0764
7	12.8	16.8	2.89	10.9	0.208	0.792	0.521	0.137
6	13.0	18.0	3.62	14.5	0.158	0.650	0.914	0.223
5	13.1	18.0	3.54	18.0	0.033	0.017	5.40	10.8
4	13.1	18.2	3.69	21.7	0.042	0.117	5.21	1.86
3	13.2	18.2	3.62	25.3	0.083	-0.0167	3.04	—
2	13.3	18.2	3.54	28.9	0.267	0.000	1.08	2.89×10^9
1	13.7	18.2	3.26	32.1	0.450	0.763×10^{-5}	0.714	4.21×10^5

T_1 and T_2 are measured temperatures on May 21, 1974 and June 6, 1974, respectively.
K_1 and K_2 are calculated vertical diffusion coefficients on these two dates.

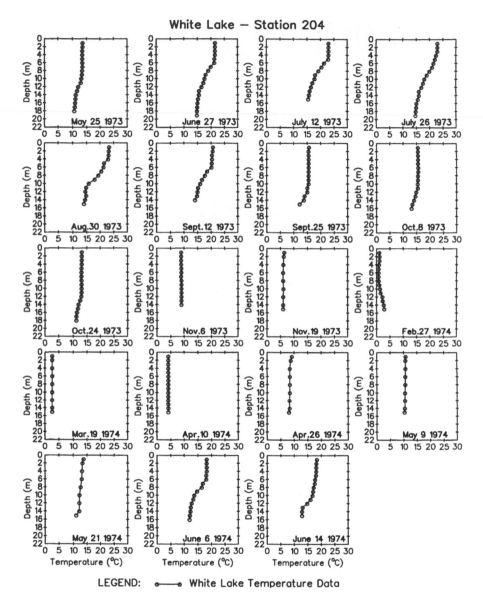

White Lake — Station 204

LEGEND: •——• White Lake Temperature Data

Figure 3-14 Vertical temperature distributions at Station 204, White Lake.

Sundaram et al. (1969) can be used. The vertical eddy diffusion coefficient can be expressed as a function of the stability effects in terms of the Richardson number:

$$K = K_o(1 + \sigma_1 R_i)^{-1} \tag{3-8}$$

where σ_1 is an empirical constant and R_i is the Richardson number, which is evaluated as follows:

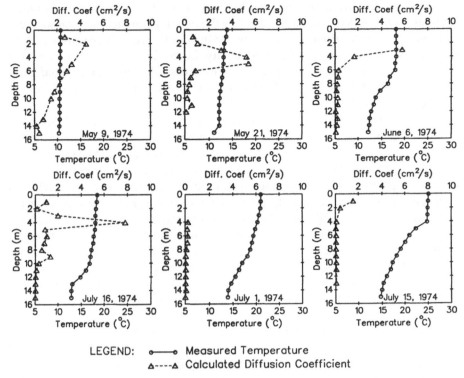

Figure 3-15   Derived vertical diffusion coefficients in White Lake using temperature data.

$$R_i = -\alpha_v g z^2 \frac{\left(\dfrac{\partial T}{\partial z}\right)}{W_w^2} \qquad (3-9)$$

where α_v is the coefficient of volumetric thermal expansion of water and W_w is the shear velocity. Advantages of this method are that it is easy to use and the computation procedure is straightforward.

3.5   TWO-DIMENSIONAL VERTICALLY INTEGRATED MASS TRANSPORT

For this type of mass transport, the hydrodynamics are not coupled to the water quality parameters. In particular, it is assumed that the salinity and temperature distributions do not affect the vertically averaged flow field. In order to quantify the horizontal advection and dispersion in this two-dimensional mass transport, we use a hydrodynamic model to generate the circulation pattern and a mass transport model to calibrate horizontal dispersion in the system. Since the circulation is not coupled with any density-driven currents, the hydrodynamic model calculation is rather straightforward. A good example of deriving the two-dimensional mass transport coefficients is presented from Lake Okeechobee in Florida. Figure 3-16 shows the horizontal distributions of specific conductivity in the lake used to calibrate the horizontal dispersion coefficient with a two-dimensional finite-segment configuration like the WASP (see Chapter 7) model. Time-variable model results of specific conductivity from WASP are compared with the measured data.

Two-dimensional mass transport models are generally used for calibrating the horizontal dispersion coefficients using a conservative substance. As shown in the following example, dye concentrations measured in a hydraulic model of the Western Delta and Suisan Bay in California (Figure 3-17) are used to determine the dispersion coefficient values. In this hydraulic model test, dye was released instantaneously in two separate locations, Grizzly Bay and Honker Bay, after the model had reached a steady state. The temporal variation of dye was measured at a number of locations at the high slack water for many tidal cycles after the releases. Therefore, a time-variable mass transport model based on the WASP modeling framework was used to simulate the dynamic behavior of dye dispersion using constant advective flows and dispersion coefficients. While the order of the magnitude of the dispersion coefficient is known for this area, the specific value was assigned on the basis of fitting both the dye and salinity data. Figure 3-17 shows the results for a net Delta outflow of 10,000 cfs. The model data are the tidally and vertically averaged values. The agreement between the calculated salinity and the observed data indicates the proper assignment of the transport coefficients in the main channel. The calculations also reproduce the dye concentrations, demonstrating that adequate lateral dispersion characteristics have been utilized between the main channel and shallow bays. Deviation between calculation and data in dye in the early tidal cycles is attributed to the relative coarseness with respect to the small time scale of the dynamic behavior (i.e., in order to simulate rapid mixing, small segments may be needed). Nevertheless, the calculation appears to reproduce the hydraulic model data correctly and the mass transport coefficients are considered properly calibrated.

3.6 TWO-DIMENSIONAL LONGITUDINAL/VERTICAL MASS TRANSPORT

In one-dimensional estuarine mass transport, a longitudinal dispersion coefficient is used to account for tidal diffusion, mixing due to velocity gradients in the vertical direction, and to compensate for other spatial and temporal averaging of tidal currents in the model configuration. In other situations, a two-dimensional analysis is necessary to characterize the mixing and mass transport in the water column, giving a better quantification of the vertical stratification of water quality constituents. The two-dimensional mass transport coefficients must be quantified in a more rigorous manner than the one-dimensional longitudinal dispersion coefficient, resulting in a greater physical insight into the water column mixing process. The advective flow in

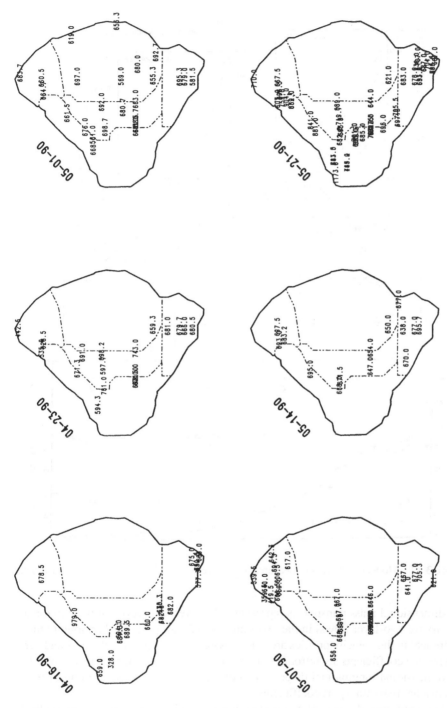

Figure 3-16 Two-dimensional distribution of specific conductivity in Lake Okeechobee, Florida, used to derive horizontal dispersion coefficients.

45

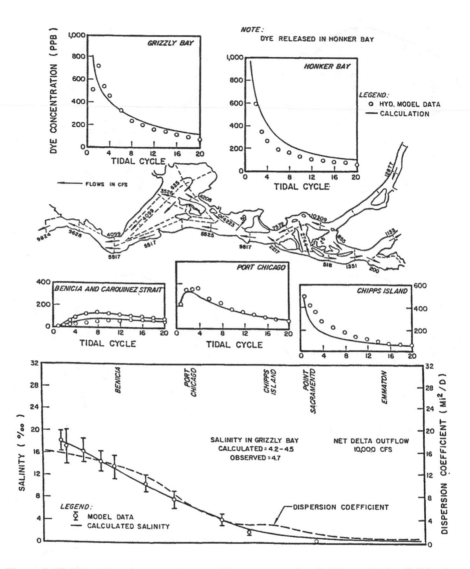

Figure 3-17 Model results versus measured dye concentrations in Western Delta, California.

one-dimensional mass transport is represented by a vertical profile of the horizontal velocity (or two-layer flows) in the two-dimensional transport. Further, longitudinal differences in the velocity profiles are compensated by the vertical velocities and vertical mixing coefficients are introduced to account for the vertical exchange. Thus the one-dimensional longitudinal dispersion coefficient is replaced by a group of two-dimensional mass transport coefficients.

In partially mixed estuaries, the driving forces for circulation are the longitudinal surface slope, acting in a downestuary direction, and the longitudinal salinity gradient, acting in an upestuary direction. These two forces are balanced by the internal and bottom frictional forces (assuming the wind stress at the surface is minimal). The resulting circulation pattern is characterized by a net landward flow in the bottom waters and a net seaward flow in the surface.

The observation of Tully (1949) in Alberni Inlet, British Columbia, featuring this two-layer flow with entrainment, represents some of the early work on mass transport in a stratified estuary. Further insight on partially mixed estuaries was due mainly to a series of investigations by Pritchard (1952, 1956, 1958), who measured tidally averaged velocity and salinity profiles in the James Estuary. Following Pritchard's analysis, Rattray and Hansen (1962) developed solutions for this type of circulation and mixing using similar techniques. In their solution, the hydrodynamic and mass transport equations are decoupled so that each may be solved separately and directly. Such direct solutions have been given by Officer (1976, 1977) and have been extended to include bottom frictional effects. In another study, Prichard (1969) presented a similar approach using a box model configuration to simplify the analysis. Officer (1980) later refined this box model approach. Similar analytical solutions have also been presented by Fisher et al. (1972) with satisfactory applications. All equations in these models are averaged over one or more tidal cycles with a correlation term for the tidal velocity. Later, a more rigorous methodology was developed by Lung and O'Connor (1984) and Lung (1986), following the path laid out by previous investigators. Through a number of successful applications, this practical engineering method is shown to be a very useful tool in estuarine wasteload allocations.

This approach to the analysis is based on an assumption that the two-dimensional (longitudinal-vertical) distribution of salinity in an estuary is given and utilized to determine the vertical profiles of velocity along the estuary under a tidally averaged condition. The density-driven circulation can be decoupled from the salt balance equation, thereby significantly simplifying the mass transport analysis. Under steady-state tidally averaged conditions, the longitudinal momentum equation for a laterally homogeneous estuary is:

$$U_o \frac{\partial U_o}{\partial x} + \frac{1}{\rho} \frac{\partial p}{\partial x} = N \frac{\partial^2 u}{\partial y^2} \tag{3-10}$$

The vertical component of the momentum equation is simply the hydrostatic pressure equation:

$$\frac{1}{\rho} \frac{\partial p}{\partial y} = g \tag{3-11}$$

The equation of state is approximated as:

$$\rho = \rho_f(1 + \alpha C) \tag{3-12}$$

The equation of continuity can be expressed as:

$$\frac{\partial(bu)}{\partial x} + \frac{\partial(bv)}{\partial y} = 0 \qquad (3\text{-}13)$$

The coordinate system is shown in Figure 3-18, in which the longitudinal x-axis is positive toward the ocean and the vertical y-axis is positive toward the bed of the channel. In these equations:

U_o = amplitude of tidal current
ρ = density of the saline water
p = pressure
N = vertical eddy viscosity (assumed constant with depth)
u = laterally averaged horizontal velocity
g = gravitational acceleration
ρ_f = density of freshwater
α = 0.000757 parts per thousand^{-1}
C = salinity (parts per thousand)
b = depth-averaged width
v = laterally averaged vertical velocity

Equation 3-10 may then be solved for the horizontal velocity, u, subject to the following boundary conditions: at the free surface ($y = -\eta$), $\partial u/\partial y = 0$, that is, there is no wind effect; at the bottom ($y = -h$), $-N(\partial u/\partial y) = k \, |u_h|u_h$, in which k is a dimensionless friction coefficient, and u_h is the velocity at the bed. The analytical solution to Eqs. 3-10 to 3-13 is presented by Lung and O'Connor (1984) as follows:

$$u = \frac{g\alpha}{N}\frac{dC_s}{dx}\left(\frac{y^3}{6} + \frac{ay^4}{24} - \frac{h^3}{24} - \frac{ah^4}{120}\right) + \frac{g}{N}\left(\frac{d\eta}{dx} + \frac{U_o}{g}\frac{\partial U_o}{\partial x}\right)\left(\frac{y^2}{2} - \frac{h^2}{6}\right) + \frac{Q}{bh} \quad (3\text{-}14)$$

The water surface gradient, $d\eta/dx$ in Eq. 3-14, must be evaluated prior to calculating u. Lung (1993) presented the derivation of the water surface gradient.

To use Eq. 3-14, one needs to assign salinity gradients and the associated freshwater flow as well as channel characteristics at selected locations throughout the saline zone. Next, the surface salinity, C_s, and the longitudinal salinity gradient, dC_s/dx, are derived from tidally averaged salinity data. These parameters and a first estimate of eddy viscosity, N, are substituted into Eq. 3-14 to obtain the vertical profile of horizontal velocity at each location. Within the reported values for estuaries, the eddy viscosity is adjusted such that the calculated vertical profile of horizontal velocity agrees with the data at a particular location. The above analysis, including Eq. 3-14, can be implemented in a spreadsheet. Table 3-2 shows the results of applying the spreadsheet calculations to the Sacramento–San Joaquin Delta in California under a Delta outflow of 4,400 cfs. The key input is the eddy viscosity, which in this case is assigned a value of 9.5 cm^2/s (= 0.00095 m^2/s). Parameters that are associated with this Delta outflow are:

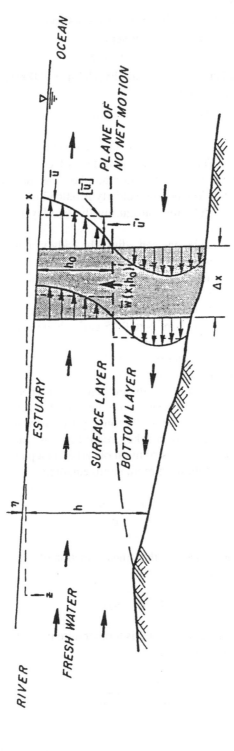

Figure 3-18 Two-dimensional estuarine circulation pattern and coordinate system.

49

1. Freshwater flow rate, $Q = 124.65$ m^3/s (= 4,400 cfs)
2. Average channel depth, $h = 8.84$ meters
3. Average channel width, $b = 2,916.79$ meters
4. Longitudinal gradient of surface salinity, $dC_s/dx = 0.000413$ ppt/m
5. Surface salinity, $C_s = 24.4$ ppt
6. Vertical salinity gradient constant, $a = 0.014043$

The calculated velocity profile is shown as the longitudinal velocity at every one-tenth of the average depth. The velocity at the water surface is 0.0876 m/s, which decreases progressively in the vertical direction and becomes negative, that is, in the landward direction, at 0.6 h. The bottom velocity is 0.0632 m/s. (Table 3-2 also displays additional analysis results for configuring the two-layer mass transport, discussed later in this section.)

If velocity data are available, one can adjust the eddy viscosity, N, to match the calculated velocity profile with the measured velocity data. Figure 3-19 shows the calculated velocity profiles matching the measured velocities at a number of locations along the tidal flume at the U.S. Army Engineers' Waterways Experiment Station in Vicksburg, MS. The calibrated eddy viscosity values that best reproduce the velocity data are shown, ranging from 0.3 cm^2/s to 0.65 cm^2/s. Note that eddy viscosity values increase in the downstream direction.

This two-dimensional mass transport can also be applied to freshwater systems such as the Genesee River in New York, discharging into Lake Ontario. During the summer months, river temperature levels are generally higher than the lake water temperature. Thus denser water associated with the lower temperature from Lake Ontario enters into the Genesee River in the bottom layer of the water column while the river water is being discharged into the lake, creating a two-dimensional temperature distribution similar to the salinity distribution observed in a partially mixed estuary. The analysis starts with the following momentum equation:

$$N \frac{\partial^2 u}{\partial z} = \frac{1}{\rho} \frac{\partial p}{\partial x} \qquad (3\text{-}15)$$

The water density is approximated with a linear function of temperature without the salt effect in freshwater systems:

$$\rho = a_1 - a_2 T \qquad (3\text{-}16)$$

where ρ is water density in gm/cm^3 and T is temperature in °C. Evaluating the derivative of Eq. 3-16 yields the following:

$$\frac{\partial \rho}{\partial x} = -a_2 \frac{\partial T}{\partial x} \qquad (3\text{-}17)$$

TABLE 3-2   Lung and O'Conner's Method for Computing Vertical Tidally Averaged Velocity Distributions (Sausolito Bay Example)

INPUT

Constants	Value
$g(m^2/s)$	9.8
alpha,α	0.000757
k	0.0025

Parameters	
Vert. sal. grad. const., a (1/m)	0.014043
Freshwater flow, Q (m³/s)	124.68
Surface sal. grad., dC_s/dx (ppt/m)	0.000413
Average depth, h (m)	8.84
Average width, b (m)	2916.79
Surface salinity, C_s (ppt)	24.4
Eddy viscosity, N (m²/s)	0.00095

OUTPUT

CALCULATED VELOCITY PROFILE

Dimensionless Depth	Depth (m)	Velocity (m/s)
0.0	0.0	0.0876
0.1	0.884	0.0827
0.2	1.768	0.0693
0.3	2.652	0.0497
0.4	3.536	0.0263
0.5	4.420	0.0013
0.6	5.304	−0.0229
0.7	6.188	−0.0438
0.8	7.072	−0.0592
0.9	7.956	−0.0664
1.0	8.840	−0.0632

TWO-LAYER MASS TRANSPORT COEFFICIENTS

Depth of zero velocity	0.505
Surface layer flow (m³/s)	702.33
Average surface velocity (m/s)	0.0539
Bottom layer flow (m³/s)	577.70
Average bottom velocity (m/s)	-0.045
Density gradient (1/s²)	0.00254
Velocity gradient (1/s²)	0.0006
Richardson number	4.273
Vertical diffusion coef. (cm²/s)	1.802

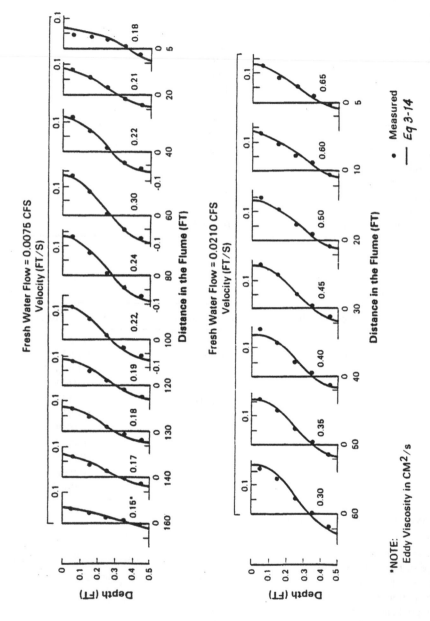

Figure 3-19 Calculated versus measured tidally averaged velocity in flume at Waterways Experiment Station, Vicksburg, Mississippi.

52

Assume:

$$\frac{\partial T}{\partial x} = \left(\frac{\partial T}{\partial x}\right)_s (1 + a_3 z) \qquad (3\text{-}18)$$

where $(\partial T/\partial x)_s$ is the longitudinal gradient of temperature at the water surface. Equation 3-18 states that the temperature gradient varies with depth in a linear fashion when the constant a_3 is derived from the temperature data. Equation 3-15 is solved with the following boundary conditions: $\partial u/\partial z = 0$ at $z = -\eta$ (water surface) and $u = 0$ at $z = h$ (river bottom). The solution to Eq. 3-15 is:

$$u = -\phi_1 \frac{5z^4 + h^4 - 6z^2 h^2}{60} - \phi_2 \frac{8z^3 + h^3 - 9z^2 h}{48} - \frac{3Q}{2bh^3}(z^2 - h^2) \quad (3\text{-}19)$$

where

$$\phi_1 = \frac{g a_2 a_3}{N a_1}\left(\frac{\partial T}{\partial x}\right)_s \quad \text{and} \quad \phi_2 = \phi_1/a_3$$

Using the temperature data at Station 14 (near the river mouth and Lake Ontario in Figure 3-20) collected in the Genesee River in August 1975 and the following attributes of $h = 6.7$ meters and $b = 78$ meters, one can derive $(\partial T/\partial x) = -6.32 \times 10^{-5}$ °C/m at the water surface and $(\partial T/\partial x) = -8.34 \times 10^{-4}$ °C/m at the bottom. As a result, $a_3 = 2.023$ m^{-1}. A velocity profile is therefore calculated using Eq. 3-19 as follows: 17.3 cm/s at $z = 0$; 15.9 cm/s at $z = 1$ meter; 11.9 cm/s at $z = 2$ meters; 6.2 cm/s at $z = 3$ meters; 3 cm/s at $z = 4$ meters; -3.9 cm/s at $z = 5$ meters; and -4 m/s at $z = 6$ meters. The negative velocities indicate the intrusion of colder water from Lake Ontario. This procedure can be repeated for other stations along the Genesee River to develop the velocity pattern, which is then used to derive the two-layer mass transport for the water quality model of the Genesee River.

A special case of the two-dimensional analysis involves configuring the water into two layers. The point of no net motion at each station where the longitudinal velocity is zero, that is, where $u = 0$, is used as the dividing point for the two layers. In the Sacramento–San Joaquin Delta example (Table 3-2), the depth of zero velocity at that location is $0.505 h$ (or $z = 4.467$ meters). The plane of no net motion is defined for the entire saline zone to form a two-layer system. The surface layer flow is 702.33 m³/s with an average velocity of 0.0539 m/s. In the bottom layer, the flow is 575.73 m³/s in a landward direction. The average bottom layer velocity is 0.045 m/s. Note that the difference between the two layer-averaged flows is approximately equal to the net Delta outflow of 124.68 m³/s. In the spreadsheet calculation, the vertical diffusion coefficient for mass transport, ε, is determined from the density gradient and velocity gradient in the vertical direction (Lung and O'Connor, 1984) to be 1.802 cm²/s. The average longitudinal velocity in each layer is then determined from the velocity profiles. Each layer of the estuary is further divided into longitudinal segments. Repeat-

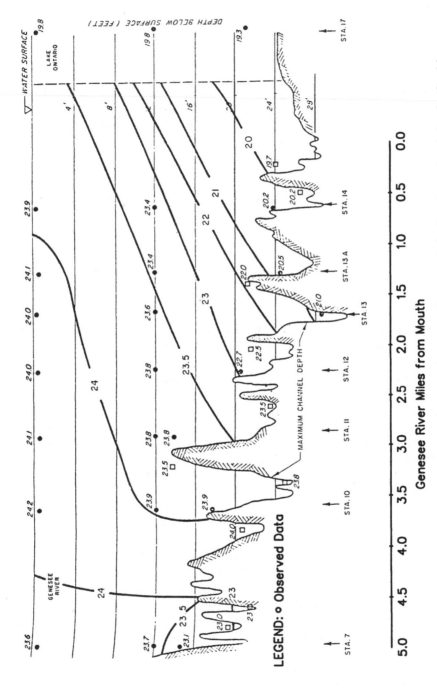

Figure 3-20 Two-dimensional temperature distribution in the Genesee River near Lake Ontario used to derive density-driven mass transport.

ing this spreadsheet calculation for a number of locations along the estuary results in a set of mass transport coefficients, which in turn are incorporated into a two-layer box model to calculate the salinity distribution.

The two-layer mass transport model is applied to the Sacramento–San Joaquin Delta and is displayed in Figure 3-21 for two Delta outflows: 4,400 cfs and 10,000 cfs. By using the measured salinity distribution and assigning values for the vertical eddy viscosity, the horizontal velocities are calculated and the plane of no net motion is determined. Subsequently, the layer-averaged horizontal velocity is obtained as shown in Figure 3-21. The vertical dispersion coefficient is calculated as plotted together with the eddy viscosity. The solid line is used to filter out the local irregularity. Finally, this two-layer mass transport pattern generates the average salinity in both layers for comparison with the data. The longitudinal dispersion due to tidal effect is no longer necessary as it is replaced by the derived two-layer mass transport that produces a salinity distribution matching the observed values. Figure 3-21 shows that the null zone moves upstream under the lower Delta outflow of 4,400 cfs because of stronger salinity intrusion. On the other hand, the higher Delta outflow of 10,000 cfs pushes the salt contents further downstream with a more pronounced vertical stratification of salinity. Note that available velocity data in Figure 3-21 provide additional checks for the velocity calculation by adjusting the eddy viscosity, N. The calibrated two-layer mass transport is subsequently used to model the total suspended solids concentrations in the Sacramento–San Joaquin Delta as described later in this section.

Observe that the vertical salinity gradient, horizontal salinity gradient, and the surface salinity concentration at mean tide conditions are assigned at each location along the estuary for this analysis. If salinity measurements are taken at slack tide, then the mean tide salinity is established by spatially translating the observations over half a tidal excursion. On the other hand, if salinity data are available over the entire tidal period, then the average of all salinity values is taken as being representative of mean tide conditions.

Ideally, the calculated velocities should be compared with observed net (residue) velocity measurements when available. If data are not available, the analysis must be repeated by adjusting the eddy viscosity to yield different velocities that eventually lead to a best match of observed salinity distribution in the estuary, as demonstrated in the modeling analysis of the Hudson River Estuary, New York. Results of the two-layer mass transport modeling analysis are shown in Figure 3-22. The two-layer model was applied to four freshwater flows at Bear Mountain, ranging from 3,200 cfs to 18,400 cfs. As expected, higher freshwater flows push the salinity profiles further downstream and result in greater salinity stratification throughout the entire estuary. Lower freshwater flows result in salinity intrusion further inland and yield less significant vertical stratification of salinity. It is important to recognize that the salinity differences between the surface and bottom layers increase with increasing flow. This phenomenon is accounted for in the net circulation pattern by decreased vertical dispersion with greater freshwater flows.

The plane of no net motion at a specific location is defined through the velocity calculations. The plane of no net motion is usually at a depth of approximately 50% of the total depth regardless of the freshwater flow or the assigned eddy viscosity without significant variation. Therefore, due to the iterative nature of the analysis, it

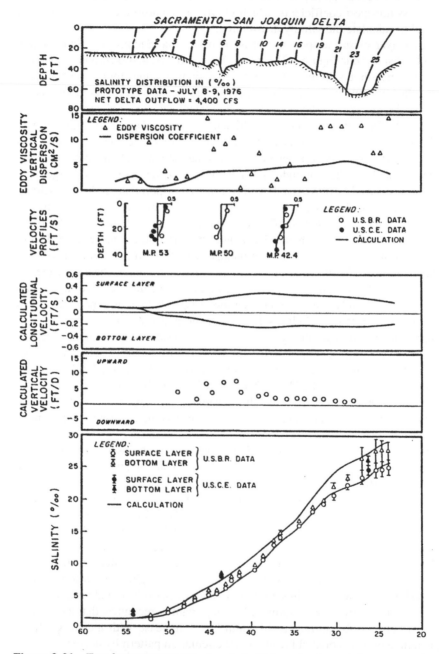

Figure 3-21 Two-layer mass transport analysis of the main channel, Sacramento–San Joaquin Delta at outflows of 4,400 cfs and 10,000 cfs. *(continues)*

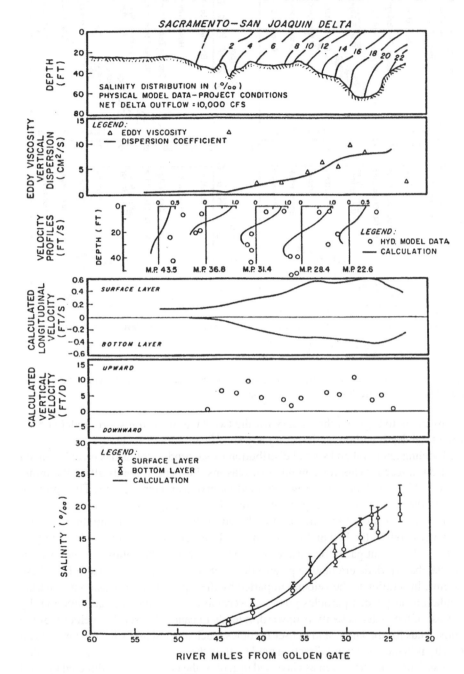

Figure 3-21   (*Continued*)

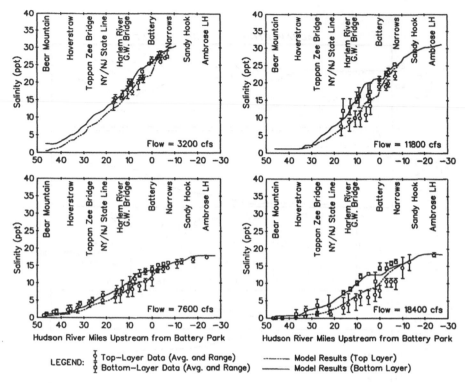

Figure 3-22   Two-layer mass transport analysis of the Hudson River Estuary, New York, under four different freshwater flow conditions.

is convenient to segment the estuary for the two-layer mass transport model at 50% of the total depth without anticipating any significant loss in accuracy.

Modeling temporal and spatial distribution of suspended solids has received much attention in recent years for a number of reasons. Suspended solids affect the transmission of light and thus, the growth of phytoplankton and other plants. They provide sites for the growth of microorganisms, which have an impact on water quality. They absorb heavy metals and pesticides and thereby influence the concentration of these substances both in the bed and in suspension. In estuaries, suspended solids are particularly significant partly because of the cohesiveness of the solids in saline waters and the characteristic circulation pattern that increases the retention of solids in these systems. In addition to the velocity imparted by the typical two-dimensional estuarine circulation, suspended particles possess a vertical flux of solids in that direction; by contrast, if the water velocity is upward, it tends to cancel the settling velocity of the solids. Thus for the larger and denser particles, such as sand and silts, the net vertical velocity is in the downward direction but less than the settling velocity. For the smaller and less dense particles, such as clays and organics, the net velocity is directed upward when the water velocity is greater than the settling velocity. Such vertical movement, in conjunction with the convergence of the landward-flowing density current and the seaward-flowing surface river current at the tail of salinity intrusion, is responsible for the solids' concentrations in the saline zone of the estuary. These solids' concentrations are greater than those of the upstream freshwater inflow and downstream density current. This phenomenon is generally referred to as turbidity maximum.

The two-layer mass transport modeling analysis of the Sacramento–San Joaquin Delta was originally developed for modeling the turbidity maxima in that system. The two-layer mass transport is incorporated with the settling velocity of suspended solids to model suspended solids concentrations in the Sacramento–San Joaquin Delta. The average settling velocity used in the analysis is 8 ft/day on a tidally averaged basis. The model reproduces the turbidity maxima reasonably well (Figure 3-23) under a Delta outflow of 4,400 cfs. The location of the turbidity maxima is very close to the location of null zone as seen in the salinity results in Figure 3-23.

Another successful application of this two-layer analysis is the mass transport and suspended solids modeling of the Norwalk Harbor in Connecticut. Figure 3-24 shows the calculated two-layer salinity matching the measured data under a tidally averaged condition. This mass transport analysis was part of the water quality modeling study to analyze the dissolved oxygen suppression in the bottom waters of Norwalk Harbor.

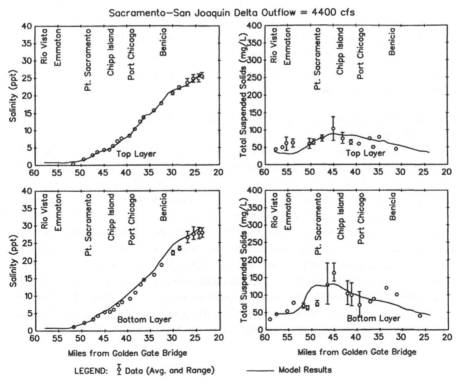

Figure 3-23 Two-layer modeling of salinity and total suspended solids in Sacramento–San Joaquin Delta.

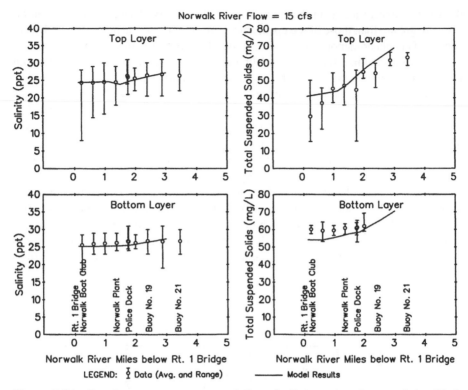

Figure 3-24 Two-layer mass transport modeling of salinity and total suspended solids in Norwalk Harbor, Connecticut.

The two-layer mass transport proved to be the key mechanism responsible for the anoxic condition in the Norwalk Harbor.

The two-layer mass transport analysis is expanded in the application to the Patuxent Estuary in Maryland, a major focal point in water quality management for well over a decade. Over 20 slack water surveys of the estuary provide salinity distributions used to generate a series of two-layer mass transport patterns. They are then incorporated into a 38-segment mass transport model to simulate the temporal and spatial distributions of salinity throughout the year. These preliminary mass transport patterns (i.e., advective flows and vertical dispersion coefficients) are subsequently adjusted to match the salinity distributions. Figure 3-25 shows the successful calibration of the mass transport for the Patuxent Estuary using the data from 1995. These results illustrate that the model reproduces seasonal variations of salinity (primarily due to freshwater flows). In addition, the pronounced vertical salinity gradients during the spring high flow months are mimicked by the model results.

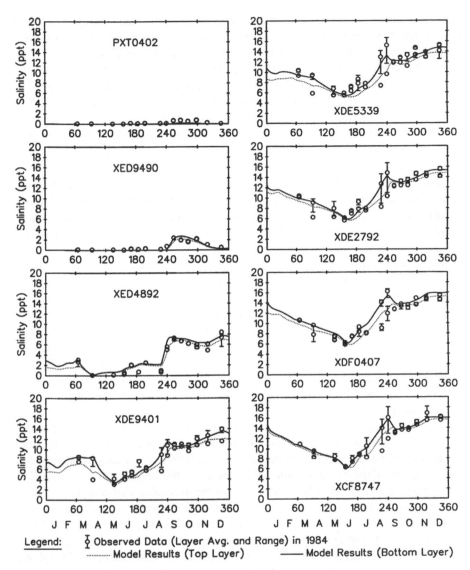

Figure 3-25 Time-variable two-layer mass transport modeling of the Patuxent Estuary, Maryland.

3.7 NEED A HYDRODYNAMIC MODEL?

The preceding analysis provides a technical basis for evaluating the net circulation in a number of estuaries, ranging from seasonal steady-state to tidally averaged time-variable computations. As was discussed, the analysis requires knowledge of the mean tide salinity distribution. Yet, because salinity distributions at future conditions are not available, an approximation is used to develop the mass transport for other freshwater flow conditions. For example, two-layer mass transport patterns have been developed for the Hudson River Estuary under freshwater flows at Bear Mountain of 3,200 cfs, 7,600 cfs, 11,800 cfs, and 18,400 cfs. These transport patterns include horizontal flow, vertical flow, horizontal dispersion, and vertical dispersion, and they provide circulation patterns that can be used in the water quality modeling analysis. If freshwater flow at Bear Mountain is similar to any of the flows for which transport is evaluated and verified, then the exact circulation patterns developed earlier can be used in the projection model. Typically, the 7-day 10-year low flow is chosen for wasteload allocations. Since the 7-day 10-year low flow in the Hudson River at Bear Mountain is 3,280 cfs, the mass transport pattern developed at 3,200 cfs can be directly substituted into a projection model. If the freshwater flow in the Hudson is different than this flow, then an interpolation scheme is needed to develop the tidally averaged mass transport. In doing this, the entire hydrodynamic calculation can be eliminated. Since the vertical dispersion changes very little from one mass transport pattern to the next, the vertical dispersion coefficients at the closest freshwater flow can be used in the projection model.

While the above procedure still lacks full projection capabilities, the methodology serves as a useful tool in calibrating the mass transport. This mass transport can be used readily to calibrate the water quality model when a hydrodynamic model is being developed concurrently. This is a timesaving step in practice, that is, to achieve the calibration of the water quality model without waiting for the hydrodynamic model. The Patuxent Estuary water quality model was first calibrated using the two-layer mass transport prior to obtaining the results from a two-dimensional hydrodynamic model (Lung, 1986).

A full hydrodynamic model is commonly used in water quality modeling studies these days, as computation limitations are becoming less of a problem with the rapid advances being made in computer hardware and software. In estuarine modeling, spatial averaging of the hydrodynamic model results to minimize the computational effort is no longer necessary when using some of the most advanced computer systems. In other words, the same spatial grid may be used for both the hydrodynamic and water quality models. In a recent modeling study of the Patuxent Estuary in Maryland, the Army Corps of Engineers Waterways Experiment Station's CE-QUAL-W2 model (Cole and Buchak, 1995) was used. While the presentation of the CE-QUAL-W2 model can be found in the model documentation (Wells and Cole, 2000), some salient features are briefly described in the following paragraphs.

CE-QUAL-W2 is a two-dimensional hydrodynamic and water quality code supported by the Army Corps of Engineers' Waterways Experiment Station (Cole and Buchak, 1995). This model has been widely applied to stratified surface water systems like lakes, reservoirs, and estuaries, and it computes water levels, horizontal and vertical velocities, and temperature, as well as 21 other water quality parameters (Wells and Cole, 2000). A predominant feature of the model is its ability to compute the two-dimensional velocity field for narrow systems that stratify. In contrast to many reservoir models that are zero-dimensional hydrodynamic models, an understanding of the mass transport can be as important as the water column kinetics in predicting water quality changes.

One limitation of CE-QUAL-W2 is its inability to model sloping river stretches. In the development of the modeling framework, vertical accelerations were considered negligible compared to gravity forces. This assumption leads to the approximation of hydrostatic pressure for the vertical momentum equation (Wells and Cole, 2000). Yet in sloping channels, this assumption is not always valid because the vertical accelerations cannot be neglected if the longitudinal and vertical axes are aligned with an elevation datum and gravity, respectively. To address this issue, models such as WQRSS (Smith, 1978), HEC-5Q, and HSPF (Donigian et al., 1984) have been developed for river basin modeling, but they too have serious limitations. One issue is that the HEC-5C (similar to WQRSS) and HSPF models incorporate a one-dimensional longitudinal river model with a one-dimensional vertical reservoir model (only one-dimensional in temperature and water quality and zero-dimensional in hydrodynamics). The modeler must choose the location of the transition from one-dimensional longitudinal to one-dimensional vertical. Besides the limitation of not solving for the velocity field in the stratified, reservoir system, any point source inputs to the reservoir section are spread over the entire longitudinal distribution of the reservoir cell.

Despite the limitations of the CE-QUAL-W2 model, its capabilities were sufficient for application to the Patuxent Estuary. The water column is divided into 42 longitudinal segments and each layer is 2 m thick. The total number of segments is 225 (Figure 3-26). Due to the sharp bottom gradients near the deepest portion of the river, the spatial resolution is increased to accommodate the bottom bathymetry in that region. The W2 model of the Patuxent Estuary has been calibrated using data from 1990 to 1994. Figure 3-27 shows the calculated temperature profiles at Station XDE2792 versus data for 20 days from January to December 1994. Station 2792 is approximately 14 miles from the mouth of the river. In general, the model results match the measured temperature closely. Note the lack of vertical temperature stratification in the water column during the entire year, primarily due to strong tidal actions at this location. Additionally, the CE-QUAL-W2 model is also capable of simulating conservative tracers such as salinity. Figure 3-28 presents the salinity results versus data in 1994. The model results again match the salinity profiles very well for the entire year, showing minimum vertical stratification during the summer months.

The CE-QUAL-W2 model can perform water quality simulations concurrently with the hydrodynamic calculations. In practice the hydrodynamic results can also be used for linking up with another water quality model, such as the EPA's WASP/EUTRO model, since it is straightforward to configure the water quality model with the same spatial grid system used in the hydrodynamic calculations. Chapter 7 presents a case study of this direct linkage in an application to a water supply reservoir (Loch Raven) in the Gunpowder River basin, Maryland.

For complicated hydrodynamics, a three-dimensional hydrodynamic model is usually required to drive the water quality calculations. A recent, successful modeling

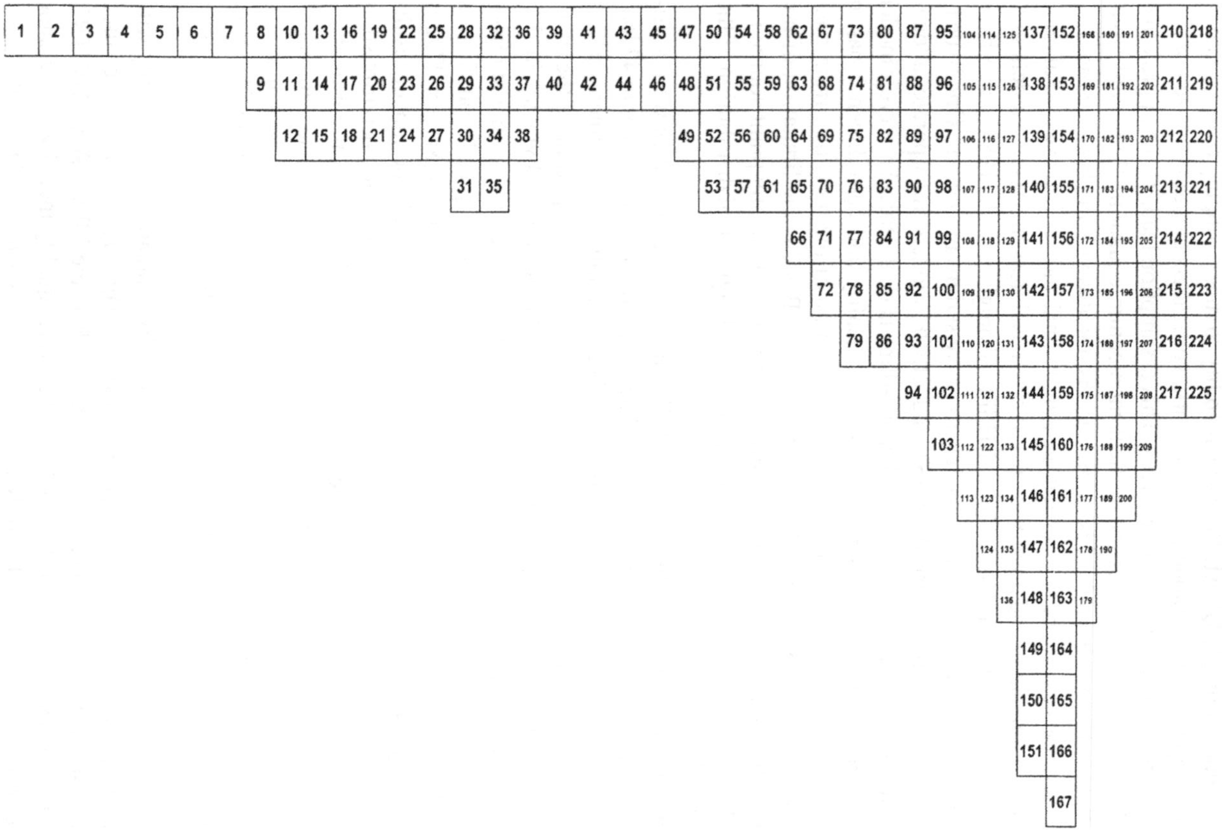

Figure 3-26 CE-QUAL-W2 model segmentation for the Patuxent Estuary.

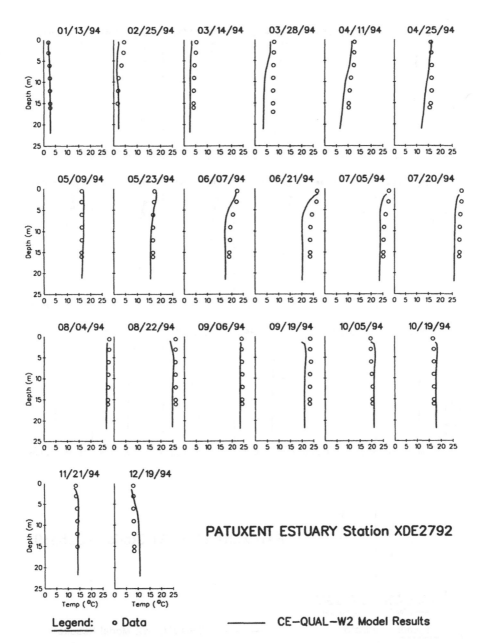

Figure 3-27 CE-QUAL-W2 model temperature results versus data at Station XDE2792 in the Patuxent Estuary, 1994.

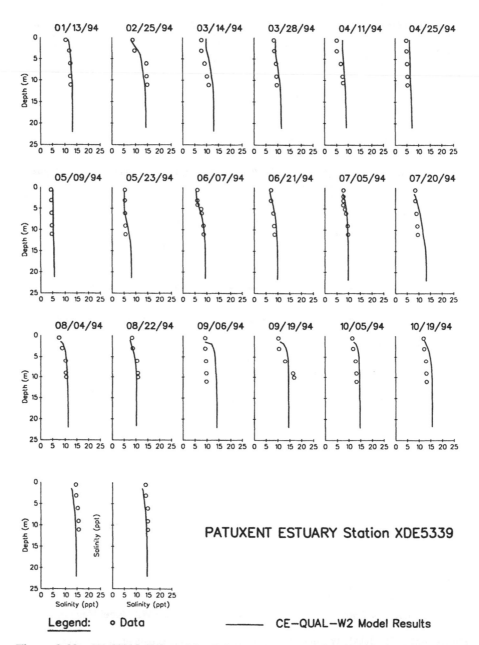

Figure 3-28 CE-QUAL-W2 model salinity results versus data at Station XDE5339 in the Patuxent Estuary, 1994.

study where the three-dimensional hydrodynamic model was developed to drive the water quality simulations is the EPA's Chesapeake Bay water quality modeling effort (Johnson et al., 1993). Although three-dimensional hydrodynamic models are continuously being developed, there are only a limited number of program codes that have been fully debugged and documented. One of the widely used three-dimensional hydrodynamic model codes is EFDC, originally developed by Hamrick (1992a) at Virginia Institute of Marine Science. It is currently being used for a wide range of water quality and ecosystem studies, including simulation of oyster and crab larvae transport, and evaluation of dredging and dredge spoil disposal alternatives (Hamrick, 1992b, 1994, 1995).

EFDC is a general-purpose three-dimensional model applicable to most water bodies. It solves the vertically hydrostatic, free surface, variable density, and turbulent-averaged equations of momentum, as well as transport equations for turbulence intensity and length scale, salinity, and temperature in a stretched, vertical coordinate system, and in horizontal coordinate systems, which may be Cartesian or curvilinear-orthogonal. Sediment transport modeling is also part of the modeling framework. Some of the useful features of the model are the wetting and drying of shallow areas, hydraulic control structures, vegetation resistance for wetlands, and Lagrangian particle tracking. The model provides output formatted to yield transport fields for water quality models such as WASP5 (Ambrose et al., 1993a) and CE-QUAL-IC (Cerco and Cole, 1993).

Another commonly used three-dimensional hydrodynamic code is ECOM, developed by Blumberg and Mellor (1980). It has been constantly refined and upgraded to include better physics and more robust numerical techniques. ECOM has become well known and respected in the hydrodynamic modeling profession because a confidence has been established that the model realistically reproduces the predominant oceanographic physics. Blumberg and Herring (1987) have described the physics of the model and the techniques employed to solve the governing equations that were given in detail by Blumberg and Mellor (1980).

The most important prognostic/diagnostic variables computed by the model are:

1. Water surface elevation
2. ε_1 component of the velocity vector (u)
3. ε_2 component of the velocity vector (v)
4. Vertical component of the velocity vector (w)
5. Salinity
6. Temperature
7. Water density
8. Turbulence kinetic energy
9. Turbulence macroscale
10. Vertical eddy viscosity
11. Vertical eddy diffusivity

12. ε_1 component of the bottom stress vector
13. ε_2 component of the bottom stress vector

This model has been used in numerous studies on a variety of natural water systems: South Atlantic Bight (Blumberg and Mellor, 1983), the Hudson-Raritan Estuary (Oey et al., 1985a,b,c), the Gulf of Mexico (Blumberg and Mellor, 1985), the Delaware Bay (Galperin and Mellor, 1990a,b), the Chesapeake Bay (Blumberg and Goodrich, 1990), the Georges Bank (Chen et al., 1995), the Oregon Continental Shelf (Allen et al., 1995), and the Gulf Stream Region (Ezer and Mellor, 1992). In all of these studies, the modeling framework's capability to mimic the circulation in the prototypes was assessed via extensive comparisons with data, establishing a confidence that the model realistically reproduces the predominant physics.

The governing equations, together with their boundary conditions, are solved by finite difference techniques. A horizontally and vertically staggered lattice of grid points is used for the computations. An implicit numerical scheme in the vertical direction and a mode splitting technique in time have been adopted for computational efficiency. The finite difference equations conserve energy, mass, and momentum, and they introduce no artificial horizontal diffusion.

3.8 LINKING A HYDRODYNAMIC MODEL WITH A WATER QUALITY MODEL

Many of the water quality modeling frameworks used in practice these days are based on the finite segment (box model) approach (Thomann and Mueller, 1987), while most of the hydrodynamic models are using finite difference schemes with either Cartesian or curvilinear coordinates. (The finite segment procedure solves a three-dimensional problem with three one-dimensional calculations, thereby significantly reducing the computation effort.) About a decade ago, a water quality model was configured with a coarser spatial grid than that used in the hydrodynamic model simply because the times scales associated with the water quality processes are much greater than those for real-time velocities, particularly in estuaries and coastal waters. Due to the limitations on the computer resources, it was not desirable to link the two models using the same temporal and spatial resolutions (Lung and Hwang, 1989). Therefore, temporal and spatial averaging of the hydrodynamic model results was desirable. Such an approach never proved satisfactory, as additional computation time is required to perform the averaging procedure. More importantly, excessive averaging loses much information from the hydrodynamic model, resulting in unrealistic mass transport coefficients. In addition, temporal averaging of estuarine hydrodynamic model results requires much caution because the residual velocity, that is, the tidally averaged velocity, could be one order of magnitude less than the tidal currents.

In recent years, the significant improvements in computer hardware have made it possible to link the two models with the same spatial grid. In many cases, it is efficient to use the hydrodynamic model grid in the water quality model. For example, several successful case studies in Chapter 7 use the CE-QUAL-W2 model to drive the

WASP/EUTRO5 model with the same spatial grid. However, there are other problems in this approach. The numerical computations in the finite segment model, such as WASP/EUTRO5, have two constraints: integration time-step and numerical dispersion, which are closely related to each other. The upwind differencing scheme used in WASP/EUTRO5 to guarantee positive solutions yields numerical dispersion. The magnitude of numerical dispersion is a function of the Courant number. Given the same spatial and temporal resolution as in the hydrodynamic model, the mass transport computations in the water quality model cannot be carried out without generating excessive numerical dispersion (Lung and Hwang, 1989). While longer time-steps will increase the Courant number and thereby reduce numerical dispersion to some extent, there is an upper bound for the time-step to maintain numerical stability.

An effective measure to reduce numerical dispersion is to use a high-order finite difference scheme in quantifying the advective transport in the box model. The QUICKEST scheme, originally developed by Leonard (1979), is a popular technique. In QUICKEST, the concentration at a segment interface is approximated by a quadratic interpolation using the concentrations in the two adjacent segments and in the next upstream segment. This approximation yields a third-order accuracy in space. Another means is to use numerical boundaries (McBride, 1987) at the physical boundaries to avoid artificially generating sources or sinks of mass. In such a scheme, the wall concentration and concentration gradient of the boundary cell are interpolated or extrapolated using the concentrations at the last two interior segments. The quadratic interpolation function is the same as that used for interior segments.

Effort is underway by the EPA to incorporate the QUICKEST scheme into the WASP/EUTRO5 model to significantly reduce numerical dispersion, thereby making it more efficient to link up with the CE-QUAL-W2 model, which already uses QUICKEST. Note that the CE-QUAL-W2 model also has a water quality module. Running the model from hydrodynamics to water quality would basically eliminate any linkage problem. Yet, over 90% of the CE-QUAL-W2 users apply the model only to hydrodynamic calculations, with very few documented case studies of using the model for water quality calculations to date. Why? The EUTRO5 module offers much more flexibility than the CE-QUAL-W2 model. For example, the EUTRO5 model allows liberal assignment of spatial and temporal variations of time functions and parameters for the water column kinetics. As a result, the combination of CE-QUAL-W2 and WASP/EUTRO5 is a popular choice in many wasteload allocation and TMDL modeling studies, taking advantage of the best of both models.

As a reminder, the WASP/EUTRO5 modeling framework has a hydrodynamic module, the one-dimensional DYNHYD module, whose output is fed into the EUTRO5 module. However, the one-dimensional hydrodynamic calculation limits its use in two- or three-dimensional wasteload allocation and TMDL modeling work. While the three-dimensional EFDC code has been expanded to include water quality simulations, there are no documented TMDL applications. An ideal approach in the near future would be to use an integrated code with hydrodynamic and water quality calculations such as EFDC or ECOM, thereby eliminating the linkage problem. The modeler would have the option to run the hydrodynamic model first and understand the mass transport prior to running the water quality model.

Using this advanced model code requires much training and skill on the part of the modeler. Furthermore, experience in modeling with the guidance of field data is essential to the success of the modeling effort. The modeler must have a full understanding of the underlying assumptions and limitations of the code and, more importantly, the extent of field data available to support the analysis. Otherwise, using this advanced code would simply be like running a black-box model.

CHAPTER 4

DERIVATION OF KINETIC COEFFICIENTS

Deriving the kinetic coefficients for a water quality model is the key step in model calibration. Since most water quality models are data driven and many water column kinetic processes are of an empirical nature, the direct approach to developing the kinetic coefficient values is independently deriving them from the field data. In water quality modeling, this approach is commonly used in deriving carbonaceous biochemical oxygen demand (CBOD) deoxygenation rates in the water column, using the measured CBOD concentrations of the receiving water (as seen later in this chapter). Note that this field data approach is also used in deriving mass transport coefficients, such as the vertical diffusion coefficient derived using the temperature data in Chapter 3.

When data from routine sampling is not available, special field studies may be performed to derive the coefficients. Using a tracer gas in the field to develop reaeration coefficients in the stream is one good example. Also the dark and light bottle method is frequently used to quantify algal growth rate and photosynthetic oxygen production rate in the water column. Further, sediment chambers are used to perform *in situ* measurements of sediment oxygen demand and nutrient release flux rates across the sediment-water interface. While these special field studies are comprehensive and labor intensive, they are invaluable in filling key data gaps and in supporting the modeling effort.

A third approach in quantifying the kinetic coefficients is using data or empirical equations from literature, for example, calculating reaeration coefficients or mass transfer coefficients across the air-water interface. An example is determining the light extinction coefficient in the water column based on the measured Secchi depth via an empirical formula.

Finally, model calibration is another alternative for properly determining kinetic coefficients, where additional model sensitivity analyses are needed to further fine-tune the coefficients. Calibrating the reaeration coefficient over a lake using the measured dissolved oxygen concentrations over a period of time is a good example. Also, in the modeling study of the Blackstone River (see Chapter 6), the CBOD deoxygenation coefficient is calibrated via model calibration using the QUAL2E modeling framework.

In summary, kinetic coefficient values can be obtained in four ways:

- Direct measurement
- Estimation from field data
- Literature values
- Model calibration

Materials presented in this chapter will assist modelers in determining kinetic coefficients using these approaches. Again, case studies are presented to demonstrate the techniques used in the data analysis.

4.1   BIOCHEMICAL OXYGEN DEMAND AND $CBOD_U$ TO $CBOD_5$ RATIO

Before this discussion gets underway, an important question must be addressed. Why is such emphasis still placed on the widely known topic of biochemical oxygen demand (BOD) when many recent water quality models track total carbon instead of BOD as a surrogate in the receiving water? The answer is simple and practical. Since discharge permits (i.e., NPDES) are still written by regulatory agencies in terms of BOD concentration in the effluent, it is necessary to address this water quality parameter in terms of wasteload allocations.

4.1.1   BOD in Effluent and Its Measurement

Looking at BOD from a historical perspective proves useful in understanding the concept of BOD in wasteload allocation water quality modeling. The evolution of BOD measurements is closely related to the treatment level of wastewater. The Metropolitan Wastewater Treatment Plant (Metro Plant) in St. Paul, Minnesota, is an excellent example illustrating this evolution.

The Metro Plant, located at UM835.1 (see Figure 4-1), is the single largest point source discharge (250 mgd) on the Upper Mississippi River. Originally constructed as a primary treatment plant in 1938, it was upgraded to a secondary treatment facility in 1966 and further upgraded to advanced secondary with nitrification in 1985. Table 4-1 shows the effluent characteristics of the Metro Plant at the three different treatment levels. The most significant change in effluent characteristics is in CBOD concentration, reduced from 101 mg/L of $CBOD_5$ (5-day CBOD) in the primary

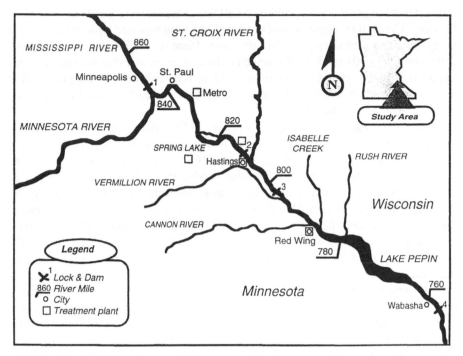

Figure 4-1   The Metro Plant and its receiving water, Upper Mississippi River between Lock & Dam No. 1 and Lock & Dam No. 2.

TABLE 4-1   Metro Plant Effluent Concentrations (mg/L) at Different Treatment Levels

Parameter	Primary Treatment[a]	Secondary Treatment[b]	Secondary with Nitrification[c]
CBOD	101[d]	40.89[d]	20.28[e]
CBOD$_u$/CBOD$_5$	1.0	2.50	——
Organic nitrogen	10.2	3.27	3.09
Ammonia nitrogen	11.68	14.95	0.80
Nitrite/nitrate	0.30	0.09	12.63
Total nitrogen	22.2	18.31	16.51
Organic phosphorus	5.50	0.89	0.19
Ortho-P	8.00	2.77	2.53
Total phosphorus	13.5	3.65	2.73
In-stream K_d (day^{-1})	0.35	0.25	0.073

[a]1964–1965 data.

[b]August 1976 data.

[c]June/July 1988 data.

[d]CBOD$_5$.

[e]CBOD$_u$.

effluent to 20 mg/L of $CBOD_u$ (ultimate CBOD) in the secondary effluent with nitrification. Although incidental CBOD removal has been reported with nitrification at many advanced secondary plants like the Metro Plant (U.S. EPA, 1995), the $CBOD_5$ concentration of 101 mg/L for the primary effluent given in Table 4-1 should be interpreted as BOD_5, as nitrification in the effluent is probably not taking place within the first 5 days of incubation of laboratory tests. In fact, measured BOD was reported as BOD_5 without significant error.

As the wastewater treatment level increases, differentiation of CBOD from nitrogenous BOD (NBOD, due to nitrification) is required. The work by Hall and Foxen (1983) explains the necessity of measuring CBOD as well as ammonia and nitrate concentrations in the wastewater. Previously, after the CBOD measurements were obtained, numerous problems in laboratory analyses were reported in accurately determining its concentration. One difficulty in the laboratory analysis, nitrification suppression, has since been eliminated. Often, anomalous results of CBOD analysis were reported, in which $CBOD_u$ exceeded BOD_u values and the CBOD versus the time curve did not follow first-order kinetics. Haffely and Johnson (1994) suggested that the source of the errors was the nitrification inhibitor, 2-chloro-6-(trichloromethyl) pyridine (TCMP), used in the laboratory analysis. In their studies, they concluded that TCMP is biodegradable and can contribute significant oxygen demand in CBOD tests. Further, TCMP degraders (bacteria) can be transferred between BOD bottles during routine $CBOD_u$ analyses. In many cases TCMP degrades were present in the wastewater and water samples, and the high TCMP dose and long incubation period enhanced the acclimation of the microbial population to TCMP. The potential for biodegradation casts doubt on the integrity of results obtained when TCMP, or perhaps any chemical inhibitor, is used in a long-term BOD test.

The laboratory protocol to quantify the $CBOD_u$ of wastewaters has improved significantly in recent years. The current practice of determining $CBOD_u$ does not call for the use of nitrification suppressors (NCASI, 1982). Instead, the total amount of oxygen consumption is recorded, along with concurrent measurements of ammonium, nitrite, and nitrate concentrations, insuring an accurate mass balance of the nitrogen components. The CBOD is then derived by subtracting the amount of oxygen used in the nitrification process from the measured total oxygen consumption. Using this protocol, Haffely (1997) has obtained excellent long-term BOD test results for the Metro Plant final effluent (Figure 4-2) and ambient water samples from the Upper Mississippi River. Results from the Metro Plant show that the $CBOD_u$ is about 14 mg/L. The NBOD is about 12 mg/L, equivalent to an ammonium concentration of 2.63 mg/L ($= 12/4.57$), and close to the nitrite/nitrate production of 2.46 mg/L. The test also tracks the amount of ammonium consumed and finds that to be 1.96 mg/L, indicating that a small amount of organic nitrogen has been converted to ammonium, which in turn is oxidized to form nitrate. The time series plots in Figure 4-2 show that the majority of ammonium oxidation (or nitrate production) takes place between days 5 and 10. In general, the mass balance between ammonium, nitrite/nitrate, and NBOD is maintained during the long-term BOD test for the Metro Plant final effluent.

Leo et al. (1984) compiled data from 144 municipal wastewater treatment plants to assess the effluent characteristics. Table 4-2 shows the 5-day BOD (BOD_5),

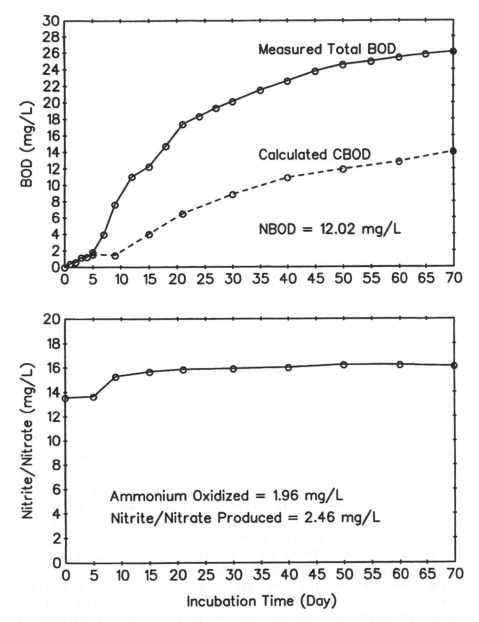

Figure 4-2   Long-term CBOD test results of the Metro Plant effluent—BOD versus incubation time and nitrite/nitrate versus incubation time.

TABLE 4-2   Mean Effluent Concentrations (mg/L) at Municipal Plants

Treatment Type	Number of Plants[a]	BOD_5	$CBOD_5$	NH_3
Primary	2	101	——	——
Trickling filter	13	41.2	——	16.6
Secondary	38	19.1	10.3	8.9
Secondary with nitrification	10	11.5	4.8	1.0
Secondary with phosphorus removal	9	16.2	14.6	7.9
Secondary with phosphorus removal and nitrification	3	13.6	——	0.9
Secondary with nitrification and filters	3	3.9	——	4.8

Source: From Leo et al. (1984).
[a]Number of plants with BOD_5 data; in some cases number with $CBOD_5$ or NH_3 data may be less.

$CBOD_5$, and ammonia concentrations in wastewaters ranging from primary effluent to secondary effluent with nitrification and filters.

4.1.2 $CBOD_u$ to $CBOD_5$ Ratio in Effluent

While NPDES permits are written in terms of 5-day CBOD ($CBOD_5$) concentrations, water quality models use ultimate CBOD ($CBOD_u$) in their calculations. So the model calculated wasteloads (in $CBOD_u$) must be converted to $CBOD_5$ values for use in the permit, thereby requiring a ratio of $CBOD_u$ to $CBOD_5$. This ratio is strongly dependent on the wastewater characteristics via the following equation:

$$\frac{CBOD_u}{CBOD_5} = \frac{1}{1 - e^{-(k_1)(5)}} \tag{4-1}$$

Equation 4-1 suggests that increased wastewater treatment (i.e., a lower k_1) tends to stabilize the wastewater, resulting in a higher $CBOD_u$ to $CBOD_5$ ratio. Such a change reflects not only the reduced impact of the effluent CBOD on the k_d rate, but also indicates the presence of highly refractory material in the well treated effluent (Lung, 1996a). Table 4-1 shows that this ratio for the Metro Plant effluent increases from 1.0 for the primary effluent to 2.50 for the secondary effluent. Results from the long-term CBOD test of the recent Metro Plant wastewater, a secondary effluent with nitrification, yield a bottle k_1 rate of 0.065 day^{-1} (see Figure 4-2), associated with a $CBOD_u$ to $CBOD_5$ ratio of 3.60 and significantly higher than 1.0 and 2.50.

The effluent data from 144 wastewater treatment plants compiled by Leo et al. (1984) have been analyzed to show the ratio in Figure 4-3. The first half of Figure 4-3 plots $CBOD_u/BOD_5$ versus BOD_5, while the second half shows $CBOD_u/CBOD_5$ versus $CBOD_5$. The data indicate that 2.47 is a better estimate of the $CBOD_u$ to BOD_5 ratio. A $CBOD_u$ to $CBOD_5$ ratio of 2.84 is also developed from the data (Figure 4-3), and the standard deviations for these two ratios are 1.52 and 1.17, respectively. To present a point of comparison, data from the Metro Plant in St. Paul,

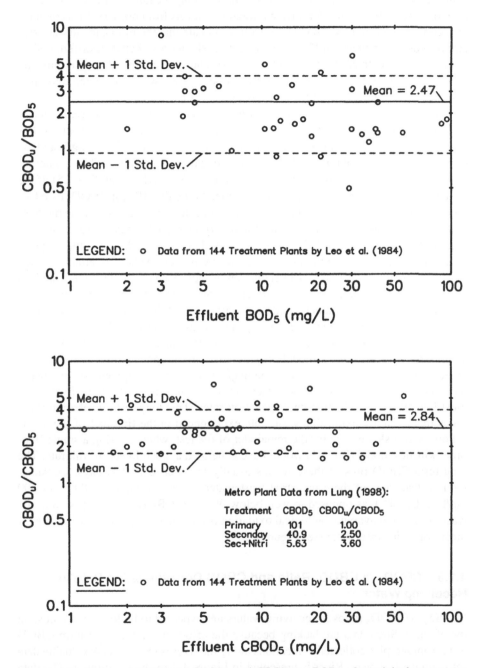

Figure 4-3   Ratios of CBOD$_u$ to CBOD$_5$ and CBOD$_u$ to BOD$_5$ of municipal wastewater effluents.

Minnesota, are also shown in Figure 4-3, demonstrating the ranges of $CBOD_5$ and the ratio of $CBOD_u$ to $CBOD_5$ for different treatment levels from primary, to secondary, to secondary with nitrification. These values indicate that the ratio can vary considerably, not only between different treatment levels but also between different sites with the same treatment level (Leo et al., 1984). Thus it is important to determine the ratio for each facility. However, this may not be possible where projected treatment conditions are significantly different from current conditions.

One consequence of using a ratio that has not been developed from field data, is understating the wastewater strength on receiving water oxygen concentrations. For example, a modeler may measure a secondary effluent $CBOD_5$ of 10.0 mg/L and assign a $CBOD_u$ to $CBOD_5$ ratio of 1.5 in a model calibration analysis. This combination would result in a calculated effluent $CBOD_u$ of 15.0 mg/L (10 mg/L × 1.5). If the actual ratio was 3.0, the modeler would be understating the effluent $CBOD_u$ by a factor of two (30.0 mg/L compared to 15.0 mg/L). In calibrating the model, the modeler will have to assign this error to another source of dissolved oxygen impact, such as nonpoint loadings. Extrapolating to wasteload allocation conditions, this nonpoint source loading may cause the modeler to require higher levels of treatment, which may not actually be necessary. Depending on the approach taken, the understated effluent $CBOD_u$ may have a variety of effects on the wasteload allocations. Because of the importance of this parameter and the observed variability in the ratio from site to site, site-specific ratios should be developed on a case-by-case basis.

The above discussions are based on first-order kinetics for CBOD, but other studies have used kinetics models besides the first-order reaction. For example, McKeown et al. (1981) showed that the majority of some biologically treated paper industry effluent oxidizes at a slow rate and that a dual first-order model would be most beneficial in evaluating deoxygenation rates. This long-term incubation illustrates that a readily oxidizable waste rate experienced in the first half of the test is followed by a slower rate for the remainder of the test, whereas using a single first-order kinetics model (Eq. 4-1) would have generated a much lower overall rate. The long-term CBOD runs of these wastes usually last well over 100 days. Consider, though, that most of these industrial wastewaters were at a high CBOD (over 300 mg/L of $CBOD_u$), as opposed to well below 30 mg/L $CBOD_u$ for today's domestic wastewaters. On average, biological oxidation is complete in about 60 to 70 days for most domestic wastewater (see Figure 4-2).

4.1.3 $CBOD_u$ to $CBOD_5$ Ratio and CBOD Deoxygenation Rate in Receiving Water

$CBOD_u$ to $CBOD_5$ ratios in receiving waters are expected to be lower than those in the effluent. Such data are lacking because the effort to conduct long-term CBOD tests from samples collected at locations along the river is extensive. A valuable data set from the Delaware River is presented in Figure 4-4 to show the ratios. The data were collected during three intensive surveys from August to October in 1979 when the major point sources, such as the Philadelphia area wastewater treatment plants,

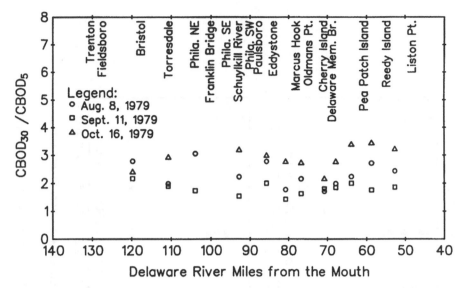

Figure 4-4 CBOD$_{30}$ to CBOD$_5$ ratio in the Delaware River Estuary, August–October 1979.

were at the secondary treatment level. Figure 4-4 shows that the ratio ranges from 1.5 to 3.5 along the river, as expected in a natural water system receiving secondary treatment effluents.

In a river or stream with no significant algal growth taking place, the concentration of CBOD is expected to decrease in the downstream direction following wastewater input. When algal biomass is significant and the unfiltered receiving water samples are analyzed for ultimate or long-term BOD, the CBOD concentration may reflect the oxygen demand to break down the algal biomass. Figure 4-5 shows the CBOD$_u$ data measured from the Upper Mississippi River near St. Paul, Minnesota. The unfiltered CBOD$_u$ concentration increases slightly immediately below the Metro Plant waste input, suggesting that the wastewater has low CBOD loads. Also note the sharp rise of CBOD$_u$ concentrations in the downstream direction toward Lock & Dam No. 2, where significant algal growth in the pool behind the dam yields much CBOD in the unfiltered samples. The filtered samples show the CBOD$_u$ concentrations decreasing in the downstream direction.

Since long-term laboratory CBOD tests are conducted for all samples to determine the CBOD$_u$ concentrations, their bottle rate (in day^{-1}) can be quantified following Eq. 4-1. Again, the unfiltered CBOD$_u$ samples usually show a progressive increase in the bottle rate in the downstream direction while the filtered samples show a small decrease. The general observation here is that both CBOD$_u$ concentration and the bottle rate of filtered samples reflect the dissolved organic carbon attenuation in the receiving water (i.e., continuing stabilization in the water column). When algae is present in the water column, extra care is needed when analyzing the CBOD data and filtered samples are required.

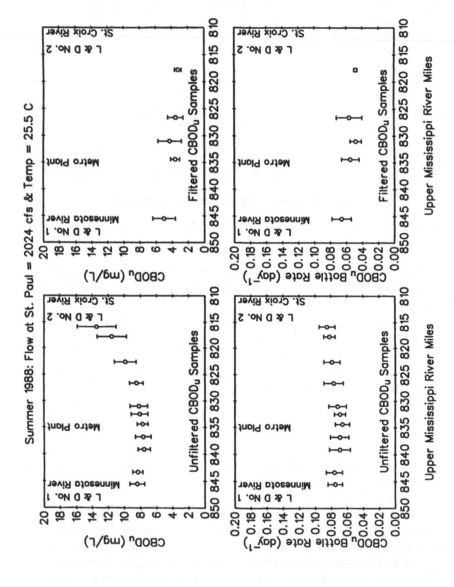

Figure 4-5 CBOD (filtered and unfiltered) concentrations and CBOD bottle rates in Upper Mississippi River.

The deoxygenation of CBOD in the receiving water is generally characterized by first-order kinetics:

$$C = C_o \exp\left(-k_d \frac{x}{U}\right) \tag{4-2}$$

where

C_o = the completely mixed CBOD concentration following the waste input
k_d = the in-stream deoxygenation rate (day^{-1})
x = distance downstream
U = average stream velocity

Note that x/U is called travel time (day). By measuring the filtered CBOD concentrations in the receiving water (as shown in Figure 4-5), one can determine the in-stream k_d rate by fitting the data points to the exponential decay in Eq. 4-2. To account for the dilution effect, the CBOD$_u$ at each station is expressed in loading rate (pounds per day), and the filtered CBOD$_u$ data are then fitted by a straight line in a regression analysis. The slope of the straight line yields an in-stream deoxygenation rate, k_d, of 0.073 day^{-1} from the semilog plot of loading rate versus travel time along the Upper Mississippi River following the Metro Plant input (Figure 4-6).

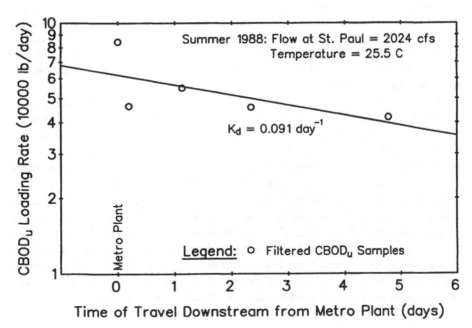

Figure 4-6   Derivation of CBOD deoxygenation rate, k_d, for Upper Mississippi River below the Metro Plant—loading rate of filtered CBOD versus travel time.

This in-stream deoxygenation rate, k_d, of 0.073 day^{-1} is a direct reflection of the wastewater characteristics and should therefore be close to the filtered CBOD bottle rate, k_1, of the wastewater (see Figure 4-2). The fact that k_d is close to k_1 is particularly true for well-treated effluents, resulting in a small impact on the receiving water with a low k_d rate. Following with the concept that the bottle rate of the effluent decreases with an increase in treatment level, the in-stream k_d should also decrease following treatment upgrade. Such a relationship is clearly demonstrated by the Metro Plant discharge and the Upper Mississippi River in Figure 4-7. When the Metro Plant discharged primary effluents into the Upper Mississippi River, the in-stream k_d rate was 0.35 day^{-1}. When the Metro Plant was upgraded to secondary treatment, the corresponding k_d rate decreased to 0.25 day^{-1}. Further upgrading, all the way up to installing the nitrification process (with incidental removal of CBOD) at the Metro Plant, dropped the k_d rate to 0.073 day^{-1}. The associated dissolved oxygen profiles shown in Figure 4-7 document the benefit of the treatment upgrades, that is, eliminating the dissolved oxygen sag and meeting the dissolved oxygen standard of 5.0 mg/L. Note that these three conditions are successfully modeled by the same water quality model with all kinetic coefficients unchanging (except the in-stream k_d rate) for each scenario.

The decrease of the in-stream deoxygenation rate following treatment upgrade has been well documented in literature (Leo et al., 1984). Figure 4-8 shows another case study of the Upper Patuxent River in Maryland. Although the rates derived from these data are slightly different than those shown in Figure 4-7 for the Mississippi River, a reduction in rates again occurs after treatment is upgraded.

Observe that the in-stream deoxygenation rates, k_d, in many U.S. rivers have decreased in recent years, due to water pollution abatement, to values below 0.1 day^{-1}. Figure 4-9 shows examples of long-term CBOD tests for the Upper Mississippi River, the Minnesota River, and the St. Croix River in Minnesota, all yielding bottle rates well below 0.1 day^{-1}. The in-stream k_d rates are also equal to the bottle rates of the receiving water samples.

This discussion strongly suggests that field data are crucial in determining the CBOD deoxygenation rate in the receiving water (Lung, 1993). In situations where no field data are available, the Environmental Protection Agency's guidance manual (U.S. EPA, 1995) recommends the use of a plot of k_d versus river depth (Figure 4-10). Note that the range of observations in Figure 4-10 is derived from data measured in the 1970s and earlier, when many wastewater treatment plants had only primary treatment. Data from 1980s and 1990s are shown for the Shirtee Creek in Alabama, the Blackstone River in Massachusetts, the Roanoke River in Alta Vista, Virginia, and the Upper Mississippi River in St. Paul and Minneapolis, Minnesota. These four data points are below the lower bound of the historical observations, suggesting that improved wastewater treatment has lowered the deoxygenation rate, k_d, in the receiving water. Figure 4-10 also indicates that the majority of k_d rates are below 0.1 day^{-1}, with the exception of extremely shallow waters such as the Shirtee Creek.

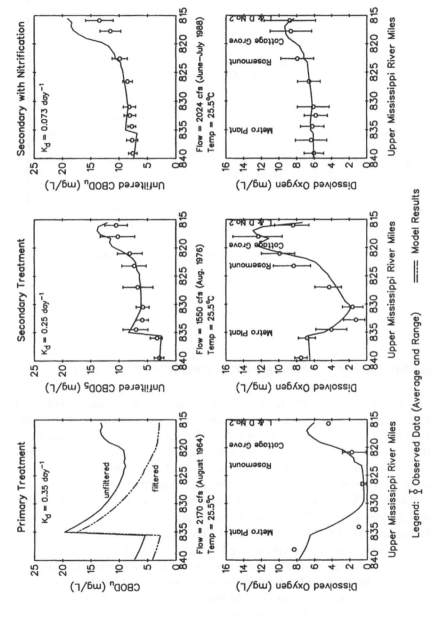

Figure 4-7 Model calculated and measured CBOD and DO concentrations in Upper Mississippi River—different treatment levels at the Metro Plant.

83

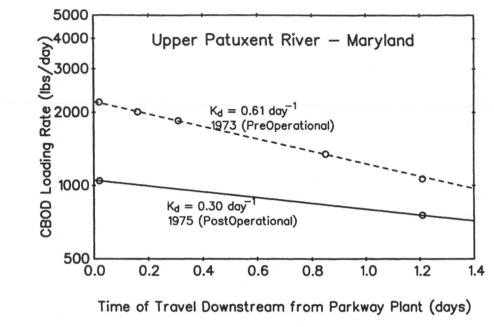

Figure 4-8   CBOD deoxygenation rates, k_d, in the Upper Patuxent River, before and after treatment upgrades at various plants.

4.2   NITRIFICATION IN WASTEWATER AND RECEIVING WATER

Biological nitrification for wastewater ammonia reduction is commonly practiced at most advanced secondary and tertiary treatment plants. Previously when the treatment level was limited to primary treatment, nitrification did not occur at the treatment plant within the first 5 days of incubation. With treatment upgrade and installation of the nitrification process, nitrification started prior to the 5th day of incubation in the laboratory analysis. The paper by Hall and Foxen (1983) points out this phenomenon. Therefore, it becomes necessary to separate the nitrification process in the long-term BOD test. The procedure presented in Section 4.1.1 would not only quantify the CBOD, but also the NBOD by measuring ammonia and nitrate concurrently with oxygen consumption (see Figures 4-2 and 4-9).

To derive the in-stream nitrification rate when modeling NBOD, the standard procedure involves plotting the NBOD loading rate (calculated from field data) with the travel time (as in the process of deriving the in-stream k_d rate for CBOD). The slope of the best-fit straight line from this regression analysis yields the nitrification rate, k_n, in the receiving water. Figure 4-11 shows such a derivation for the Upper Patuxent River in Maryland. Note that the k_n rate also changes from the pre-operational value to a lower value following the treatment upgrade at the Parkway treatment plant.

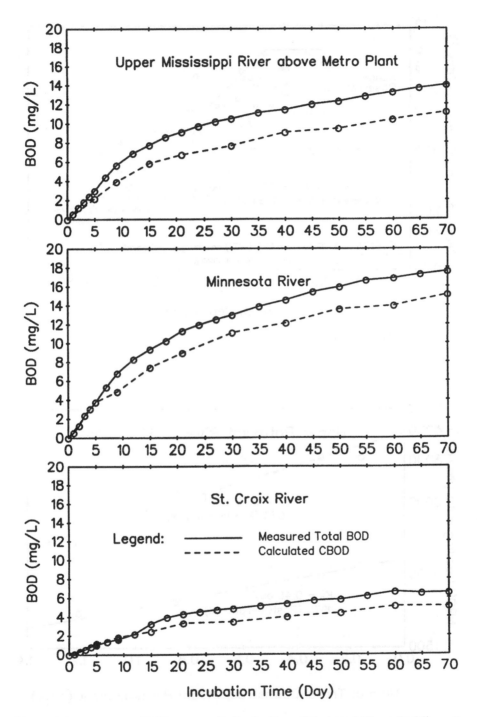

Figure 4-9 Long-term CBOD test results for the Upper Mississippi River, the Minnesota River, and the St. Croix River, 1996.

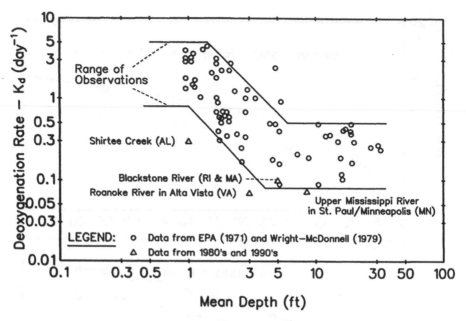

Figure 4-10 CBOD deoxygenation rate, k_d, versus river depth.

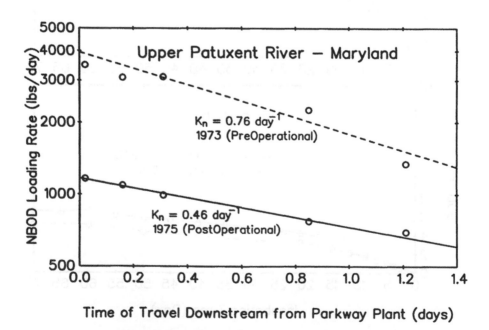

Figure 4-11 Nitrification rates, k_n, in the Upper Patuxent River, before and after nitrification process installed at treatment plants.

Several issues related to k_n in BOD/DO (dissolved oxygen) modeling are:

1. The variability of k_n as a function of the waterbody physical characteristics,
2. Algal effects on k_n, and
3. Seasonal variations of k_n and ammonia removal requirements.

In general, the availability of surfaces for nitrifier attachment can affect k_n rates. These surfaces include the stream bottom and suspended particles in the water column. Consequently, shallow streams with rocky bottoms favor the growth of nitrifying bacteria with associated high k_n rates, while deep rivers composed of sands, silts, or clays generally have fewer attached nitrifiers. Although some deep streams may support significant populations of nitrifiers in the water column, they tend to have lower k_n values than shallow streams. And since deeper streams usually have lower k_n rates and provide more dilution than shallow streams, the need for ammonia removal is less definitive. Thus large and deep streams require site-specific data and a rigorous determination of the k_n rate in the modeling analysis to support wasteload allocation studies.

k_n rates determined for shallow, fast moving streams are usually highly variable, ranging from 0.1 to 1.5 day^{-1} (U.S. EPA, 1984). Since small and shallow streams tend to be effluent dominated, the DO impact from warm weather nitrification is usually significant even with relatively low k_n rates. Thus the selection of an appropriate k_n rate should also be based on site-specific nitrogen series data.

Much like the derivation of the in-stream k_d rate for CBOD, high concentrations of algae, either suspended or attached, significantly affect the k_n rate derivation. Plotting the ammonia loading rate in the receiving water versus time of travel reflects only the ammonia loss. Such an approach could result in the overestimation of k_n where algal effects are significant because algae consume ammonia as a key nutrient (Lung, 1993). A k_n rate derivation based on the total loss of ammonia would include ammonia uptake by algae as well as ammonia oxidation. In many cases, observing a concurrent increase of nitrate proves a better approach for estimating k_n because a nitrate increase results directly from ammonia oxidation in the stream. As a cautionary note, under some conditions, algae can uptake nitrate as well as ammonia. Therefore, the k_n rate derived from nitrate increase would represent the minimum k_n. Figure 4-12 shows relatively constant nitrate concentrations throughout a stream reach. As a result, the 1.0 day^{-1} k_n rate based on the reduction of ammonia concentrations alone exceeds by five times the rate derived from the nitrate data. As such, algal uptake of ammonia instead of nitrification may have caused the reduction in ammonia concentrations. Results from Figure 4-12 serve as the basis to reject nitrification at the point sources along the river.

Figure 4-13 shows the impact of nitrification process at the Metro Plant on the ammonium and nitrate concentrations in the Upper Mississippi River. Data from a water quality survey in 1976, prior to the installation of the nitrification process at the Metro Plant, show much higher ammonium concentrations in the Upper Mississippi River below the Metro Plant. In fact, the receiving water ammonium concentrations were so high in 1976 that ammonia standards were violated while the concurrent nitrate concentrations were low. By 1988, when the nitrification process at the Metro Plant was in full operation, the low effluent ammonium concentrations resulted in

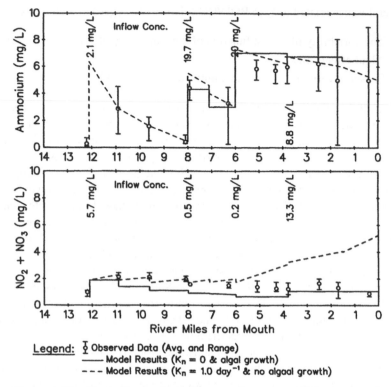

Figure 4-12   Algal effect on the derivation of in-stream nitrification rate, k_n.

much lower receiving water ammonium levels, thereby meeting the ammonia standard. Yet the nitrate concentration in the Upper Mississippi River sharply increased. These data indicate that, while the total inorganic nitrogen concentrations in 1976 and 1988 are about the same, their composition is much different. Also, it is not surprising that dissolved oxygen concentrations in 1988 are much improved over those in 1976, further confirming the water quality benefit of nitrification.

Another critical issue in determining the k_n rate involves the time of the year these rates are determined and applied. Although ammonia and nitrate data may indicate relatively high k_n rates during July and August, k_n rates for the same river may be negligible during the winter months, and even during "transitional periods" such as April through June and September through November (U.S. EPA, 1984). Seasonal adjustments in k_n, like using the temperature correction relationships, may be appropriate during months when wastewater temperatures are above 20°C. During months when temperatures fall below this level, nitrogen series data collected may be necessary when selecting appropriate cool weather nitrification rates.

Finally, a large body of literature exists for case studies of k_n rates in streams and rivers. The U.S. EPA (1995) guidance manual on BOD/DO modeling of rivers and streams summarizes nitrification rates measured in the field and uses parameter values for models from a number of investigations.

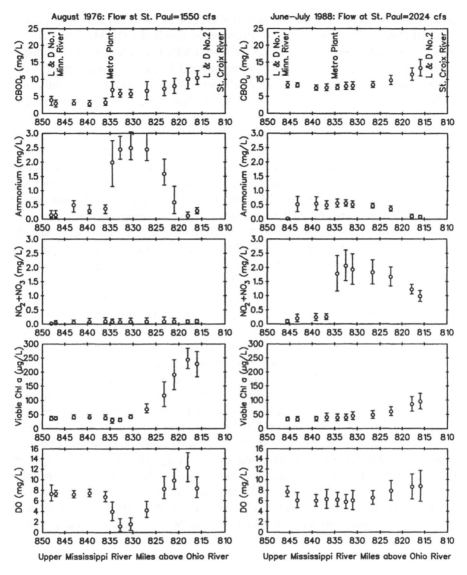

Legend: ⚲ Observed Data (Average and Standard Deviation)

Figure 4-13 Water quality of Upper Mississippi River (1976 data versus 1988 data), before and after Metro Plant nitrification.

4.3 REAERATION COEFFICIENT

4.3.1 A Case Study of Stream Reaeration Coefficient

Quantifying correct reaeration coefficients for the receiving water is a crucial step in BOD/DO modeling for wasteload allocations and is illustrated in the following case study. In a wasteload allocation modeling study of Shirtee Creek in Alabama in 1991, the Alabama Department of Environmental Management (ADEM) selected the Langbien and Durum (1967) equation for stream reaeration in the QUAL2E model. The QUAL2E model of Shirtee Creek was calibrated with two sets of data collected in August 1989 and October 1990. The calibrated model was then run under the 7-day 10-year low flow conditions to develop effluent CBOD$_5$ limits for two point source dischargers: the City of Sylacauga wastewater treatment plant and Avondale Mills (a textile mill).

To meet the 5-mg/L daily average DO standard for Shirtee Creek, ADEM developed stringent CBOD$_5$ limits for the point sources. To challenge these limits, the consultant for Avondale Mills used the STREAM (see Chapter 6) model and conducted two stream surveys in the winter of 1991 and August 1991 to support the modeling analysis. The consultant's STREAM model was able to reproduce the four sets of field data (two collected by ADEM and two by the consultant). However, the consultant used the Tsivoglou equation (Tsivoglou and Neal, 1976) to calculate the stream reaeration coefficient, while ADEM used the Langbien and Durum equation. In using the calibrated STREAM model, the consultant predicted that much less stringent effluent CBOD$_5$ limits were needed for the point source discharges.

The reaeration coefficients used in these two modeling analyses are summarized in Table 4-3. The reaeration coefficients differ significantly between the two equations, thereby leading to contrasting assessments of the effluent CBOD limits. The foremost question then is: Which equation is more appropriate for Shirtee Creek? First, a literature review on surface mass transfer and reaeration coefficients in streams indicated that the Tsivoglou equation is more suitable for small and shallow streams like Shirtee Creek. The U.S. EPA (1971) summarized reaeration coefficient as a function of depth, as shown in Figure 4-14. In the August 1989 survey, the water depth was less than 2 ft, often even below 1 ft in the upstream reaches, making the average stream velocity close to 0.5 ft/s. The reaeration coefficients, ranging between 4 day^{-1} to 10 day^{-1}, are reasonable based on the average stream velocity (Figure 4-14). The reaeration coefficient values calculated by the Tsivoglou equation are either within or very close to this range, with the exception of one value in the reach between mile 0.75 and 2.02, where the steep channel slope contributes to a high value. (Note that this reach is not a critical reach as far as the DO levels are concerned.) On

TABLE 4-3 Comparison of Reaeration Coefficients (day^{-1}) at 20°C

Mile Point	Using Tsivoglou Equation			Using Langbien and Durum Equation	
	Aug. 1991	March 1991	Aug. 1989	Aug. 1989	Oct. 1990
0.0–0.53	5.074	6.266	5.392	9.77	1.82
0.53–0.75	9.886	11.79	14.62	0.76	0.10
0.75–2.02	14.94	17.05	17.45	0.76	2.25
2.02–3.95	8.427	9.222	9.181	0.76	1.44
3.95–6.07	4.897	3.598	3.072	0.76	1.44

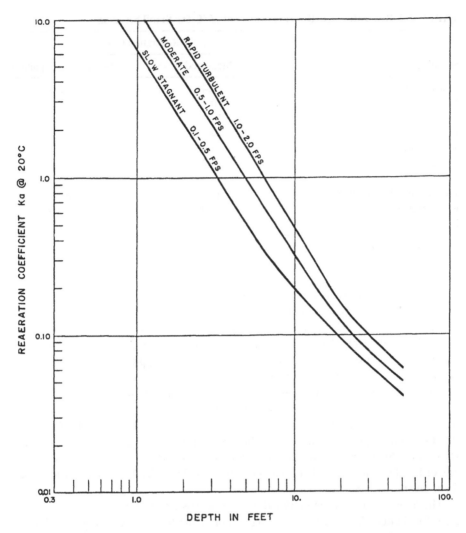

Figure 4-14 Stream reaeration coefficient, k_a, versus river depth (U.S. EPA, 1971).

the contrary, the values calculated by Langbien and Durum equation are considerably lower than the literature values; they are off by an order of magnitude.

A review of the Langbien and Durum equation by Covar (1976) questioned the merit of using considerably dissimilar data to derive a single equation. Langbien and Durum (1967) used field data of O'Connor and Dobbins (1958) and of Churchill et al. (1962) with the laboratory data of Krenkel and Orlob (1963) and of Streeter (1936). These data cover a wide range of conditions and should have been used separately to develop equations for streams with conditions similar to those used in Langbien and Durum's derivation.

TABLE 4-4 Reaeration Coefficients (day⁻¹) at 25°C for Small Streams

	Reaeration Coefficient	
Stream	Measured	Tsivoglou's Equation
Black Earth Creek	8.46	7.8
Mud Creek Tributary	10.7	4.2
Dodge Branch	33.1	34.6
Isabelle Creek	14	——
Madison Effluent Channel	2.06	4.1
Mill Creek	3.31	2.2
Honey Creek	18.4	27.4
West Branch Sugar River	42.5	36.4
Koshkonong Creek	6.09	4.8
Badger Mill Creek	7.98	9.1

Source: From Grant (1976).

There are numerous documentation that state that the Tsivoglou equation is particularly suitable for small and shallow streams (McCutcheon, 1989; St. John et al., 1984; Thomann and Mueller, 1987). In several wasteload allocation guidance manuals, the U.S. EPA (1985, 1990, 1991a, 1995) also recommends the use of the Tsivoglou equation for small and shallow streams. The equation is most appropriate for streams with depths of up to 2 or 3 ft and velocities between 0.3 ft/s and 0.6 ft/s (U.S. EPA, 1984). The hydraulic geometry conditions in Shirtee Creek fall into this category.

Table 4-4 presents a comparison of predicted (using the Tsivoglou equation) and observed reaeration coefficients on small streams in Wisconsin (Grant, 1976). Again, the Tsivoglou equation is accurate when compared with the data. Table 4-4 also shows that the reaeration coefficients for small Wisconsin streams are high, at least one order of magnitude higher, than the values used by ADEM in their wasteload allocation study for Shirtee Creek.

The above analysis eventually led to a consensus on the correct reaeration coefficient for Shirtee Creek. Both parties agreed on using the Tsivoglou equation for that system, which eventually resulted in developing correct wasteloads for the dischargers.

4.3.2 Assigning Stream Reaeration Coefficient

An excellent discussion of predicting volatilization coefficients for surface water by Rathbun (1998) is summarized in Table 4-5. Rathbun classifies streams into three categories for the determination of reaeration coefficients: shallow-flow, intermediate-depth, and deep-flow. Each category of waters is characterized with an empirical equation to calculate the reaeration coefficient: Owens et al. (1964), Churchill et al. (1962), and O'Connor and Dobbins (1958). These three equations are commonly used for stream reaeration, and the values provided by these equations in Table 4-5 are for 20°C water. Rathbun (1998) provides a chart (Figure 4-15) showing various combinations of water velocity and flow depth appropriate for each of these equations.

TABLE 4-5   Summary of Stream Reaeration Equations

Equation	Range of Velocity, V (m/s)	Range of Depth, D (m)
Shallow-flow (Owens et al., 1964): K_a (day^{-1} at 20°C) = 6.92 $V^{0.73}$ $D^{-1.75}$	0.040–0.558	0.12–0.75
Intermediate-depth (Churchill et al., 1962): K_a (day^{-1} at 20°C) = 5.01 $V^{0.969}$ $D^{-1.673}$	0.564–1.52	0.646–3.48
Deep-flow (O'Connor and Dobbins, 1958): K_a (day^{-1} at 20°C) = 3.93 $V^{0.50}$ $D^{-1.50}$	0.058–1.28	0.274–11.3

Source: From Rathbun (1998).

Each equation is valid within the database's limits in terms of water velocity and flow depth. If the water velocity and/or flow depth values fall outside the three rectangles in Figure 4-15, the following options are recommended (Rathbun, 1998):

1. Use whichever equation in Table 4-5 is closest to the desired values and extrapolate, with the understanding that prediction errors are likely to be larger when an equation is used outside the range of data upon which the equation was based.
2. Search the literature—for example, Rathbun (1977)—for an equation based on data that encompass the values of interest.

There are many other empirical equations for stream reaeration. The EPA guidance manual on BOD and DO modeling for rivers and streams alone lists over 20 empirical equations reported by various investigators (U.S. EPA, 1995). Whittemore (1986) has compiled a database of stream reaeration measurements obtained over a 30-year period under a wide range of environmental and hydraulic conditions. More recently, Melching and Flores (1999) put together a number of reaeration equations from the U.S. Geological Survey (USGS) data base. These sources are excellent references on stream reaeration.

4.3.3   Dam Reaeration

A commonly used equation to quantify the oxygen input from dam reaeration is given by Butts and Evans (1983):

$$D_a - D_b = \left[1 - \frac{1}{1 + 0.116abH(1 - 0.034H)(1 + 0.046T)}\right]D_a \qquad (4\text{-}3)$$

where
D_a = dissolved oxygen deficit above the dam (mg/L)
D_b = dissolved oxygen deficit below the dam (mg/L)

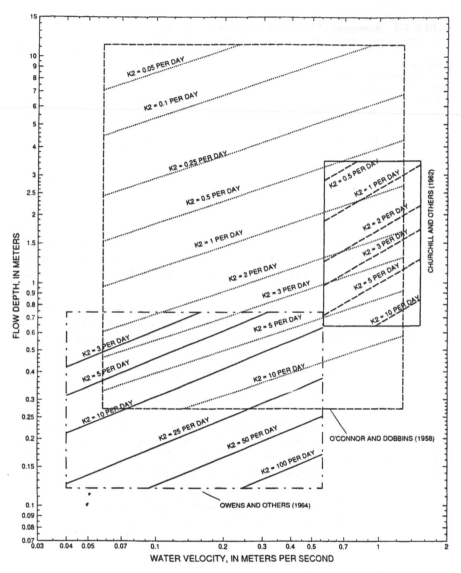

Figure 4-15 Stream reaeration coefficient as a function of river velocity and flow depth (Rathbun, 1998).

T = water temperature (°C)
H = height through which water falls (ft)
a = 1.80 in clean water
 = 1.60 in slightly polluted water
 = 1.00 in moderately polluted water
 = 0.65 in grossly polluted water

b = 0.70 to 0.90 for flat, broad crested weir
= 1.05 for sharp crested weir with straight slope face
= 0.80 for sharp crested weir with vertical face
= 0.05 for sluice gates with submerged discharge

The parameters H, a, and b must be assigned for each dam.

4.3.4 Reaeration on Lakes and Estuaries

For relatively insoluble gases such as O_2, N_2, and CO_2, the resistance to mass transfer across the air-water interface is localized in the liquid side of the interface according to the two-film theory. In streams and rivers, as discussed in the preceding section, water motion is primarily responsible for the transfer. Therefore, velocity plays a major role in determining the reaeration coefficient. In lakes and impoundments where water motion is less significant, wind becomes the dominating factor in reaeration.

One of the most used empirical equations for reaeration over a lake surface is by Banks and Herrera (1977):

$$K_L = 0.728W^{0.5} - 0.317W + 0.0372W^2 \qquad (4\text{-}4)$$

where K_L is mass transfer coefficient (m/day) and W is wind speed measured 10 meters above the water surface (m/s).

In the eutrophication model for the Potomac Estuary, Thomann and Fitzpatrick (1982) used the following equation for the reaeration coefficient:

$$K_a = 3.93\frac{V^{0.5}}{D^{1.5}} + \frac{0.728W^{0.5} - 0.317W + 0.0372W^2}{D} \qquad (4\text{-}5)$$

in which

V = tidally averaged, longitudinal velocity (m/s)
D = depth (meters)
W = wind speed (m/s)

Note that the first part of Eq. 4-5 is the O'Connor–Dobbins equation (see Table 4-5) and the second part is Eq. 4-4. Other empirical equations for lakes and estuaries include Harleman et al. (1977), Hartman and Hammond (1985), Broecker et al. (1978), and Wanninkhof et al. (1991).

As stated earlier, using models to backcalculate, that is, to calibrate a kinetic coefficient, is an alternative to determining the coefficient value. Applying this technique to reaeration coefficient determination involves the use of a simple DO model for the lake by tracking the DO budget in a time-variable fashion. A good example is shown in Muskegon Lake, Michigan. The lake water column is divided into two layers (epilimnion and hypolimnion) due to strong vertical stratification of temperature and dissolved oxygen during the summer months. The processes in the epilimnion are inflow from the watershed, outflow at the outlet, vertical exchange with the hypolimnion, and surface reaeration. The processes in the hypolimnion include exchange with the epilimnion and oxygen consumption in the hypolimnion due to sediment oxygen demand (SOD) and algal respiration. Note the hypolimnetic oxygen consumption is determined using the methodology of Chapra and Canale (1991). The DO mass loading rates via inflow and outflow are independently quantified using measured flows and DO concentrations. The vertical diffusion coefficient, which characterizes the exchange process between the two layers, is quantified using the gradient method (described in Chapter 3) with temperature data. The only coefficient left to be determined is reaeration because all other processes have already been independently calculated.

This DO model is then used to calibrate the reaeration coefficient by reproducing the DO concentrations in the epilimnion and hypolimnion. Figure 4-16 shows the model results versus data in both layers for 1991. This approach is similar to that used by Gelda et al. (1996) for Onondaga Lake in Syracuse, New York, with the exception that a two-layer model is used for Muskegon Lake to account for the significant vertical stratification of temperature and DO. While the calculation for Onondaga Lake is only for a period of 35 days following the fall overturn, the Muskegon Lake simulation is performed for an entire year. The calibrated reaeration coefficient for Muskegon Lake is 0.03 day^{-1} for the year until the fall overturn. A higher value of 0.15 day^{-1} is needed following the fall overturn to account for the sharp rise of DO concentrations in the lake. [Note that Gelda et al. (1996) calibrate a reaeration coefficient of 0.055 day^{-1} for Onondaga Lake following the fall overturn.] Additional fine-tuning with wind speed dependent reaeration rates on a time-variable basis would improve the match with the data.

The above calculation, called the whole lake technique, is based on measured DO concentrations in the water column over time. This analysis is particularly appealing and useful because it is nonobtrusive (Chapra, 1997); that is, it does not require using tracers and dyes, characteristic of a comprehensive field study. Most often the reaeration process is the dominating process in the DO budget of the water column (in Muskegon Lake, the epilimnion). Because the model results are sensitive to varying the reaeration coefficient, this estimating technique becomes even more attractive.

For the sake of completeness, total phosphorus model results versus data are also shown in Figure 4-16 because the oxygen consumption rate in the hypolimnion is related to the total phosphorus concentration (see Chapra and Canale, 1991). However, total phosphorus model calculations are decoupled from the DO calculations. Note that this total phosphorus and DO model for Muskegon Lake is a hybrid model derived from the two-layer phosphorus model in the water column for White Lake (Lung and Canale, 1976) and the total phosphorus and hypolimnetic DO model by Chapra and Canale (1991).

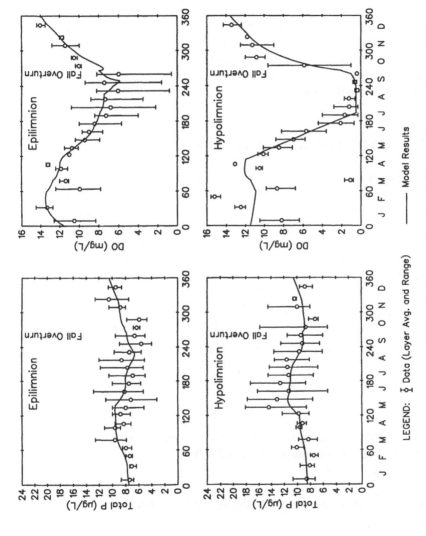

Figure 4-16 Total phosphorus and DO in Muskegon Lake, 1991. Calibrating the reaeration coefficient using the DO data.

4.3.5 How Is the Reaeration Process Incorporated in the Models?

The STREAM model has the O'Connor–Dobbins and Tsivoglou equations in the code. The user can select either one for appropriate application. The QUAL2E model offers eight empirical equations for stream reaeration plus a user override; that is, the user can assign reaeration coefficients without using any of the equations provided. Dam reaeration is also included in the QUAL2E model. Finally, the WASP/EUTRO model offers reaeration in streams, rivers, lakes, impoundments, and estuaries. Wind-induced reaeration is also incorporated, including a useful feature called ICE COVER, which simply turns off the reaeration process when the surface is frozen during the winter months. Finally, temperature effect on the reaeration coefficient is also accounted for in the models.

4.4 SATURATION DISSOLVED OXYGEN LEVEL

The solubility of oxygen in water is dependent on water temperature, atmospheric pressure (i.e., partial pressure of oxygen in the air), and the concentration of dissolved solids (or specific conductivity or salinity) in the water. Oxygen saturation concentration decreases with an increase in temperature and salt concentration increases with increasing atmospheric pressure. One of the most commonly used equations to establish the dependence of oxygen saturation on temperature is the following (APHA, 1992):

$$\ln[C_{sf}] = -139.344411 + \frac{1.575701 \times 10^5}{T_a} - \frac{6.642308 \times 10^7}{T_a^2}$$
$$+ \frac{1.243800 \times 10^{10}}{T_a^3} - \frac{8.621949 \times 10^{11}}{T_a^4} \qquad (4\text{-}6)$$

where

C_{sf} = saturation concentration of DO (mg/L) in fresh water at 1 atm
T_a = absolute temperature in °K = $T + 273.15$
T = temperature in °C

A more compact equation is recommended by the U.S. EPA (1995):

$$C_{sf} = \frac{468}{31.5 + T} \qquad (4\text{-}7)$$

where C_{sf} is freshwater DO saturation concentration in mg/L and T is water temperature in °C. This equation is accurate to within 0.03 mg/L, as compared with the Benson–Krause equation on which the Standard Methods tables are based (McCutcheon, 1985).

For saline waters, the following equation is recommended (APHA, 1992):

$$\ln C_{ss} = \ln C_{sf} - S\left(1.764 \times 10^{-2} - \frac{1.0754 \times 10^1}{T_a} + \frac{2.1407 \times 10^3}{T_a^2}\right) \quad (4\text{-}8)$$

where C_{ss} is saline water DO saturation concentration in mg/L and S is salinity in ppt. Equation 4-8 is incorporated into the WASP/EUTRO5 model code.

For high elevation streams or lakes, the barometric pressure effect is important. The following equation is used to quantify the pressure effect on saturated DO concentrations:

$$C_{sp} = C_{s0}P\left[\frac{[1 - (P_{wv} / P)](1 - \theta P)}{(1 - P_{wv})(1 - \theta)}\right] \quad (4\text{-}9)$$

where

$$
\begin{aligned}
C_{sp} &= \text{DO saturation (mg/L) at pressure } P \\
Cs_0 &= \text{DO saturation (mg/L) at sea level} \\
P &= \text{nonstandard pressure in atm} \\
P_{wv} &= \text{partial pressure of water vapor in atm} \\
\ln P_{wv} &= 11.8671 - 3840.70/T - 216961/T^2 \\
\theta &= 0.000975 - (1.426 \times 10^{-5}T) + (6.436 \times 10^{-8} \, T^2)
\end{aligned}
$$

4.5 SEDIMENT OXYGEN DEMAND

4.5.1 Field Measurement of SOD

Although this topic is often discussed in technical documents and one could easily obtain the SOD from the literature, the preferred approach for obtaining model input data is direct field measurement of sediment oxygen consumption upstream and downstream of the discharge. Consistent field techniques for determining SOD in natural waters are evolving, with a specific approach developed by Murphy and Hicks (1986). The two basic measurement techniques are *in situ* chambers, and sediment core extraction and laboratory measurement (Hatcher, 1986; Whittemore, 1986). The *in situ* method requires submersion of a chamber on the waterbody bottom with periodic measurements of oxygen to determine the uptake rate in the chamber. Laboratory measurements are based on a sample core from the sediment, preferably placed in a well-aerated column with oxygen measurements taken over time to determine the uptake rate. Different investigators have varying opinions on the relative merits of each technique; however, the use of *in situ* chambers, with minimal disturbance of the sample sediments, appears to be the preferred technique (Murphy and Hicks, 1986). [See Vigil (1992) for a comprehensive review of SOD measurement techniques.]

4.5.2 Determining SOD with Model Calibration

In many stream water quality modeling studies, the SOD values are obtained through model calibration. Calibrating the SOD values requires that other processes in the DO budget be accurately quantified via an independent data analysis. Figure 4-17 shows the SOD and its impact on the receiving water DO profiles in the Lower Minnesota River, the Upper Mississippi River, and Lake Pepin (all in the state of Minnesota). The first row in Figure 4-17 for each of the three study sites is the model calculated DO profile versus field data collected in 1988. The same modeling framework, WASP/EUTRO, is used for all three cases. Figure 4-17 shows that model results of DO match the data well. The calibrated SOD values in each case are displayed in the second row of panels. In general, SOD values between 2 and 3 gm O_2/m^2/day are commonly observed following the discharge of municipal wastewaters. Note the excessively high value of 5 gm O_2/m^2/day immediately below the Shakopee Plant discharge on the Minnesota River. Also note that the SOD progressively increases in the Upper Mississippi River toward Lock & Dam No. 2 as a result of yearly deposits accumulating behind the dam. On the other hand, the SOD values in Lake Pepin decrease slightly in the downstream direction. Because Lake Pepin is a natural impoundment on the Upper Mississippi River, there is no dam at the lake outlet to trap oxygen consumption material. The third row of panels in Figure 4-17 presents the DO deficit in mg/L resulting from the SOD along each of the three systems. Over 2 mg/L of DO deficit is calculated in the Upper Mississippi River behind Lock & Dam No. 2. Data in Figure 4-17 provide a perspective to the significance of SOD in these three areas.

4.5.3 SOD Reduction Following Treatment Upgrade

In wasteload allocations, the traditional approach in assigning the SOD values for model projections is a conservative one, that is, using the calibrated values for future conditions. As waste treatment increases to secondary and beyond, the occurrence of gross, noxious sludge deposits of raw sewage origin decreases (Thomann and Mueller, 1987). However, transient treatment plant bypasses and combined sewer overflows still can contribute to the benthal demand. In addition, algal biomass resulting from excessive nutrient inputs may contribute to sediment deposits and to the SOD. A good example of such a slow or near-zero SOD change over a long time period is seen in the Upper Mississippi River following successive treatment upgrades at the Metro Plant (see Figure 4-1). From 1964 to 1988, the SOD values stayed approximately the same (Figure 4-18) even when the Metro Plant was upgraded from primary treatment to secondary, and then to secondary with nitrification. The SOD values measured in the Upper Mississippi River in 1995 show a small reduction (see the open circles in Figure 4-18). In another case, HydroQual (1987) demonstrated that a reduction of total organic carbon loading to the Potomac Estuary from 92,540 lb/day in 1969 to 57,800 lb/day in 1985 resulted in a reduction of the mean SOD from 2.2 to 1.8 gm O_2/m^2/day. The relationship used to infer the long-term coupling

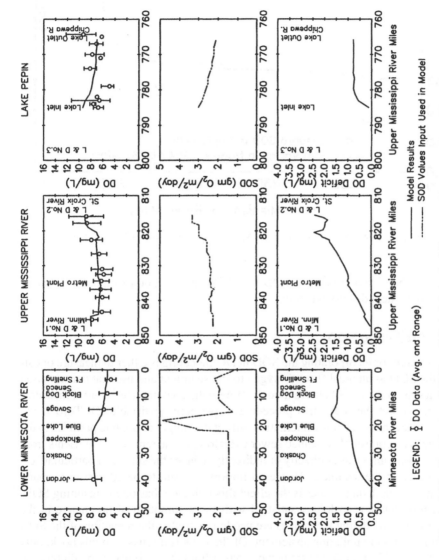

Figure 4-17 Contribution of SOD to DO deficit in the Minnesota River, Upper Mississippi River, and Lake Pepin.

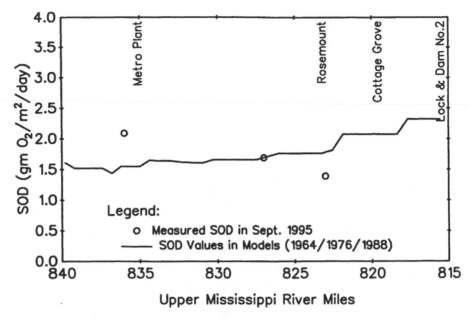

Figure 4-18 SOD rates in Upper Mississippi River from 1964 to 1995.

between carbon loading rate and SOD in the Potomac is not a simple formulation, leading to the next discussion topic.

4.5.4 Predicting SOD

The procedure for incorporating the SOD in DO modeling is either to measure or calibrate the SOD as an areal flux of oxygen to the sediment and use that consumption rate as a sink in the mass balance models. Where the contemplated control measures would not affect the SOD, this procedure is perfectly justifiable (Di Toro et al., 1990). However, most control alternatives affect the supply of organic particles to the sediment. The control of combined sewer overflows or the removal of nutrients from point and nonpoint sources directly or indirectly reduces the supply of particulate organic matter to the sediment. Hence an important issue to be addressed in many water quality modeling studies is the effect this reduction has on the resulting SOD. Today, modelers are no longer satisfied with the assumption of no change in the SOD. As such, modeling techniques are needed to predict the reduction of the SOD over time following pollution abatement. Of the models discussed in this book, only the WASP/EUTRO modeling framework has a limited capability of predicting the SOD under future conditions by configuring a sediment system dynamically interacting with the water column. The other models, such as STREAM and QUAL2E, lack this feature, and the user must assign the SOD as an external parameter.

The difficulty in SOD prediction is that SOD is not linearly proportional to the organic carbon loading in natural waters. For projecting future water quality conditions, it is advisable to use the same SOD values that were used in the model calibration and verification analysis. Such an approach will result in a somewhat conservative projection of future oxygen levels, since SOD is likely to be reduced following treatment upgrades. A more aggressive approach is to use a modeling framework that explicitly couples particulate organic carbon deposition and SOD (see Section 4.9), an approach based on the landmark paper by Di Toro et al. (1990). An approximating method is outlined below to assist modelers in quantifying the effects of nutrient removal on SOD. Note that the following steps to derive a range of SOD reductions are valid only for steady-state calculations:

1. Compute particulate and dissolved fluxes across the sediment-water interface.
2. Establish an aerobic/anaerobic nutrient flux switch (at DO = 1 mg/L).
3. Calculate direct flux subtraction. A change in nutrient and carbon flux to the sediment due to point source reduction would result in an equivalent reduction in sediment fluxes. For every gram/m^2/day reduction of particulate flux delivered to the bed, SOD delivered back to the water column would be reduced by a gram/m^2/day.
4. Allow for proportional flux reduction, assuming some degree of attenuation. In other words, the local sediment fluxes are reduced in proportion to the reduction of particulate flux to the bed averaged over the entire bottom surface area of the water system. This provides a lower bound for expected sediment nutrient flux reductions, since sediment nutrient fluxes would be reduced on a less than gram for gram basis.
5. Establish lower limits for sediment fluxes. For some systems, zero is not a reasonable lower limit for nutrient or SOD fluxes. There are always nutrients from other sources.

4.6   PHYTOPLANKTON AND DISSOLVED OXYGEN

4.6.1   Algal Oxygen Production and Respiration

In many BOD/DO models, algal photosynthetic oxygen production and respiration consumption are formulated with zero-order kinetics. In other words, average gross photosynthetic production of DO and respiration are represented as mg O_2/L/day. Three estimation methods are commonly used (Thomann and Mueller, 1987):

1. Estimation from observed chlorophyll levels
2. Light and dark bottle or chamber measurements of DO
3. Measurements of diurnal DO range

In the first method, given the algal chlorophyll a concentration measured in the field, one estimates the average daily oxygen production. A technique for performing this estimation, developed by Di Toro (1975), can be found in Thomann and Mueller (1987). The following equations are used:

$$P = 0.25 \text{ Chl } a \tag{4-10}$$

and

$$R = 0.025 \text{ Chl } a \tag{4-11}$$

in which

P = daily average oxygen production rate (mg O_2/L/day)
R = daily average respiration rate (mg O_2/L/day)
Chl a = measured chlorophyll a concentration (μg/L)

The light and dark bottle method is described in detail by Standard Methods (APHA, 1992). Clear glass (light) and foil-wrapped glass (dark) bottles are positioned and suspended at various fixed depths in the water column and filled with water collected at their respective depths. Usually, an attempt is made in deep waters to suspend the bottles at least to the depth of the euphotic zone, taken to be the 1% light level of the surface light intensity. Measurements of DO are made at regular time intervals. Since the light bottles receive solar radiation, net photosynthetic oxygen production $(P - R)$ is measured. The dark bottles, in the absence of light, measure gross respiration (R). Note that this method does not measure the photosynthetic activities of benthic algae and macrophytes, only the activities of the suspended algae. If there is significant attached algae or rooted plants, their photosynthetic contribution is not included.

The estimate of respiration (R) made from the dark bottles includes both algal respiration and bacterial respiration from oxidation of carbonaceous and nitrogenous compounds, that is, CBOD and NBOD. If the sample contains significant BOD, special care must be followed to determine the oxygen consumption due to algal respiration, as illustrated in the following calculation. The initial DO in a light and dark bottle test is 7 mg/L. After one day the DO in the dark bottle is 2 mg/L and the DO in the light bottle is 9 mg/L. The CBOD$_5$ of the water sample without algae (i.e., filtered sample) is 10 mg/L and the CBOD bottle rate, k_1, is 0.30 day^{-1}. Using Eq. 4-1,

$$10 = \text{CBOD}_u [1 - e^{-(0.30)(5)}]$$

yielding CBOD$_u$ equal to 12.9 mg/L. Thus CBOD$_1$ can be calculated as

$$\text{CBOD}_1 = 12.9[1 - e^{-(0.30)(1)}] = 3.34$$

In other words, the amount of oxygen consumed by bacteria for CBOD decay is 3.34 mg/L. The algal respiration is then equal to $7.0 - 2.0 - 3.34$ or 1.66 mg/L.

The productivity varies with depth and the results from different depths are synthesized to determine the depth-averaged productivity rate. The extent to which it is time-averaged depends on the period of the day covered by the measurements. Because of the significant variations in P with depth and time, care must be taken to ensure that light and dark bottle test results are interpreted correctly. It is also essential that the dissolved oxygen concentration in the light bottle does not reach its saturation level during the test. The maximum hourly increase in dissolved oxygen in the light bottle can be computed as follows:

$$\Delta C = \frac{a_{oc}a_c G_{max} 1.066^{(T-20)}}{(1,000)(24)} \text{Chl } a \qquad (4\text{-}12)$$

where

ΔC = maximum hourly increase in dissolved oxygen (mg/L/hr)
a_{oc} = stoichiometric ratio of oxygen to carbon = 2.67 (mg O_2/mg C)
a_c = stoichiometric ratio of carbon to chlorophyll a (mg C/mg Chl a)
G_{max} = maximum algal growth rate (day^{-1})
T = water temperature (°C)
Chl a = instantaneous chlorophyll a concentration (µg/L)

Equation 4-12 can be used to plan sampling intervals and maximum duration of light bottle tests.

For the third method Di Toro (1975) has developed an analytical equation, called the Delta method, to calculate P based on the measured diurnal dissolved oxygen range, Δ. As seen in the next section, the Delta method can also be used in reverse to estimate the diurnal range of dissolved oxygen with an estimate of P from the first two methods described in this section (Thomann and Mueller, 1987).

4.6.2 Diurnal DO and the Delta Method

As secondary treatment has become commonplace in this country, the emphasis of many wasteload allocation studies has shifted from bacterially mediated decomposition to the effect of aquatic plants on stream oxygen reserves (Chapra and Di Toro, 1991). As such, diurnal DO fluctuations in streams need to be quantified.

An analytical expression for the DO fluctuation (Δ) during the day is difficult to obtain since finding the exact time of maximum and minimum DO levels requires the solution of a transcendental equation (Di Toro, 1975). However, Di Toro derived a numerical solution by using an important property of Δ that is relatively insensitive to the reaeration coefficient, K_a, for $K_a < 2.0$ day^{-1}. The solution is a relationship between Δ/P_{av} and K_a, where P_{av} is daily average photosynthetic DO production (mg/L/day). The time-variable gross photosynthetic oxygen production can be estimated from diurnal DO data simply by observing Δ, the length of daylight, f, and Δ versus P_{av}. This provides a quick estimate of the algal oxygen production in terms of

P_{av}. Conversely, for a P_{av} calculated from chlorophyll a measurements, an estimate of the range of DO variation over the day can be made without further analysis. If K_a is not known precisely but is small, the estimate is still relatively good since the maximum algal oxygen production does not vary too much with varying K_a. Hence, this method provides a simple relationship between the range of diurnal DO fluctuation and the average daily algal oxygen production rate. Di Toro (1975) presented the following equation to estimate the diurnal DO variation due to algal photosynthesis:

$$\Delta = P_{av} \frac{\left(1 - e^{-K_a fT}\right)\left(1 - e^{-K_a T(1-f)}\right)}{fK_a\left(1 - e^{-K_a T}\right)} \tag{4-13}$$

where

Δ = range of diurnal DO (mg/L)
K_a = stream reaeration coefficient (day^{-1})
f = photoperiod (0–1.0)
T = 1 day
P_{av} = daily average photosynthetic DO production (mg/L/day)

Equation 4-13 is referred to as the Delta method. The value of Δ approaches a constant value independent of K_a as K_a decreases. On the other end, the special case for small K_a corresponds to the approximation:

$$e^{-K_a T} = 1 - K_a T$$

for $K_a < 0.2$ day^{-1} so that

$$\frac{\Delta}{P_{av}} = T(1 - f)$$

which is independent of K_a.

Applying the Delta method to Shirtee Creek in Alabama generates a plot of Δ/P_{av} for four different values of fT ranging from 8 hours to 14 hours (Figure 4-19). For a P_{av} calculated from chlorophyll a measurements, an estimate of the range of DO variation over the day can be made from Eq. 4-11. Figure 4-19 also indicates that for K_a values greater than 10 day^{-1}, Δ/P_{av} decreases and the fT value becomes less important. Thus for the K_a values in Shirtee Creek, the diurnal DO variations are not that significant as high reaeration coefficients release oxygen from the water column into the atmosphere as quickly as the oxygen is produced from algal photosynthesis. More specifically, a very high P_{av} value of 20 mg/L would yield a Δ of 2 mg/L in the water column.

The Delta method outlined above is useful in calculating the range of DO fluctuation under a steady-state condition. To calculate the real-time DO concentrations throughout the day requires a more elaborate analysis. In a total maximum daily load

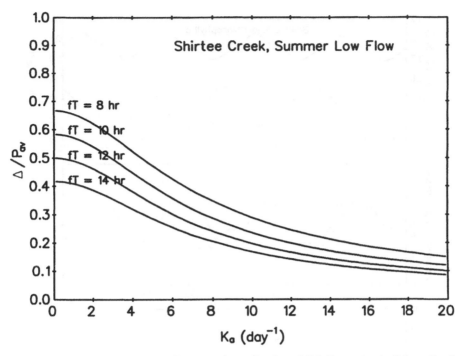

Figure 4-19 Using the Delta method to estimate the diurnal DO fluctuation in Shirtee Creek.

(TMDL) modeling study for the Santa Fe River in New Mexico, Lung et al. (2000) modified the EPA's WASP/EUTRO5 model to calculate real-time DO and pH levels in the river in response to the significant growth of attached, benthic algae in the water column. To support the modeling calculation, light intensity and water temperature values on a 15-min basis were incorporated into the model. Therefore, the light effect on the algal growth is instantaneous, that is, not being averaged over a day. Since the light level needed for the calculation is the value reaching the bottom of the water column, no depth-averaged light effect is required. The light effect on algal growth is simply as follows:

$$r = \frac{I}{I_s} \exp\left[-\frac{I}{I_s} + 1\right] \qquad (4\text{-}14)$$

where

> r = light reduction factor for algal growth rate
> I = real-time light intensity reaching the bottom of the water column
> I_s = saturated light intensity for optimum algal growth

The modified WASP/EUTRO5 model for the Santa Fe River performs well, producing real-time results of DO and pH, matching the measured values (Figure 4-20).

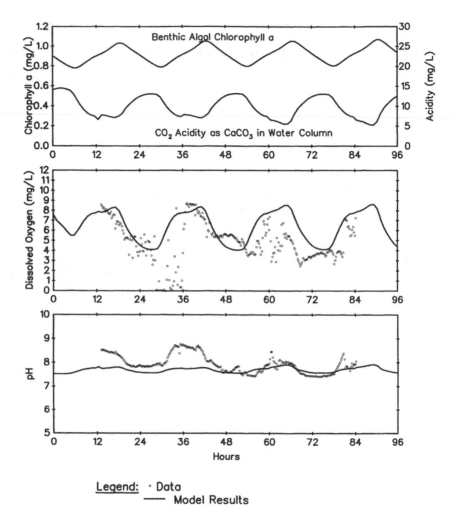

Figure 4-20   Modeling diurnal DO and pH fluctuations in the Santa Fe River, New Mexico.

4.6.3   Some Helpful Guidelines

The following observations prove to be helpful in BOD/DO modeling:

- Algae can affect the CBOD data used to calculate k_d. Algal respiration and decay in the BOD bottle can cause higher measured CBOD values and thus higher k_1 rates (in the lab) compared to samples without algae.
- If the concentration of suspended algae (i.e., algae that gets collected into the BOD bottle) is not constant in the stream below the outfall, then the measured CBOD will not indicate a defined decay rate, k_d.
- Algal impacts on k_d exist wherever high concentrations of chlorophyll a or large diurnal DO fluctuations occur.

- 10 µg/L of chlorophyll *a* will increase the $CBOD_u$ concentration by 1 mg/L above that without algae.
- If the stream is effluent dominated with most of the CBOD coming from the discharge rather than coming from algae, filtering the samples is not necessary. On the other hand, if the stream is not effluent dominated and most of the CBOD is from algae, both filtered and unfiltered samples should be run to provide data for the modeling effort.

4.7 DISSOLVED OXYGEN IMPACT

A typical DO response to point source BOD input is shown in Figure 4-21 for the Delaware River Estuary from Trenton, New Jersey, to the Delaware Bay. The data shown were collected in a low water slack survey during August 1975 under dry weather conditions. At the time of the water quality survey, not all Philadelphia area wastewater treatment plants were providing secondary treatment. The combined impact of these point source BOD loads results in a sharp decline of DO (Figure 4-21),

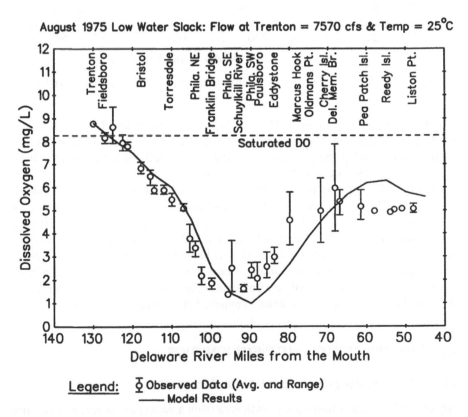

Figure 4-21 Model results versus DO data of the Delaware River Estuary, August 1975 low water slack.

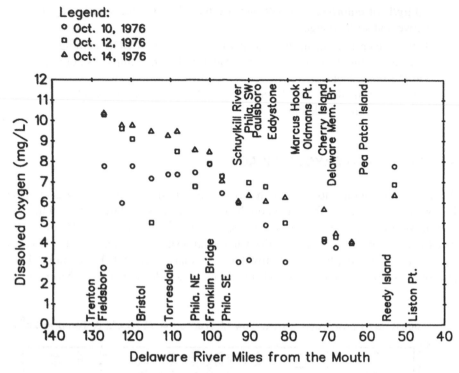

Figure 4-22 Wet weather DO response in the Delaware River Estuary, October 1976.

with a minimum DO near River Mile 90. Subsequent recovery of DO follows primarily due to reaeration. Many steady-state BOD/DO models are capable of mimicking this profile and quantifying the cause-and-effect relationship between the BOD loads and the DO response in the receiving water.

Figure 4-22 shows the DO response in the Delaware River Estuary following a storm event. The wet weather DO profile observed from October 10 to October 14, 1976, is quite different from that observed during the dry weather. During the first wet weather survey, the DO levels are much depressed following the combined sewer overflow discharges. By October 14, the DO levels begin to recover slowly. To reproduce such a transient response, a time-variable model must be used not only with the freshwater flow changes, but also with the pollutant input loads. For the stormwater related DO modeling, fine temporal resolution is required in the modeling analysis.

4.8 PHYTOPLANKTON/NUTRIENT KINETICS

4.8.1 Temperature-Dependent Algal Growth Rate

Eppley (1972) summarized algal growth data from a variety of sources as a function of temperature and developed the following equation:

$$G_T = G_{max}\theta^{T-20} \tag{4-15}$$

where

$\quad G_T$ = temperature-adjusted growth rate (day^{-1})
$\quad G_{max}$ = maximum growth rate at 20°C (day^{-1})
$\quad \theta$ = constant for temperature adjustment
$\quad T$ = temperature (°C)

Both G_T and G_{max} are specific growth rates under optimum light and nutrient conditions. Reported ranges for G_{max} and θ are G_{max} = 1 to 3 day^{-1} at 20°C and θ = 1.01 to 1.18.

4.8.2 Light Effect on Phytoplankton Growth

A depth- and time-averaged effect of available light energy on phytoplankton growth rate can be obtained (Di Toro et al., 1971) by integrating the light intensity relationships over depth and time. This reduces to

$$r_L = \frac{ef}{HTK_e}(e^{-\alpha_1} - e^{-\alpha_2}) \tag{4-16}$$

and

$$\alpha_1 = -\frac{I_a}{I_s}e^{-K_eH}$$

$$\alpha_2 = e^{-I_a/I_s}$$

where

$\quad r_L$ = light limitation factor
$\quad e$ = 2.718
$\quad f$ = photoperiod—daylight fraction of averaging period (day)
$\quad T$ = averaging period = 1 day
$\quad K_e$ = light extinction coefficient (m^{-1})
$\quad H$ = average depth of water column layer (m)
$\quad I_a$ = average of incident light over photoperiod (langley/day)
$\quad I_s$ = saturated light intensity (langley/day)

The complete derivation of Eq. 4-16, starting with Eq. 4-14, is presented in Table 4-6. Solar radiation is measured routinely at selected weather stations in the United States and is usually reported as daily values in langley (ly), which is a measure of the total radiation of all wavelengths that reach the surface of the earth. Since visible light energy is measured in various ways, the following light unit conversions are useful when relating one measurement to another:

TABLE 4-6 Calculating Average Algal Growth Rate by Integrating the Light Intensities Relationships over Depth and Time

$$r_l = \frac{1}{H} \int_0^H \int_0^f \frac{I(z,t)}{I_s} \left[e^{-\frac{I(z,t)}{I_s}+1} \right] dt \, dz$$

$$I_o(t) = I_a \qquad 0 < t < f$$

$$I_o(t) = 0 \qquad f < t < 1$$

$$r_l = \frac{1}{H} \int_0^H \frac{1}{T} \int_0^f \frac{I_a e^{-K_e z}}{I_s} \left[e^{-\frac{I_a e^{-K_e z}}{I_s}+1} \right] dt \, dz$$

$$= \frac{ef}{TH} \int_0^H \frac{I_a e^{-K_e z}}{I_s} \left[e^{-\frac{I_a}{I_s} e^{-K_e z}} \right] dz$$

Let

$$y = \frac{I_a}{I_s} e^{-K_e z}$$

$$y = \frac{I_a}{I_s} \qquad at \qquad z = 0$$

$$y = \frac{I_a}{I_s} e^{-K_e H} \qquad at \qquad z = H$$

$$dy = \frac{I_a}{I_x} (-K_e) e^{-K_e z} \, dz = -K_e y \, dz$$

Thus

$$y \, dz = -\frac{1}{K_e} dy$$

$$r_l = \frac{ef}{HT} \int_{I_a/I_s}^{(I_a/I_s)e^{-K_e H}} \left(-\frac{1}{K_e} e^{-y} \right) dy$$

$$= \frac{ef}{HTK_e} \int_{I_a/I_s}^{(I_a/I_s)e^{-K_e H}} -e^{-y} \, dy = \frac{ef}{HTK_e} \int_{(I_a/I_s)e^{-K_e H}}^{I_a/I_s} e^{-y} \, dy$$

$$= \frac{ef}{HTK_e} (-1) e^{-y} \Big|_{(I_a/I_s)e^{-K_e H}}^{I_a/I_s} = \frac{ef}{HTK_e} e^{-y} \Big|_{I_a/I_s}^{(I_a/I_s)e^{-K_e H}} = \frac{ef}{HTK_e} [e^{-(I_a/I_s)e^{-K_e H}} - e^{-I_a/I_s}]$$

---

1 langley = 1 g cal/cm^2

1 g cal = 4.185 watt-sec

1 foot candle ≈ 10.76 lux

1 einstein ≈ 52 × 10^3 g cal

The light limitation factor, r_L, which represents a percentage of the optimum growth rate, is sensitive to HK_e, a dimensionless number, also referred to as the light extinction factor.

The key parameter in Eq. 4-16 is the light extinction coefficient, K_e. It can be determined from direct measurements of light intensity in the water column as light intensity versus depth follows this relationship:

$$I(z) = I_o e^{-K_e z} \tag{4-17}$$

where I_o is surface light intensity and z is depth in the water column. The slope of $\ln(I/I_o)$ versus depth, z, provides an estimate of K_e. Figure 4-23 shows the results of quantifying K_e with light intensity data from White Lake, Michigan. The light measurements in the water column, that is, the light measurements at various depths, are recorded as a percentage of the water surface light intensity. Plotting the percentage versus depth shows an exponential decrease of light intensity. Results in Figure 4-23 also show seasonal variations of light extinction, primarily due to the composition of the suspended solids (i.e., more algal biomass during the growing season).

When light intensity data are not available, an approximation can be made with a regression analysis between the light extinction coefficient and Secchi depth:

$$K_e = \frac{C}{H_s} \tag{4-18}$$

where C is a constant ranging from 1.2 to 5.0 (Holmes, 1970) and H_s is the Secchi depth. The large range of C is due to the poor correlation of the light extinction coefficient and Secchi depth in many waters.

Di Toro (1978) has provided a theoretical and empirical basis for estimating the light extinction coefficient as a function of nonvolatile suspended solids, detritus (nonliving organics), and phytoplankton chlorophyll a:

$$K_e = 0.052N + 0.174D + 0.031 \, \text{Chl} \, a \tag{4-19}$$

where

$\quad N =$ nonvolatile suspended solids concentration (mg/L)
$\quad D =$ detritus concentration (mg/L)
$\text{Chl} \, a =$ chlorophyll a concentration (µg/L)

The nonvolatile suspended solids (the fixed inorganic particulate) both absorb and scatter light, whereas the organic detritus and phytoplankton biomass mainly absorb light. Equation 4-19 yields K_e values in m^{-1} and is valid for $K_e < 5$ m^{-1} (Di Toro, 1978). The last term in Eq. 4-19 is referred to as the algal self-shading effect.

Because there were no direct measurements of light intensity in the water column, Eq. 4-17 could not be used to calculate the light extinction coefficients in a modeling study of the Upper Mississippi River by Lung (1996a). Instead, Eq. 4-19 is used with

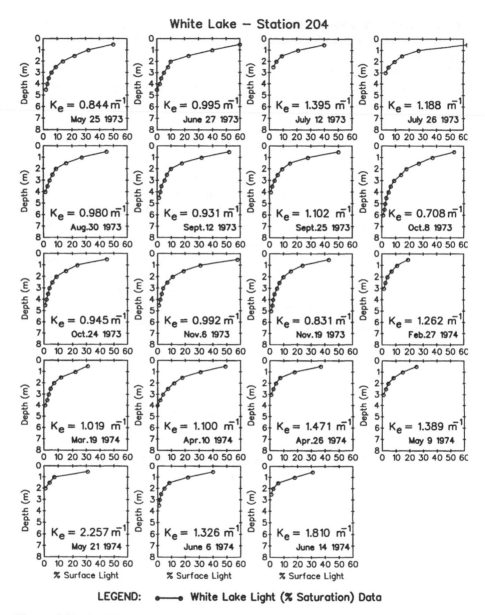

Figure 4-23 Deriving the light extinction coefficient, K_e, in White Lake using water column light intensity measurement data in 1974.

the measured total suspended solids and volatile suspended solids concentrations from a water quality survey. First, the detritus concentration is derived by subtracting the dry weight of the algal biomass from the volatile suspended solids. As an approximation, the ratio of algal biomass dry weight to algal chlorophyll a is 100–1. That is, a 100 µg/L chlorophyll a concentration is equivalent to 10,000 µg/L, or 10 mg/L, of suspended solids. With this conversion, one can calculate the detritus concentration for

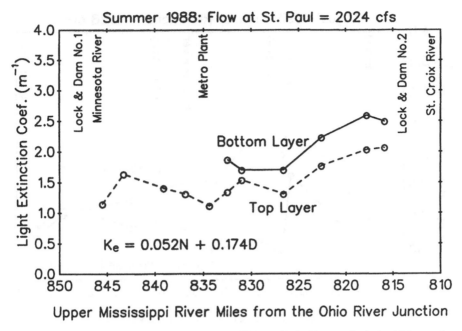

Figure 4-24 Deriving the light extinction coefficient, K_e, in Upper Mississippi River using measured water column total and volatile suspended solids concentrations.

use in Eq. 4-19. The calculated K_e values based on the first two terms of Eq. 4-19 and used as model inputs, are shown in Figure 4-24. Note the progressive increase in the solids concentration toward Lock & Dam No. 2. The algal self-shading effect is then added to this K_e value by the water quality model to yield the total light extinction.

The situation is slightly different in the James Estuary, Virginia, where only the Secchi depth data are available. First, Eq. 4-18 is used to calculate the total K_e in the water column. The algal self-shading effect is then subtracted from this value for input into the water quality model. When applying Eq. 4-18 to the James Estuary, the C value is 1.2 (Cerco, 2000). Figure 4-25 shows the calculated K_e for model input, characterizing a spatial variation of light extinction coefficients in the system. Note the maximum K_e is located at mile point 77, which is where the turbidity maxima is observed in the James Estuary under that freshwater flow condition. In other words, suspended solids concentration at this location is higher than at any other location along the estuary.

4.8.3 Nutrient Effect on Phytoplankton Growth

The phytoplankton growth rate is also a function of nutrient concentrations up to a saturating condition, after which it remains constant with increasing nutrients. At zero nutrient concentration, there is no growth. As the nutrient level is increased, the growth rate is linearly proportional to the availability of nutrients. However, as nutrient levels continue to increase, the effect on the growth rate of the phytoplankton

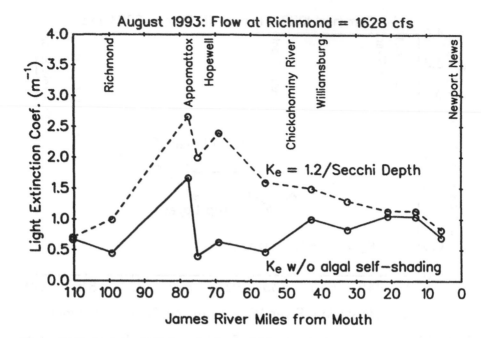

Figure 4-25 Deriving the light extinction coefficient, K_e, in James Estuary using water column Secchi depth data.

becomes saturated. Such an overall nonlinear relationship is described by a Michaelis–Menton formulation where the nutrient reduction factor, or nutrient effect, for algal growth, r_N, is:

$$r_N = \frac{Nut}{K_m + Nut} \tag{4-20}$$

where *Nut* is the inorganic nutrient concentration (μg/L) and K_m is the half-saturation constant (μg/L). K_m varies with nutrients, that is, with orthophosphate, ammonium plus nitrite/nitrate, and silica. The K_m values can be found in the literature with a wide variation range (U.S. EPA, 1995). With more than one nutrient accounted for in the model, the nutrient limitation effect is the minimum of the three individual r_N factors. By combining the temperature, light, and nutrient effects, the full expression for algal growth is given:

$$G_p = G_{max}(1.066)^{T-20}\left[\frac{ef}{HTK_e}(e^{-\alpha_1} - e^{-\alpha_2})\right]$$

$$\min\left[\frac{DIN}{K_{mn} + DIN}; \frac{DIP}{K_{mp} + DIP}; \frac{Si}{K_{Si} + Si}\right] \tag{4-21}$$

where

DIN = sum of ammonium, nitrite, and nitrate concentrations
DIP = dissolved inorganic phosphorus concentration
 Si = dissolved inorganic silica concentration
K_{mn} = Michaelis–Menton constant for nitrogen
K_{mp} = Michaelis–Menton constant for phosphorus
 K_{si} = Michaelis–Menton constant for silica

4.8.4   Algal Death Rate

The endogenous respiration rate, D_{p1}, is given by:

$$D_{p1} = \mu_R(1.08)^{T-20} \qquad (4\text{-}22)$$

where μ_R varies from 0.05 to 0.25 day^{-1} (Thomann and Mueller, 1987). A value of 0.15 day^{-1} is usually used as a first approximation.

4.8.5   Phytoplankton and Organic Nutrient Settling Rate

Phytoplankton settling rate is estimated by dividing the settling velocity by the stream depth. Phytoplankton settling velocities can be found in U.S. EPA (1995). Additional data are available in a review by Smayda (1970). Some phytoplankton such as blue-green algae develop gas vacuoles, which result in buoyancy and subsequent aggregation at the water surface. The proliferation of such species is a particular problem since the settling velocity may be zero or even negative, causing phytoplankton to remain in the water column or at the surface.

Note that algal biomass is part of the particulate organic matter that settles in the water column. That is, organic phosphorus and nitrogen in the algal biomass settle at the same velocity as algae. In many eutrophic systems, this particulate organic matter is the primary source of organic material to the sediment.

While algal settling velocity is usually formulated as a spatial and temporal constant in many water quality models, modeling total nutrients such as total phosphorus and total nitrogen should allow temporal variations of the settling rate. This is particularly justified for water systems where organic phosphorus in the algal biomass is the dominating portion of total phosphorus in the water column. An interesting example is Platte Lake, in Michigan, where total phosphorus concentrations are relatively low, usually below 20 µg/L throughout the year. Figure 4-26 shows that the model results match the observed data quite well in a two-layer model configuration with a seasonal variable settling rate for total phosphorus in the water column. The seasonally variable settling rate follows the pattern of algal biomass throughout the year quite closely. On the other hand, an overall average settling velocity of 23 m/yr does not give acceptable results (Figure 4-26).

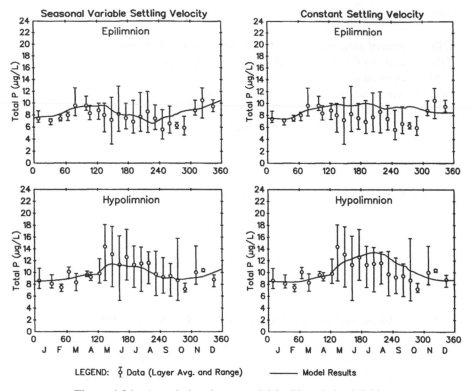

Figure 4-26 A total phosphorus model for Platte Lake, Michigan.

4.8.6 Algal Biomass Stoichiometry

Dry weight biomass is related to major nutrients (carbon, nitrogen, and phosphorus) and chlorophyll *a* through stoichiometric ratios that give the ratio of each nutrient to the total biomass. Typical algal nutrient compositions are summarized in U.S. EPA (1995). Ratios for different algal groups or species (blue-green, diatom, etc.) can be found in the literature.

4.8.7 Phosphorus Mineralization Rate

The rate of transformation from particulate phosphorus to orthophosphate in the water column ranges from 0.02 to 0.10 day^{-1} (Bowie et al., 1985). As a first approximation, a value of 0.03 day^{-1} may be used. Another source of information is U.S. EPA (1995).

4.8.8 Sediment Nutrient Release Rates

Like SOD, sediment nutrient release rates such as ammonium and orthophosphate fluxes are mostly measured or model calibrated. One extensive set of reported field work is on the Anacostia River in Washington, D.C. Combined sewer overflows (CSO) draining the District of Columbia have been the primary sources of pollution to the Anacostia River (Nemura, 1992). Two field surveys of the sediment along an 11-mile length of the river were conducted by Sampou (1990) in May and August 1990, measuring SOD, methane, ammonium, nitrate, and orthophosphate fluxes at nine locations (Figure 4-27). In Figure 4-27, positive values represent a flux from the sediment to the overlying water, that is, a source of nutrients to the Anacostia River. Negative values represent a flux to the sediment, that is, a sink. In addition to the fluxes, total carbon, nitrogen, and phosphorus concentrations in the sediments, and porewater nutrient concentrations were also measured.

Thomann and Mueller (1987) show some reported nutrient fluxes from the sediments under both aerobic and anaerobic conditions. When the overlying water is anaerobic, the flux of phosphorus from the sediment increases significantly as a result of increased diffusion between the sediment and the water. Such increased diffusion results from changes in the iron complexes at the water-sediment interface.

Table 4-7 lists SOD and nutrient fluxes observed in other systems. Note that measurements near CSO outfalls are higher than those away from the CSOs. Also note that point source pollution is insignificant when compared with nonpoint source pollution and that CSO is currently one of the most sizable nonpoint loads to many receiving waters. Particulate organic material discharged by CSOs can settle to the sediment of the receiving water. The reactive component of this organic material decomposes within the sediment through anaerobic bacterial activity in a process referred to as diagenesis. In freshwater systems the end products of diagenesis are methane and ammonia, both of which are subject to oxidation as they diffuse into the aerobic layer of the sediment near the sediment-water interface. This oxidation is SOD.

4.9 MODELING SEDIMENT DIAGENESIS PROCESSES

As part of the program to abate the CSO impact on the Anacostia River, studies were made to evaluate the water quality benefits resulting from CSO controls. A modeling framework was developed by HydroQual (1992) to quantify the expected changes in SOD and ammonia flux following the implementation of potential CSO control measures. The combination of measured fluxes, sediment production rates (Figure 4-27), and standing concentrations in the sediment was used to support this modeling effort.

The key kinetic processes in this modeling framework for SOD and ammonia flux are displayed in Figure 4-28. The diagenesis reactions are assumed to convert particulate organic carbon (POC) and particulate organic nitrogen (PON) to methane and ammonia, respectively. The diagenesis reaction can be represented as follows:

$$(CH_2O)_a NH_3 \rightarrow \frac{a}{2} CH_4 + \frac{a}{2} CO_2 + NH_3 \qquad (4-23)$$

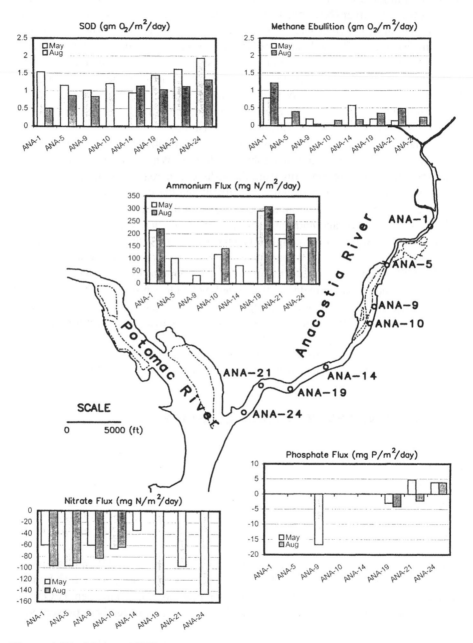

Figure 4-27 Measured SOD, methane, ammonium, nitrate, and phosphate flux rates across the sediment–water interface in the Anacostia River, 1990.

TABLE 4-7   Comparison of SOD and Nutrient Flux Rates in Natural Water Systems

Sediment Oxygen Demand	Gram O_2 m^{-2} d^{-1}	Reference
Tidal Anacostia River, 1990	0.5 to 1.9	Sampou (1990)
Tidal James River, summer 1983–84	1.0	Cerco (1985a)
Upper Potomac Estuary, 1986	1.8	HydroQual (1987)
Hunting Creek, summer 1986	4.1	HydroQual (1987)
Patuxent Estuary, 1980	2.0 to 4.0	Boynton et al. (1980)
Gunston Cove, summer 1984	2.3	Cerco (1985b)

Sediment Ammonium Release	mg N m^{-2} d^{-1}	Reference
Tidal Anacostia River, 1990	33 to 309	Sampou (1990)
Tidal James River, near CSO	740	Cerco (1985a)
Tidal James River, not near CSO	28	Cerco (1985a)
Gunston Cove	125	Cerco (1988)
Gunston Cove and Potomac	4 to 6	Seitzinger (1985)
Patuxent Estuary, 1980	70 to 530	Boynton et al. (1980)
Upper Potomac Estuary	50 to 70	HydroQual (1987)

Nitrate Flux to Sediment	mg N m^{-2} d^{-1}	Reference
Tidal Anacostia River, 1990	33 to 145	Sampou (1990)
Tidal James River, near CSO	50	Cerco (1985a)
Tidal James River, not near CSO	76	Cerco (1985a)
Gunston Cove	40	Cerco (1988)
Gunston Cove and Potomac	35 to 49	Seitzinger (1985)
Patuxent Estuary, 1980	25 to 100	Boynton et al. (1980)

Sediment Phosphate Release	mg P m^{-2} d^{-1}	Reference
Tidal Anacostia River, 1990	–5 to 17	Sampou (1990)
Tidal James River, near CSO	63	Cerco (1985a)
Tidal James River, not near CSO	4	Cerco (1985a)
Gunston Cove	10	Cerco (1988)

where a is the reactive organic carbon to nitrogen ratio. This anaerobic reaction consumes no net oxygen. It produces a more reduced (CH_4) and a more oxidized (CO_2) carbon end product and does not affect the oxidation state of nitrogen.

The reactions that determine the magnitude of the oxygen flux to the sediment, are the oxidation of methane and ammonia, occurring in the aerobic zone of the sediment. The stoichiometry for the oxidation of methane is given by:

$$CH_4 + 2O_2 \rightarrow CO_2 + 2H_2O \qquad (4\text{-}24)$$

Equation 4-24 indicates that 5.33 grams of oxygen are required for each gram of methane-carbon oxidized. As Eq. 4-23 shows, one half of a mole of methane carbon is produced for each mole of sedimentary organic carbon that is decomposed. Therefore,

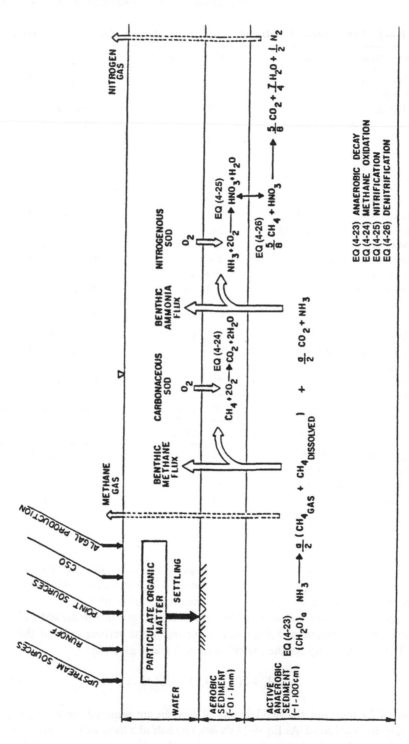

Figure 4-28 Modeling sediment–water interactions with carbon and nitrogen diagenesis kinetics.

the overall oxygen consumption stoichiometry of sedimentary organic material is 2.67 gm O_2/gm carbon.

The oxidation of ammonia proceeds in two steps. Ammonia is oxidized to nitrate by the nitrification reaction:

$$NH_3 + 2O_2 \rightarrow HNO_3 + H_2O \qquad (4\text{-}25)$$

If no further reaction occurred, the sediment would provide a flux of nitrate to the overlying water. Experimental data indicate that most if not all of the nitrate produced is denitrified to nitrogen gas since the amount of nitrogen gas present in the gas produced by the sediments cannot otherwise be explained (HydroQual, 1992). It is assumed, therefore, that the nitrate produced in the aerobic zone of the sediment is denitrified to nitrogen gas via the reaction:

$$\frac{5}{8}CH_4 + HNO_3 \rightarrow \frac{5}{8}CO_2 + \frac{1}{2}N_2 + \frac{7}{4}H_2O \qquad (4\text{-}26)$$

where methane is the electron donor. Hence the denitrification reaction is coupled to the methane reactions, yet it would be convenient to sidestep this complexity in some way. The focus of the denitrification reaction—whether it occurs in micro-anaerobic zones in the aerobic layer or just below in the anaerobic layer—is uncertain. One way of avoiding the coupling is to assume that the methane sink reduces the quantity of methane that is available for direct oxidation via Eq. 4-24. Hence, if the equivalent oxygen not consumed is substituted for the methane consumed in Eq. 4-26, the overall nitrification-denitrification reactions (Eqs. 4-25 and 4-26) become:

$$NH_3 + \frac{3}{4}O_2 \rightarrow \frac{1}{2}N_2 + \frac{6}{4}H_2O \qquad (4\text{-}27)$$

which is simply the oxidation of ammonia to nitrogen gas directly with an oxygen consumption stoichiometry of 1.714 gm O_2/gm N. For a complete development of these equations, refer to Di Toro et al. (1990).

In a recent study to develop a TMDL modeling framework for the Anacostia River, Lung (2000) implemented such an analysis using the EPA's WASP/EUTRO5 model. The resulting model is capable of calculating the internal fluxes across the sediment–water interface on a real-time basis for multiyear model runs. Figure 4-29 shows the model calculated carbon and nitrogen diagenesis rates in the sediment as well as SOD, methane, and nitrogen flux rates across the sediment–water interface. Note that methane gas is produced when its aqueous concentration exceeds its solubility in the pore water of the sediment. It is seen from the results that the diagenesis rates, SOD, and nutrient fluxes strongly depend on the seasonal pattern of water temperature in the system, reaching maximum levels during the summer months.

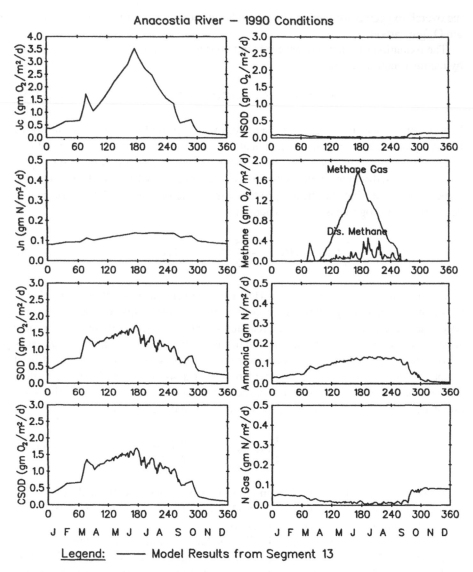

Figure 4-29 Model calculated SOD, methane, and nitrogen fluxes from the sediment in the Anacostia River.

4.10 CALIBRATING THE NUTRIENT/EUTROPHICATION KINETICS

Compared with the BOD/DO kinetics, the nutrient/eutrophication kinetics have far more parameters and coefficients that need to be quantified for model input. In practice, rarely are all these coefficient values independently determined from field and laboratory data. Often budget constraints limit the field effort on special studies to support the modeling analysis. In well-studied systems, data from other investigators may be available to guide the modeler on quantifying individual kinetic processes. For example, phytoplankton productivity study data can be used to check the algal growth rate used in the model. Nevertheless, model calibration is the common approach toward finalizing the nutrient/eutrophication coefficients. When field data are not available to independently derive the coefficient values, literature data are used as preliminary estimates for model input. Subsequent model calibration and sensitivity analyses will fine tune these coefficients by reproducing the observed data. Also, multiple data sets or multiple-year model runs are commonly used in the model calibration effort. Every attempt is designed to minimize the uncertainty with respect to the kinetic coefficients. Note that model calculations are not intended to curve fit the data. Rather the calibrated coefficients (e.g., algal growth rate) are obtained through a series of model sensitivity runs with reasonable and narrow ranges of their values derived from literature and other modeling studies. In many model sensitivity analysis cases, adjusting the kinetic coefficients and constants (within their narrow ranges) to improve the match with the observed concentration of a given water quality constituent often results in an adverse outcome of matching other water quality constituents. These constraints in the model calibration process eventually lead to the development of an acceptable set of model coefficients.

In the recent Chesapeake Bay modeling study, the phytoplankton growth rates (day^{-1}) selected for the model calibration analysis consistently yielded primary production rates that are appreciably lower than most of the *in situ* rates measured by scientists working on the Bay system. When the algal growth rate is increased to obtain the primary productivity rate matching the observed values, the model results significantly overpredict the phytoplankton biomass, that is, the chlorophyll *a*, and underpredict the nutrient concentrations in the water column. In other words, nutrients are soon depleted from the water column due to the strong growth of phytoplankton. The modeler must be careful in utilizing the data and interpreting the model results.

4.11 POST AUDIT OF BOD AND EUTROPHICATION MODELS

Post audit is the evaluation of system performance following actual implementation of environmental control facilities. In the DO models, three questions are addressed (Thomann, 1987):

1. Does the actual DO data following a treatment upgrade generally reflect the basic principles of DO models; that is, does the DO go up when the BOD goes down?
2. To what degree are the DO models successful in predicting quantitatively the observed DO?
3. Does the accuracy of the DO models really matter in the decision regarding the installation of treatment facilities?

The case study on the Upper Mississippi River (see Figure 4-1) represents an excellent post audit analysis of a BOD/DO model. The Metro Plant's 1982 NPDES permit, based on a wasteload allocation study in 1981, required a river water quality assessment study to provide additional information on these areas and to evaluate river water quality following the plant upgrade to advanced secondary treatment (including nitrification) in 1985. The study was completed at the end of 1985 except for one key element: an intensive low flow water quality survey. Low flow conditions did not occur until 1988 at which point the survey was conducted.

The post audit of the BOD/DO model of the Upper Mississippi River is summarized in Figure 4-30. Hydroscience, Inc. (1979) developed a BOD/DO model for the Upper Mississippi River from Lock & Dam No. 1 to Lock & Dam No. 2 (Figure 4-1) using four data sets. Data sets of August 1973 and August 1976 are shown in Figure 4-30, along with their model results, matching the data well. The Minnesota Pollution Control Agency (MPCA) used another data set collected in 1980 (Figure 4-30) in their wasteload allocation study, setting the requirement for nitrification. Lung (1996a) used the 1988 data set to post audit the BOD/DO model when the nitrification process was in full operation. The first question above is fully satisfied. Concentrations of DO have improved to above the 5.0 mg/L standard even during the summer 1988 low flow condition. To answer the second question, the DO model is successful in predicting the observed DO with one exception. The CBOD deoxygenation rate, k_d, is decreased from 0.25 day^{-1} prior to 1981 to 0.073 day^{-1} in 1988. This decrease is an important result from the post audit analysis. In the current construct of the BOD/DO model, the lower k_d rate is necessary to reflect the stabilization of the organic carbon matter from the Metro Plant.

Next, the Patuxent Estuary is examined in an eutrophication post audit analysis of nutrient controls. The top two panels in Figure 4-31 show the total phosphorus and nitrogen loads from point sources to the Patuxent Estuary from 1985 to 1995. The impact of the Maryland phosphate detergent ban is clear, reducing the phosphorus loads by 50% in 1985–1986. Continuing improvements in wastewater treatment have further reduced the phosphorus loads by another 50%, to about 100 kg/day by 1995. Nitrogen load reductions took place in 1991 when biological nutrient removal was installed at the wastewater treatment facilities. Note the seasonal trend in the nitrogen loads due to the seasonal nature of the nitrogen removal process. A factor of four in nitrogen load reduction is obtained by this process.

The algal biomass (chlorophyll a) peak was consistently observed at Nottingham prior to the nutrient reduction and the bottom layer dissolved oxygen concentrations were close to anoxic near Broomes Island in the Patuxent Estuary. The third and fourth panels of Figure 4-31 show these two water quality constituents from 1985 to 1996, covering the period of phosphate detergent ban, treatment improvement, and biological nutrient removal. The data show no significant improvement in these two parameters. High chlorophyll a levels are still being measured from year to year following the nutrient controls, and the anoxic condition near Broomes Island has also not improved.

Lung (1992) has developed a water quality model for the Patuxent Estuary based on the data collected in 1983, 1984, and 1985, that is, during a period prior to nutrient

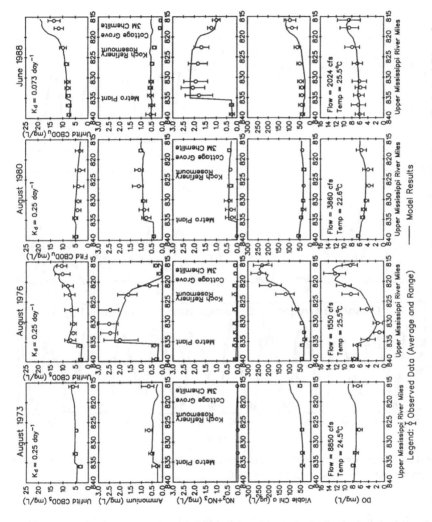

Figure 4-30 Results of a BOD/DO model post audit analysis of the Upper Mississippi River from Lock & Dam No. 1 to Lock & Dam No. 2 following treatment upgrades.

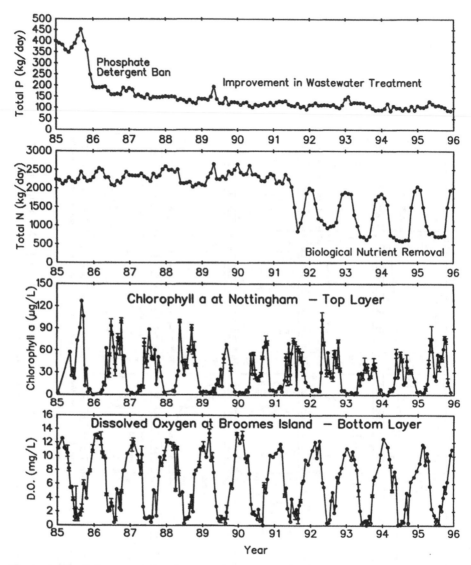

Figure 4-31 Point source phosphorus and nitrogen load reductions to the Patuxent Estuary and receiving water response from 1985 to 1996.

controls. It was interesting to see whether the model could predict the water quality response following the nutrient controls. Results from the model post audit analysis are shown in Figure 4-32 for 1985 and 1995, before and after the nutrient controls, respectively. The model results, like the field data, do not show any significant water quality improvement over this 10-year period. These results suggest that eutrophication in the Patuxent Estuary is controlled by light attenuation in the water column rather than by nutrients (Lung, 1998). Such a result from the Patuxent Estuary is not

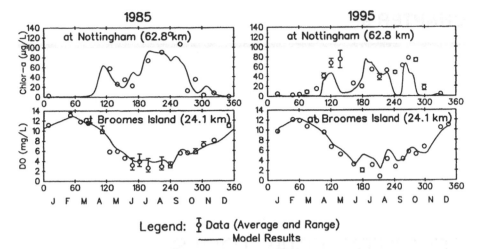

Figure 4-32 Results of a eutrophication model post audit analysis of the Patuxent Estuary following point source nutrient reductions.

unique. Studies from many eutrophication analyses have concluded that the water quality benefit from nutrient controls usually takes a long period of time to appear, if indeed it ever does.

To conclude this chapter, a perspective by Thomann (1987) provides an excellent reflection on both BOD/DO and nutrient/eutrophication modeling:

1. The aquatic plant/nutrient problems are the most difficult models with which we have worked because of the complexity of the plant biology, the non-linear interactions between nutrients and aquatic plants and the interactions of the sediment.

2. Dissolved oxygen, in spite of its long studied history, tends to be considerably more complex than generally believed.

3. Sediment interactions are important to all water quality problem contexts and credible interactive sediment models are now appearing for wasteload allocations

CHAPTER 5

COMPUTATIONAL TOOLS
AND ACCESSORIES

This chapter discusses a number of computational tools that support water quality modeling for wasteload allocations and total maximum daily loads (TMDLs). A computer system, including hardware and software, is a must-have. An important accessory in TMDL work is the geographical information system (GIS), which is particularly useful in watershed modeling. Many modeling TMDL studies require a watershed model to generate nonpoint source flows and pollutant loading rates to the receiving water. Finally, the post processing and visualization tool, which is an essential part of a modeling study, is discussed.

5.1   COMPUTER HARDWARE AND CAPABILITY

A wide range of computer systems is being used in water quality modeling. For example, the Chesapeake Bay water quality model, which has 12,000 cells in the three-dimensional water column, is run on a Cray supercomputer system due to its intensive computation requirement. (See Figure 5-1 for the two-dimensional horizontal grid of the Chesapeake Bay model. The water column is divided into layers to form the three-dimensional grid.) The computationally intensive tasks of hydrodynamic and water quality modeling of the Chesapeake Bay justify the use of supercomputers. No other computer systems are capable of handling the 10-year time-variable runs that are frequently made for the Bay study.

While supercomputer systems are powerful, they are not readily accessible to water quality professionals in general. Neither are they needed in many wasteload allocation and TMDL studies. In fact, most TMDL water quality modeling applications can be performed on Pentium-based personal computers (PCs). The Chesapeake Bay

10   0   10   20 Miles

Figure 5-1   The Chesapeake Bay hydrodynamic and water quality model grid system.

watershed model, which is based on the HSPF modeling framework, is run on PCs. Using PCs in hydrodynamic and water quality modeling has come a long way since 1980s (Lung, 1987). Due to the rapid advances that have been made in computer hardware, PCs are now used in many water quality modeling studies, providing levels of computational effort that were not even viewed as possible 5 years ago.

The biggest advantage of running hydrodynamic and water quality models on PCs is that PCs represent the most widely used computer platform in service today. Models developed on a PC system (in the form of an executable image) can be transferred to another PC system without much difficulty. Workstations are also popular and powerful, but their performance has been equaled by PCs in the last year or two. Another reason that PCs are so popular in water quality modeling these days is their low prices. The trend of increasing computation power at a decreasing price is continuing in the PC industry.

5.2   MODELING FRAMEWORKS AND FORTRAN COMPILER

Many widely used models, such as QUAL2E, WASP/EUTRO, CE-QUAL-W2, HSPF, and BASINS, are constantly being updated or enhanced. It is important to obtain the latest version of the computer code and monitor any announcements or bulletins from the model developer. Although it is rare, it is still possible that bugs exist in the code. For example, in early 1999, a bug was detected and fixed in the HSPF code regarding the yield-based uptake on cropland. Workshops on a particular modeling framework usually offer the latest information on the model as well. For example, the annual workshop on the CE-QUAL-W2 model held at Portland State University provides updates and upcoming new enhancement of the model.

There have been several versions of the WASP/EUTRO model in addition to the official version offered by the Environmental Protection Agency (EPA). For example, some modelers disconnect the run-time screen display subroutines in the WASP model to shorten the overall runtime. Others have modified the code for special uses, for example, by adding water quality constituents in numerical tagging of the fate and transport of nutrients, or, by modifying the CE-QUAL-W2 output formats to fit the hydrodynamic input for the WASP model. Extreme caution must be exercised to insure the integrity of the code through thorough testing.

FORTRAN77 and FORTRAN90 are still the main computer languages used in watershed, hydrodynamic, and water quality modeling. A FORTRAN compiler is essential in many wasteload allocation and TMDL modeling studies, as recompiling the source code to fit the configuration of the receiving water system is needed to produce the executable image of the watershed, hydrodynamic, and water quality model. FORTRAN compilers are available from a number of software manufacturers, offering many user friendly features to assist the modeler in compilation, debugging, and execution. While the language itself remains the same, different FORTRAN compilers may have different characteristics. Some are very forgiving in compiling while others may be very stringent. As part of the quality assurance/quality control (QA/QC) process of developing the source code for distribution, the Center for Exposure and Assessment Modeling (CEAM) at U.S. EPA, Athens, Georgia, has been using a variety of FORTRAN compilers to compile, debug, and execute the programs, making sure that the source code passes these tests under a number of different compilers. Following the QA/QC process, the final executable image released to the public by CEAM, along with the source code, is generated by the Lahey compiler (McCutcheon, 2000).

5.3 GEOGRAPHICAL INFORMATION SYSTEM

Over the last century it has became increasingly apparent that the effects of human activities are extremely pervasive, and that the recognition and prediction of causes and effects are highly complex. As a consequence, the technology to inventory resources and methods to model processes from the local to the global scale are of critical importance. Geographical information systems (GISs), as well as mathematical models, are a powerful resource for analyzing the interrelated systems involving environmental resources and processes. In the past, representation of spatial data was costly and time consuming, and thus development of mathematical models using spatial information was limited to small areas or very simple processes. Today the development of computer models is a more approachable task when using analytical tools such as GISs. Using GISs allows us to identify the spatial relationships among databases in a particular region. However, the greatest impediment to model development is the availability and sufficiency of data.

GISs typically link data from different sets, using space as the common key. Each data set, or layer, stores particular spatial features of a region such as land use, watershed boundaries, streams, bathymetry, and so on. These features are stored as spatial and descriptive information and can be displayed using two-dimensional maps. Spatial data representation can be conducted through vector, raster, and triangulated formats. Such displays cover a variety of environmental applications, including the development of watershed and water quality models.

This chapter uses the Gunpowder watershed and the Lock Raven reservoir in Maryland to illustrate the importance of a GIS in developing applications of a watershed and a reservoir model, and to show some of the determining factors in choosing the type of spatial data representation. Figure 5-2 shows a group of sub-watersheds within the Gunpowder River Basin, which are represented as segments in the mathematical model. The spatial representation is achieved by using discrete features in vector format, to produce enclosed polygons in which common characteristics specified by the user are met. Vector data represent features as points, lines, and polygons and are best applied to discrete objects with defined shapes and boundaries. The Gunpowder River is represented as part of a geometric network, a network that also depicts a number of other geographic features.

Figure 5-3 is an example of raster data representing land cover information by model segment or sub-watershed. This information, in conjunction with other watershed data, is used in the mathematical model to determine nutrient loads by land use. Rasters are two-dimensional arrays of cells (or pixels) in which each cell has a fixed

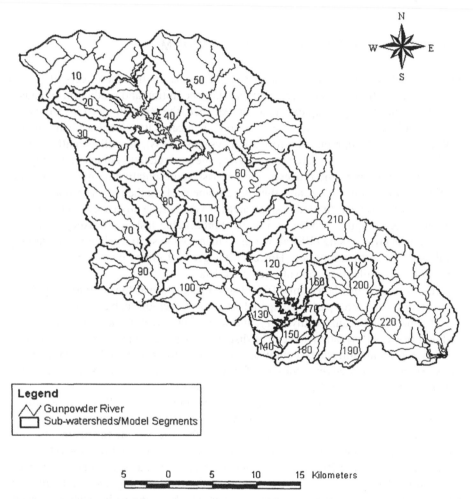

Figure 5-2 The Gunpowder River watershed and Pretty Boy and Loch Raven Reservoirs in Maryland.

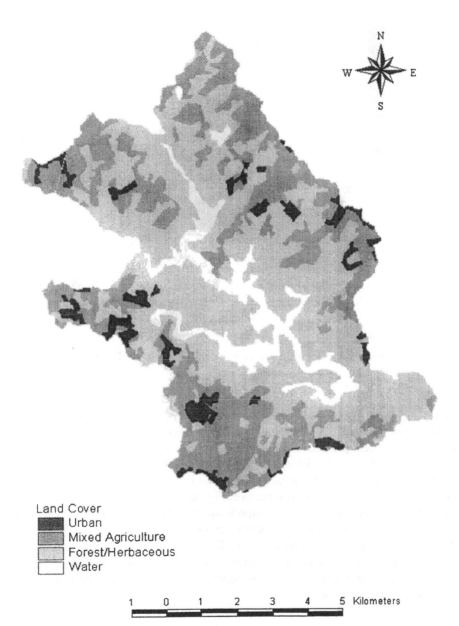

Land Cover
- Urban
- Mixed Agriculture
- Forest/Herbaceous
- Water

1   0   1   2   3   4   5 Kilometers

Figure 5-3   GIS generated land use in the Pretty Boy watershed.

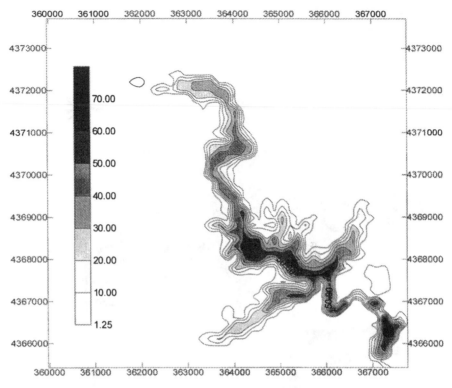

Figure 5-4   GIS generated bathymetry of Loch Raven Reservoir.

height and width, and represent the best method for presenting images and continuous features. The value associated with a cell defines the class, group, category, or measure at the cell position, and it applies to the whole area of the cell. Sources for raster data include satellite images or aerial photographs. Triangulated irregular networks (TINs) are perhaps the most useful format for representing elevation, slope, or aspect in a three-dimensional perspective. Figure 5-4 shows the bathymetry for the Loch Raven Reservoir. This information is used for the development of mathematical models to simulate hydrological and water quality processes within the reservoir. The benefits brought to environmental modeling by use of a GIS are immense, not only because the analysis and data manipulation is more efficient, but also because the information provided to the environmental models (e.g., watershed and reservoir models) is more accurate.

5.4   POST PROCESSING AND VISUALIZATION

The first step in post processing is comparing model results with observed data, followed by quantitative measures of the goodness of fit. Most model runs generate a large amount of information; synthesizing the results and presenting them in a logical and efficient fashion is key to post processing. The commonly employed display method is a graphical format. Although there are a number of displaying software programs to plot model results and data, very few are found to fit the needs of water quality modeling. The following paragraphs present a number of examples to display and visualize model results and data in an efficient manner using software programs developed by the author.

A unique feature of water quality modeling is the extent of the data to be presented. For example, temperature levels in a reservoir could vary significantly in the vertical direction. The vertical profile of temperature also varies during the year. It is necessary to display the spatial and temporal variations of reservoir temperature in one plot for a comprehensive picture. Figure 5-5 shows such variations of temperature at Station GUN0142 at the dam in Loch Raven Reservoir in 1991. Twenty-four vertical profiles of temperature are displayed on one page showing the spatial and temporal changes. Results from the CE-QUAL-W2 model run are shown as solid lines and the open circles present the measured temperatures. The reservoir hydrodynamics are displayed in these 24 temperature profiles. Such a graphical presentation requires a plotting program specially written for that study. This program not only provides screen display of the plots but also offers hardcopy prints that are report quality. The program also allows the user to change the thickness of the lines and size of the characters for labeling, putting the modeling analyst in full control of the visual presentation.

Figure 5-6 presents another example of a visual display, using the model results and data from the Patuxent Estuary, in Maryland. In this figure, seven key water quality constituents—total nitrogen, ammonium, nitrite/nitrate, total phosphorus, orthophosphate, chlorophyll a, and dissolved oxygen—at Station XDE5339 (Broomes Island) are shown in temporal plots for the top layer and bottom layer of the water column in 1984. The water quality data are shown as open circles for the layer-averaged values. The attached bars represent the maximum and minimum values in each layer. Model results are shown in solid lines. It is important that model results and data from one station for the entire year be succinctly presented in one figure, allowing a complete display of the cause-and-effect relationship between nutrient loads and response in the Patuxent Estuary. For example, nutrient uptake from algal growth during the growing season can be seen from the plots.

The two-layer display presents the dissolved oxygen (DO) stratification during the summer months, leading the anoxic condition in the bottom waters. The anaerobic release of ammonium and orthophosphate from the sediment to the water column during the summer season is seen in the plots. Another nine pages are required to display the model results and data for the other nine stations along the Patuxent Estuary.

Results for a steady-state modeling analysis of one-dimensional rivers can be presented in longitudinal profiles of water quality constituents. Figure 5-7 shows the model results and data for 5-day carbonaceous biochemical oxygen demand ($CBOD_5$), organic nitrogen, ammonium, nitrite/nitrate, organic phosphorus, orthophosphate, chlorophyll a, and DO in the James Estuary, Virginia, from Richmond to the mouth of the river near the Chesapeake Bay. The horizontal axes of these plots

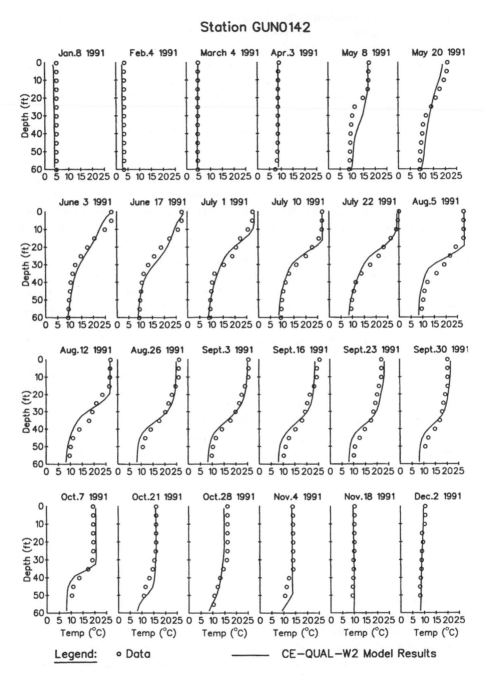

Figure 5-5   CE-QUAL-W2 model calculated temperature profiles versus measured data near the dam at Loch Raven Reservoir.

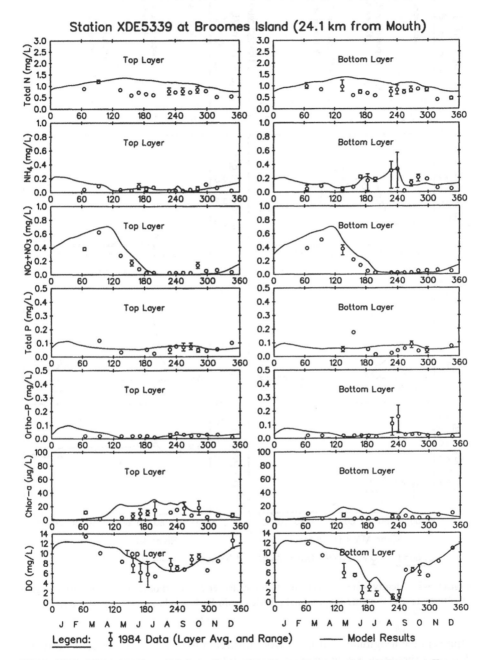

Figure 5-6 Water quality model results versus data at Broomes Island, Patuxent Estuary, Maryland, 1984 simulations.

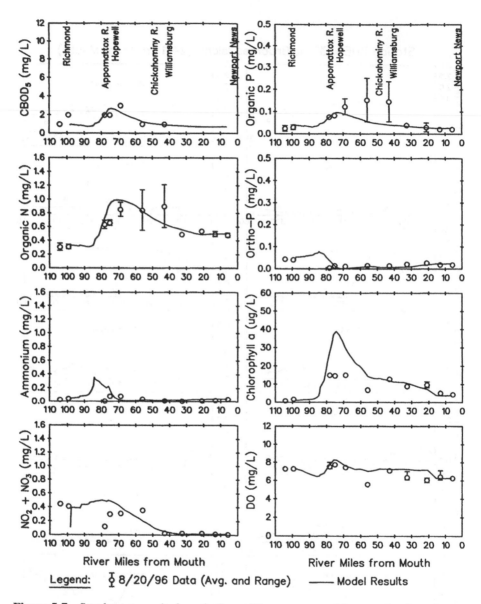

Figure 5-7 Steady-state results from the James River estuary model versus data from August 20, 1996 survey, showing longitudinal profiles from Richmond to the mouth of the river.

represent the longitudinal distance from upstream to downstream along the estuary. The data show water column average concentrations with maximum and minimum values to display the range of the data.

To show model results in a horizontal two-dimensional configuration requires a different display. Figure 5-8 shows steady-state model results and field data for DO

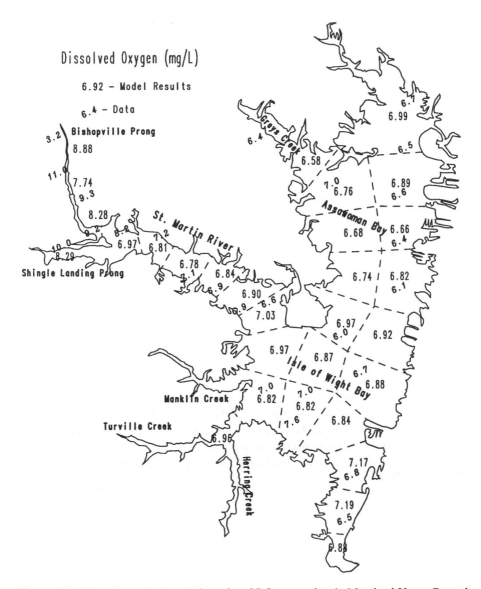

Figure 5-8 Two-dimensional model results of DO versus data in Maryland Upper Coastal Bay under steady-state conditions.

in Maryland Upper Coastal Bay near Ocean City. The model calculated DO concentrations and data are shown in different fonts. Dashed lines represent interfaces between model segments. It should be pointed out that a finer segmentation might provide sufficient spatial resolution to develop concentration contour isopleths for comparison with the DO concentration values in discrete points.

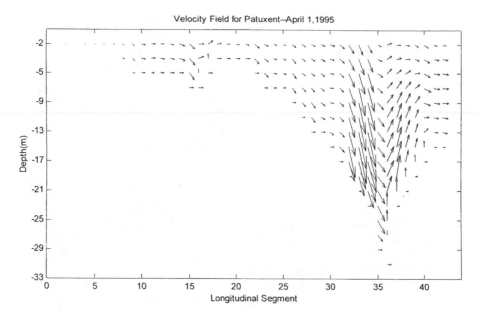

Figure 5-9 CE-QUAL-W2 calculated tidal currents in the Patuxent Estuary, April 1, 1995.

Hydrodynamic model results are usually presented in velocity vectors. Figure 5-9 shows the tidal currents (in a snapshot) calculated by the CE-QUAL-W2 model for the Patuxent Estuary from the fall line to the mouth of the river. A sequence of these plots may be combined to produce an animation of the dynamic interaction between the freshwater inflow from the head of the estuary and the tidal action from the Chesapeake Bay.

CHAPTER 6

USING THE STREAM AND QUAL2E MODELS

Modeling of rivers and streams in terms of biochemical oxygen demand and dissolved oxygen (BOD/DO) has become an integral component of water pollution control and management over the past 25 years. The DO concentration has been one of the most significant criteria in stream sanitation. The discharge of organic impurities, such as municipal sewage and industrial wastes, into a body of water presents a problem of primary importance in this regard (Eckenfelder and O'Connor, 1961). Bacteria consume oxygen to break down the organic matter. Oxygen is resupplied via the reaeration process. The analysis of such a natural purification capacity of a stream, therefore, is of fundamental and practical value. The first attempt to quantify the oxygen balance in a stream was by Streeter and Phelps (1925) for the Ohio River. O'Connor (1960) expanded this analysis by including other processes such as photosynthesis, respiration, and sediment oxygen demand (SOD) in a classic paper to present a comprehensive modeling analysis of oxygen balance in streams. His work represents the foundation of modern water quality modeling of natural waters.

More than three decades have passed since O'Connor's classic work on stream BOD/DO modeling. One-dimensional stream BOD/DO modeling has become a standard practice in the water quality modeling profession. Most one-dimensional models are run on personal computers (PCs) at a great speed, making efficient calculations and proving that running these models is not difficult. Yet performing correct analyses by assigning reasonable model coefficients is still the key to success with simple one-dimensional BOD/DO stream modeling.

One-dimensional BOD/DO stream modeling was widely used during the 1970s when the NPDES permit process, construction program, and the 208 program were beginning. In a recent article, Lung and Sobeck (1999) demonstrated the renewed interest in the use of these models in water quality management. More importantly, additional information on wastewater characteristics and new protocol on long-term BOD measurements have demonstrated new aspects of this modeling analysis that was once thought "straightforward." The purpose of this chapter is to present several examples of one-dimensional BOD/DO stream modeling with up-to-date information concerning this modeling technology.

The case studies presented in this chapter focus on two stream models. The first one, STREAM, is based on the original work by O'Connor and his colleagues at Manhattan College (1976). This modeling program is not unique, as many other workers have developed their own code to implement the governing equations for oxygen balance in streams. The second model, QUAL2E, is presently the most popular stream water quality model for the analysis of wasteload allocations and total maximum daily loads (TMDLs) for one-dimensional streams and rivers.

6.1 EQUATIONS BEHIND THE STREAM MODEL

STREAM is a steady state, one-dimensional BOD and DO model for streams and rivers. Three water quality parameters are modeled: ultimate carbonaceous BOD ($CBOD_u$), nitrogenous BOD (NBOD), and DO. For $CBOD_u$, Eq. 6-1 takes the form:

$$0 = -\frac{Q}{A}\frac{dL}{dx} - K_r L \qquad (6\text{-}1)$$

where L is the $CBOD_u$ concentration and K_r is the $CBOD_u$ removal rate (see Chapter 4).
The equivalent equation for NBOD is:

$$0 = -\frac{Q}{A}\frac{dN}{dx} - K_n N \qquad (6\text{-}2)$$

where N is the NBOD concentration.
The distribution of DO may be formulated in a similar manner by including all DO sources and sinks described earlier:

$$0 = \frac{Q}{A}\frac{dC}{dx} + K_a(C_s - C) - K_d L - K_n N + P - R - \frac{SOD}{H} \qquad (6\text{-}3)$$

If $CBOD_u$ is removed only by direct oxidation, the deoxygenation rate coefficient, K_d, reflecting actual oxygen reduction in the system, is equal to the $CBOD_u$ removal rate coefficient, K_r. The terms on the right side of Eq. 6-3 represent, respectively: the downstream transport of oxygen with the stream flow; atmospheric reaeration; biological oxidation of $CBOD_u$; biological oxidation of NBOD; algal photosynthesis; algal respiration; and the biological oxidation of sediment materials. The $CBOD_u$ concentration, L, and NBOD, N, in Eq. 6-3 may be replaced by the functional forms of Eqs. 6-1 and 6-2. The resulting expression may be integrated with given boundary conditions and expressed in terms of DO.

Further, the DO deficit, D, instead of DO itself, is used to formulate the DO profile, expressing Eq. 6-3 as:

$$0 = -\frac{Q}{A}\frac{dD}{dx} - K_aD + K_dL + K_nN - P + R + \frac{SOD}{H} \qquad (6\text{-}4)$$

The solution to Eq. 6-4 is:

$$D = \frac{K_dL_o}{K_a - K_r}\left(e^{-K_r\frac{x}{U}} - e^{-K_a\frac{x}{U}}\right) \qquad \text{CBOD}_u \qquad (6\text{-}5a)$$

$$+\frac{K_nN_o}{K_a - K_n}\left(e^{-K_n\frac{x}{U}} - e^{-K_a\frac{x}{U}}\right) \qquad \text{NBOD} \qquad (6\text{-}5b)$$

$$+D_oe^{-K_a\frac{x}{U}} \qquad \text{Initial Deficit} \qquad (6\text{-}5c)$$

$$-\frac{P}{K_a}\left(1 - e^{-K_a\frac{x}{U}}\right) \qquad \text{Algal Photosynthesis} \qquad (6\text{-}5d)$$

$$+\frac{R}{K_a}\left(1 - e^{-K_a\frac{x}{U}}\right) \qquad \text{Algal Respiration} \qquad (6\text{-}5e)$$

$$+\frac{SOD}{HK_a}\left(1 - e^{-K_a\frac{x}{U}}\right) \qquad \text{SOD} \qquad (6\text{-}5f)$$

Note that $U = Q/A$ and is the average stream velocity. D_o represents an initial DO deficit existing at the origin, if any. The initial $CBOD_u$ concentrations, L_o, in Eq. 6-5 must be expressed in terms of the ultimate oxygen demand. The DO concentration C may be determined from the computed deficit using the equation:

$$C = C_s - D$$

Because of zero-order and first-order kinetics formulated in the model, the DO deficit terms due to varying sources and sinks may be added (i.e., superimposed).

6.2 IMPLEMENTING THE GOVERNING EQUATION IN STREAM CODE

Equations 6-1 to 6-4 are developed for steady-state conditions using constant parameters. Equation 6-5 is the analytical solution of the BOD/DO equations with spatially constant hydraulic geometry, kinetic coefficients, and wasteloads. In practice, however, the hydraulic geometry varies along the river. To implement Eq. 6-5 for a real river, it is necessary to divide the river into a number of reaches, each of which has spatially constant parameters. Mass balance at the end of each reach must be maintained prior to starting a new reach. The STREAM model code is designed to accommodate numerous reaches in the one-dimensional configuration.

One of the key parameters of STREAM is the reaeration coefficient, K_a. In the original version of the STREAM model, the O'Connor–Dobbins (1958) equation was used to quantify K_a:

$$K_a(20°C) = \frac{12.9U^{0.5}}{H^{1.5}} \tag{6-6}$$

where U is average stream velocity (ft/s), and H is average stream depth (ft).

The Tsivoglou equation (Tsivoglou and Neal, 1976) has been added to quantify reaeration coefficients for small rivers and streams:

$$K_a (20°C) = CVS \tag{6-7}$$

where

V = stream velocity (ft/s)
S = stream slope (ft/mile)
C = proportionality constant:
   = 1.8 for stream flow rate between 1 cfs and 10 cfs
   = 1.3 for stream flow rate between 10 cfs and 25 cfs
   = 0.88 for stream flow rate between 25 cfs and 300 cfs

Saturated DO concentrations in each reach are calculated using the following equation (U.S. EPA, 1995):

$$C_s = \frac{468}{31.6 + T} \tag{6-8}$$

where T is water temperature in °C and C_s is saturated DO concentration in mg/L. This equation is accurate to within 0.03 mg/L compared with the Benson–Krause equation on which the Standard Methods tables are based (U.S. EPA, 1995).

6.3 STREAM MODEL INPUT STRUCTURE

The STREAM model input structure indicates a simple design (Table 6-1). Following the entry of the number of reaches and the starting mile point, each reach is characterized by two lines of input. Note that the tributary flows and wastewater treatment

TABLE 6-1 Input Data Structure for the STREAM Model

Line	Column	Variable	Description	Units	Format
1	(5)	SW1	= 1 for suppressing the listing of input data = 0 for printing the input data		I1
	(10)	SW2	= 0 for O'Connor's reaeration equation = 1 for Tsivoglou's reaeration equation		I1
2	(1–80)	Title	To identify the run		A80
3	(1–10)	NOR	Number of reaches in the river/stream		F10.1
	(11–20)	Starting X	River mile of the location to start the model	mile	F10.0
	(21–30)	PINT	Print interval of model results	mile	F10.0
	(31–40)	Ratio	$CBOD_u$ to $CBOD_5$ ratio[a]		F10.0
4	(1–10)	CBOD	$CBOD_5$ concentration of the tributary inflow	mg/L	F10.0
	(11–20)	WCBOD	Direct $CBOD_5$ load to the head of the reach	lb/day	F10.0
	(21–30)	NBOD	NBOD concentration of the tributary inflow	mg/L	F10.0
	(31–40)	WNBOD	Direct NBOD load to the head of the reach	lb/day	F10.0
	(41–50)	DOD	DO deficit concentration of the tributary inflow	mg/L	F10.0
	(51–60)	WDOD	Direct DO deficit load to the head of the reach	lb/day	F10.0
	(61–70)	BN	Benthic oxygen demand	$gm\ O_2\ m^{-2}\ day^{-1}$	F10.0
	(71–80)	PNET	Net algal photosynthetic oxygen production	$mg\ O_2\ L^{-1}\ day^{-1}$	F10.0
	(81–90)	SL	Slope of river channel bottom[b]	ft/mile	F10.0

(*Continues*)

147

TABLE 6-1 (*Continued*)

Line	Column	Variable	Description	Units	Format
5	(1–10)	Q	Incremental flow to the reach[c]	cfs	F10.0
	(11–20)	DEPTH	Average depth of the reach	ft	F10.0
	(21–30)	AREA	Average cross-sectional area of the reach	ft^2	F10.0
	(31–40)	TEMP	Average water temperature of the reach	°C	F10.0
	(41–50)	XR	Mile point that the reach ends	mile	F10.0
	(51–60)	AKR	Instream CBOD removal rate in the reach	day^{-1}	F10.0
	(61–70)	AKD	Instream CBOD deoxygenation rate in the reach	day^{-1}	F10.0
	(71–80)	AKN	Instream NBOD decay rate in the reach	day^{-1}	F10.0
6[d]					
7[e]					

[a]This ratio is used if the user prefers to generate results in CBOD$_5$ for comparison with data. However, in many cases, it is desirable to have the results in CBOD$_u$ by simply setting the ratio to 1.0; all model input parameters and results will be in CBOD$_u$.

[b]For Tsivoglou equation only.

[c]Positive flow for input and negative flow for withdrawal.

[d]Repeat lines No. 4 and No. 5 for each subsequent reach.

[e]A blank line is needed to indicate the end of input.

plant flows are entered as incremental flows to the total stream flow. Boundary conditions for CBOD, NBOD, and DO deficit concentrations are also needed. Water column kinetic coefficients, K_r, K_d, K_n, along with SOD and net algal photosynthetic oxygen production, are entered for each reach. Because all of these coefficients are temperature dependent, average water temperature for each reach is also required. Note that the user can choose either the O'Connor–Dobbins or the Tsivoglou equation for the reaeration coefficient.

6.4   APPLICATION OF STREAM TO ROCK CREEK

Rock Creek (Figure 6-1) in Gettysburg, Pennsylvania, was experiencing poor water quality in the late 1970s, resulting from the Gettysburg wastewater treatment plant discharge, which failed to meet the NPDES permit limits. Water quality and aquatic biology surveys in 1970s indicated that severe degradation in water quality was found below the treatment discharge in terms of key water quality parameters. During a 1969 water survey, dissolved oxygen concentration was down to 2 mg/L near the outfall of the plant discharge. Nutrient concentrations also were high in the Creek. Under the Construction Program of the Clean Water Act of 1977, federal grants and state support were secured by the local government to upgrade the Gettysburg treatment plant. As part of the facility planning process, a water quality model was developed in 1979 to evaluate various wastewater management alternatives designed to alleviate these water quality problems. The STREAM BOD/DO model was chosen for this study. To support the modeling work, a water quality survey was conducted in August 1979 to collect data for the model calibration analysis. Historical data from surveys in 1973 and 1969 were also used for model verifications.

6.4.1   Stream Hydraulic Geometry and Time of Travel

The physical parameters of hydraulic geometry include width, depth, cross-sectional area, and velocity in different reaches of the stream. In this study, the depth profile across the stream at each sampling station was surveyed. The cross-sectional area was planimetered from the survey data. The average depth was calculated by dividing the cross-sectional area by the width. Finally, the average flow was calculated by multiplying the measured velocity by the cross-sectional area. Figure 6-2 shows the cross-sectional area, depth, and velocity measured along Rock Creek during the August 1979 survey.

Tributary flows in the study area include the upstream flow in Rock Creek, White Run, Littles Run, Stevens Run, and Plum Run (Figure 6-1). Since there is no permanent U.S. Geological Survey (USGS) gaging station in the watershed, flows from the nearest gaging station (Evitts Creek near Centerville, PA) with similar watershed characteristics were used to approximate the flows for Rock Creek and its tributaries. Table 6-2 lists the drainage areas and derived flows at Rock Creek and the tributaries during different times. Additional flows to the stream system are from point source wastewater treatment plants (Cumberland and Gettysburg).

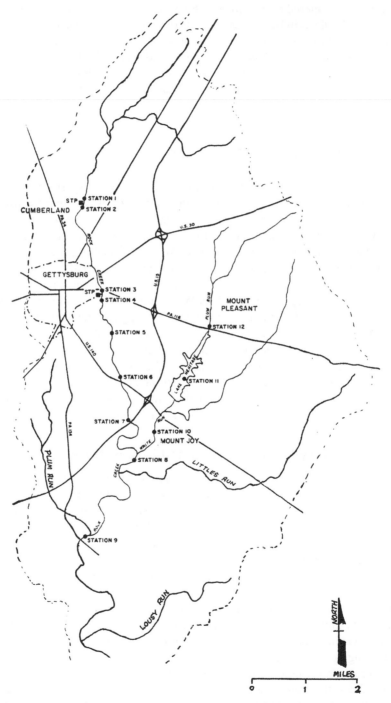

Figure 6-1 Rock Creek in Gettysburg, Pennsylvania (including the sampling stations).

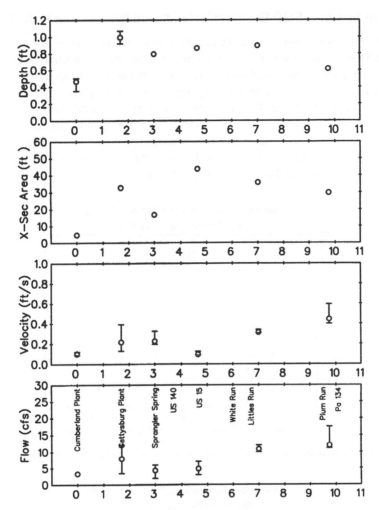

Figure 6-2 Cross-sectional area, depth, and velocity measured in Rock Creek.

TABLE 6-2 Drainage Areas and Derived Flows of Rock Creek Watershed

Location/ Tributary	Drainage Area (mi^2)	Aug. 1979 Flow (cfs)	Oct. 1973 Flow (cfs)	Aug. 1969 Flow (cfs)	7Q10 Flow (cfs)
Rock Creek above					
Gettysburg Plant	20.00	3.8	7.5	6.3	1.13
White Run below					
Heritage	13.00	2.5	5.2	4.0	0.73
Littles Run	7.67	1.5	3.1	2.4	0.42
Plum Run	3.84	0.8	1.5	1.2	0.22
Lousey Run	2.93	0.6	1.2	0.9	0.17
Stevens Run	1.15	0.2	0.5	0.4	0.07

6.4.2 In-Stream CBOD Removal Rate, K_r and Nitrification Rate, K_n

The overall decay phenomenon, as measured in Rock Creek, reflects the disappearance of CBOD from the water column due to a combination of settling and biological oxidation. Plotting the CBOD data versus time of travel provides a first estimate of the removal coefficient, K_r (Figure 6-3). Since the effluent quality from the wastewater treatment plants, particularly the Gettysburg plant, is poor and has a significant concentration of suspended solids, the CBOD removal rate immediately below the treatment plants is much higher than the biological oxidation rate, K_d (Figure 6-3).

The in-stream nitrification rate, K_n, is determined by plotting the NBOD concentration versus time of travel along the stream (Figure 6-4).

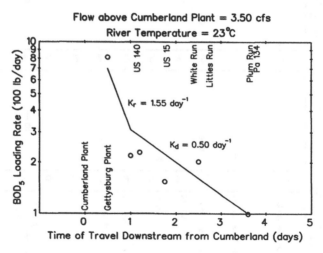

Figure 6-3   Estimation of CBOD removal and biological oxidation rates in Rock Creek.

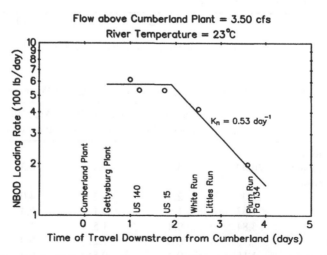

Figure 6-4   Estimation of in-stream nitrification rate in Rock Creek.

6.4.3   Deriving Other Kinetic Coefficients

The Tsivoglou equation is used to calculate the reaeration coefficients for Rock Creek, as the stream flow rates are small and well within the range of flow rates that the Tsivoglou equation deems appropriate. Since there was no available data on sediment oxygen demand (SOD) in Rock Creek, literature values were used. They range from 2 gm O_2/m^2/day to 10 gm O_2/m^2/day with an approximate average of 4 gm O_2/m^2/day for municipal wastewater sludge near the outfall (Thomann and Mueller, 1987). The SOD values decrease to between 1 gm O_2/m^2/day and 2 gm O_2/m^2/day for aged sludge downstream of the outfall. In this study, a value of 2 gm O_2/m^2/day was initially used in the reach from the Gettysburg wastewater treatment plant outfall to 2.3 miles downstream from the plant. Subsequent model calibration and verification runs would eventually fine-tune the SOD values (4.0 and 3.5 gm O_2/m^2/day) as shown in Table 6-3.

TABLE 6-3   Input Data File for the STREAM Model of Rock Creek

0 1[a]								
Rock Creek in Gettysburg, Pennsylvania (August 1979)								
11.0[b]	−1.0	0.20	1.50[c]					
1.50	0.00	5.00	0.0	1.7	0.0	0.0	0.0	5.435
1.50	0.40	15.0	25.0	0.0	0.50	0.50	0.00	
0.00	20.00	0.00	40.0	0.0	5.60	1.4	0.0	9.235
0.30	0.80	17.0	25.0	1.7	1.00	1.00	0.0	
0.00	400.0	0.00	525.0	0.0	15.5	4.0[d]	0.0	25.689
1.30	0.90	23.08	25.0	3.0	1.00	0.80	0.0	
1.00	0.0	2.00	0.0	4.0	0.0	3.5[d]	0.0	14.345
1.0	1.00	32.31	25.0	4.0	0.90	0.70	0.0	
1.00	0.0	2.0	0.0	3.0	0.0	0.7	0.0	21.789
0.4	1.10	46.00	25.00	5.0	0.5	0.5	0.63	
1.00	0.0	2.00	0.0	3.0	0.0	0.0	0.0	13.981
0.0	1.20	46.0	25.0	6.0	0.5	0.5	0.63	
1.00	0.0	2.0	0.0	3.0	0.0	0.0	0.0	6.678
2.50	1.30	44.38	25.0	6.7	0.5	0.5	0.63	
1.00	0.0	2.0	0.0	3.0	0.0	0.0	0.0	5.234
1.50	1.30	45.26	25.0	8.0	0.5	0.5	0.63	
1.00	0.0	2.00	0.0	3.0	0.0	0.0	0.0	6.129
0.0	1.40	45.26	25.0	9.0	0.5	0.5	0.63	
1.00	0.0	2.00	0.0	3.0	0.0	0.0	0.0	4.978
0.0	1.40	45.26	25.0	9.5	0.5	0.5	0.63	
1.00	0.0	2.00	0.0	3.0	0.0	0.0	0.0	3.457
0.8	1.50	44.76	25.0	11.0	0.5	0.5	0.63	

[a]The Tsivoglou's reaeration equation is selected.

[b]Rock Creek is divided into 11 reaches.

[c]$CBOD_u$-to-$CBOD_5$ ratio is 1.50.

[d]SOD values in gm O_2/m^2/day.

6.4.4   BOD/DO Model Results

The BOD/DO model of Rock Creek is first calibrated using the data collected in August 1979. Subsequent model verification runs are made using the data sets of October 1973 and August 1969. CBOD, NBOD, and DO deficit loads from point sources are developed from the data for each data set. In addition, boundary conditions for these three model constituents are also derived from the data. No algae have been observed in the study area and therefore, net production of oxygen due to algal photosynthesis is not considered in the model.

Figure 6-5 presents the BOD/DO model results compared to the field data. Also shown in Figure 6-5 are the stream flows along Rock Creek and the average water temperature for these three surveys. In general, the model results match the field data very well throughout these three data sets. In the August 1979 data, the DO standard of 5 mg/L is violated in a portion of the stream. Specifically, significant CBOD loads from the Gettysburg plant contribute to the DO sag of 2 mg/L immediately below the outfall. Due to such a low minimum DO, nitrification in the stream is lagging behind, as shown in the horizontal lines between mile point 1.7 and 4. Only after the DO concentration starts to recover does nitrification begin to take place. The October 1973 data were collected during a higher stream flow, resulting in much lower CBOD and NBOD concentrations along Rock Creek. DO concentrations during this survey are much higher and do not violate the DO standard. No NBOD data are available from the October 1973 and August 1969 surveys for comparison with the model calculated concentrations. However, nitrification is not lagging during these two model runs.

To better understand the significance of each DO consuming source, a unit response analysis is conducted using the BOD/DO model of Rock Creek. The four major oxygen-consuming sources are: Gettysburg and Cumberland plants, upstream loads, and SOD. Figure 6-6 shows the DO deficit contributed by these individual sources. It is evident that the Gettysburg plant and the SOD contribute most to the DO deficit, while the other sources are insignificant in the DO budget of Rock Creek.

6.4.5   Modeling Total Phosphorus and Fecal Coliform Using STREAM

Phosphorus compounds, like nitrogen compounds, can contribute to aquatic life, but in excessive amounts cause eutrophication of a stream. While phosphorus enters Rock Creek primarily from the Gettysburg plant, a number of processes are responsible for the removal of phosphorus from the water column. Suspended solids in the wastewater eventually settle to the bottom of the stream. Any phosphorus bound to the solids, including the algal biomass, would be incorporated, at least temporarily, into bottom deposits. Soluble phosphorus may combine chemically with metallic cations to form precipitates. Phosphorus sorption by particulate materials plays another role in soluble concentration reduction. Subsequent settling and deposition of this particulate material will reduce the total phosphorus mass in the water column. These processes all contribute to phosphorus removal and can be approximated by a first-order removal process in a similar fashion to the CBOD removal.

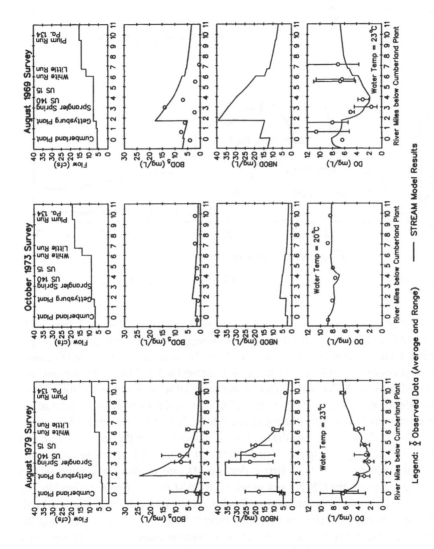

Figure 6-5 STREAM model calibration results for Rock Creek (1979, 1973, and 1969 data).

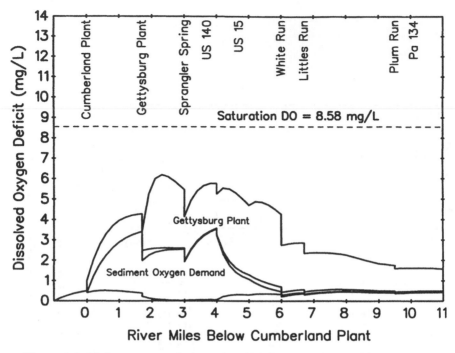

Figure 6-6 Unit response analysis results of Rock Creek using the STREAM model.

Two first-order phosphorus removal rates are estimated: 0.44 day^{-1} from mile point 1.7 to 3.0 and 0.24 day^{-1} for the rest of the stream in a manner similar to the spatial variable CBOD removal rate (see Figure 6-3). These removals rates and the measured total phosphorus loading rates from the Cumberland and Gettysburg plants are incorporated into the STREAM model, using the CBOD slot to form a total phosphorus model of Rock Creek. The model calculated total phosphorus concentrations match the measured data quite closely (Figure 6-7). Note that the removal rate considered in this analysis represents the net loss of phosphorus in the water column and inherently includes any resuspension and release of phosphorus from the sediment.

For fecal coliform bacteria, a first-order die-off rate is also considered in this analysis. Again, using the CBOD slot in the STREAM model, one can simulate the fecal coliform bacteria concentrations in Rock Creek. For freshwater, the fecal coliform die-off rate has been cited in the range from 0.5 day^{-1} to 3.5 day^{-1}. Plotting the measured fecal coliform bacteria concentration versus time of travel along Rock Creek yields a die-off rate of 2 day^{-1}, consistent with the literature-reported values. Figure 6-8 shows the comparison between the model calculated fecal coliform levels and the data.

The above applications to total phosphorus and fecal coliform bacteria further demonstrate the versatility of the STREAM model. It not only can be used to model BOD/DO in a one-dimensional stream, but is also capable of simulating the fate and transport of other constituents following simple, first-order kinetics. Obviously, the

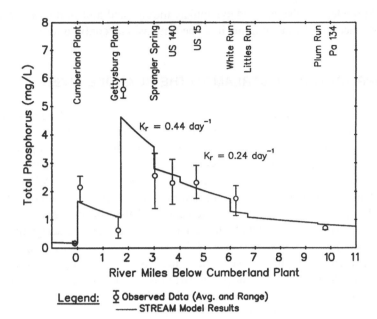

Figure 6-7   Modeling total phosphorus concentrations in Rock Creek using the STREAM model.

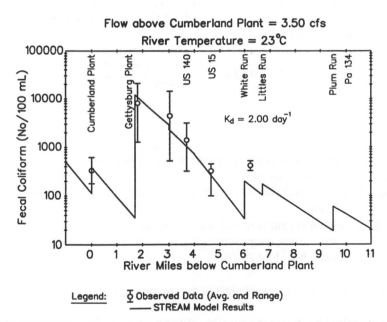

Figure 6-8   Modeling fecal coliform bacteria levels in Rock Creek using the STREAM model.

STREAM model can also be used to simulate the fate and transport of any conservative substance in the water column in a one-dimensional configuration.

6.5 APPLICATION OF STREAM TO THE ROANOKE RIVER

The Roanoke River near the Town of Alta Vista flows from the Smith Mountain–Leesville Dam complex in central Virginia (Figure 6-9). In the vicinity of Alta Vista, there are a number of water supply intakes and wastewater discharges into the Roanoke River. Yet the waters are classified as a "Public Water Supply" resource in Virginia. Downstream waters in the study area are also classified as Virginia "State Scenic Waters." Two major wastewater treatment facilities, the Town of Alta Vista Plant and Burlington Industries Plant, discharge final effluents into this section of the river at Alta Vista. When the Alta Vista plant flows approached the design limit of 1.8 mgd in 1988, the Virginia Department of Environmental Quality (VDEQ), using an enforcement action, required Alta Vista to increase the capacity of the plant and modified their NPDES permit. The permit writer issued an NPDES permit for the expanded facilities at Alta Vista and developed the BOD limits using a model known as the Tennessee Valley Authority (TVA) flat water equation.

6.5.1 What Is the TVA Flat Water Equation?

The TVA flat water equation is a multiple regression formula based on the data from 15 assimilative studies in the TVA region (Krenkel and Ruane, 1979). The equation was developed to calculate the allowable 5-day BOD (BOD$_5$) loading rate for a point source using a minimum amount of field data:

$$Y = 10,138 \frac{(DO_{mix})^{1.094} Q^{0.864} S^{0.06}}{T^{1.423} (DO_{sag})^{1.474}} \tag{6-9}$$

where

$$
\begin{aligned}
Y &= \text{Assimilative capacity of the stream (lb BOD$_5$/day)} \\
10,138 &= \text{regression constant} \\
DO_{mix} &= \text{in-stream dissolved oxygen concentration (mg/L) following complete} \\
&\quad \text{mixing} \\
Q &= \text{sum of stream flow and waste flow (cfs)} \\
T &= \text{stream water temperature (°C)} \\
S &= \text{stream bed slope (ft/ft)}
\end{aligned}
$$

$$DO_{sag} = \text{minimum allowable DO of the stream (mg/L)}$$

VDEQ applied Eq. 6-9 to the Roanoke River near Alta Vista using the following data:

$$DO_{mix} = 6.4 \text{ mg/L}$$
$$Q = 220 \text{ cfs}$$
$$T = 30°C$$
$$S = 0.0013$$
$$DO_{sag} = 5.4 \text{ mg/L}$$

The total assimilative capacity of the Roanoke River in the vicinity of Alta Vista, cal-culated with the TVA flat water equation, equals 3,600 lb/day of BOD_5. With an up-stream background BOD_5 load of 1,675 lb/day, the available BOD_5 allocation for the point source discharges equals 1,925 (= 3,600 − 1,675) lb/day. Note that the NPDES permit BOD_5 loads prior to the VDEQ enforcement action were 2,332 lb/day and 675 lb/day for the Burlington and Alta Vista Plants, respectively. The total available ca-pacity of 1,925 lb/day is less than the sum of the loads (2,332 + 675 = 3,007 lb/day) allowed in the original NPDES permits. Observe that the DO levels in the river below the Alta Vista and Burlington Industries Plants are well above the standard of 5.4 mg/L.

There are a number of technical issues concerned with using Eq. 6-9 in wasteload allocations. First, waste characteristics, like the $CBOD_u$ to $CBOD_5$ ratio of the waste-water, are not considered in this equation. Although this ratio is closely related to the in-stream deoxygenation rate of waste, K_d (Lung, 1998), Eq. 6-9 does not account for this factor, that is, for how rapidly the waste is being stabilized in the stream. Con-sider two waste discharges, one with a high K_d and the other with a low K_d rate. With all other factors being equal between these two discharges, should the waste loading rates from these two sources be the same? They should not, yet, Eq. 6-9 would pro-ject equivalent demands. Second, reaeration in the stream is not considered in Eq. 6-9. And finally, time of travel is another important factor in stream BOD/DO mod-eling but it is neglected in the TVA equation.

6.5.2   BOD/DO Modeling of the Roanoke River Using STREAM

The TVA equation calculation indicates that the current BOD wasteloads permitted are exceeding the ability of the Roanoke River to assimilate those loads and therefore impairing water quality, specifically DO. The action taken by the permit writer would result in more stringent BOD limits than secondary limits for the Alta Vista plant. The permit writer intended to reduce BOD limits for the Burlington Industries Plant based on the TVA equation, although Burlington was not intending to expand their production capacity.

The permit writer's legal basis for using the TVA equation is that it is the regula-tory part of the Roanoke River Basin Water Quality Management Plan (RRBP) and is considered law in Virginia. On the other hand, the RRBP also declared that the state of water quality in the Roanoke River near Alta Vista meets water quality stan-dards for DO and classifies the reach as effluent limited. The Town of Alta Vista and Burlington Industries decided to investigate the TVA equation and to pursue a water quality modeling study due to the potential treatment costs. They also hoped to define the economic conditions under which future growth would occur.

TABLE 6-4 **Hydraulic Geometry of Roanoke River near Alta Vista, Virginia**

Reach Mile Points	Flow (cfs)	Cross-Sectional Area (ft²)	Velocity (ft/sec)	Depth (ft)
September 1995 Survey				
131.0–129.7	294.0	267	1.10	2.55
129.7–128.7	300.0	400	0.75	3.03
128.7–128.0	302.8	481	0.62	4.00
128.0–126.9	302.8	511	0.59	3.50
126.9–125.6	312.9	502	0.62	3.00
125.6–124.8	434.9	806	0.54	4.12
124.8–117.6	457.9	501	0.94	2.35
October 1993 Survey				
131.0–129.7	831.0	704	1.18	3.55
129.7–128.7	835.8	1,116	0.75	5.03
128.7–128.0	836.0	981	0.85	6.00
128.0–126.9	838.4	811	1.03	5.50
126.9–125.6	848.4	562	1.51	3.00
125.6–124.8	931.9	556	1.68	3.12
124.8–117.6	944.2	541	1.74	2.35
7-day 10-year Low Flow				
131.0–129.7	225.0	234	0.96	2.32
129.7–128.7	232.0	365	0.64	2.95
128.7–128.0	232.4	445	0.52	3.77
128.0–126.9	238.0	474	0.56	3.36
126.9–125.6	241.0	455	0.53	2.73
125.6–124.8	263.5	745	0.35	4.03
124.8–117.6	266.8	434	0.61	2.25

The study area of 13 miles is divided into 7 reaches. Average depth and velocity in each reach were measured through stream surveys and incorporated into the model to calculate stream reaeration coefficient in each reach (Table 6-4).

6.5.3 Data to Support the Modeling Analysis

Two water quality surveys were conducted with river flows at Alta Vista of 836 cfs and 294 cfs, in October 1993 and September 1995, respectively. Stream flows were obtained at various locations along the Roanoke River from USGS gaging stations at Alta Vista, APCO records on Goose Creek and discharges over Leesville Dam, and VDEQ records on Goose Creek and the Big Otter River entering the Roanoke River below Alta Vista (see Figure 6-9). Hydraulic geometry parameters such as width,

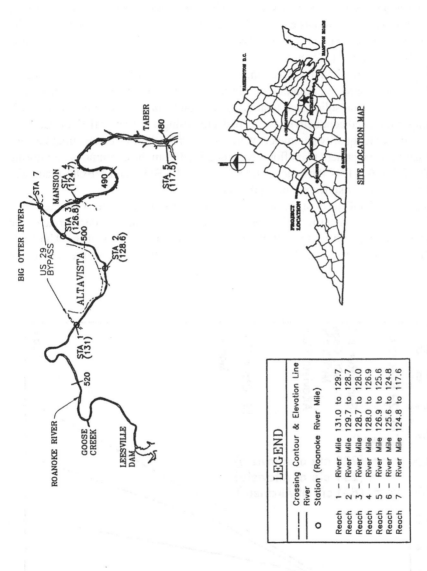

Figure 6-9 The Roanoke River near Alta Vista, Virginia (including sampling stations).

LEGEND

---‥— Crossing Contour & Elevation Line
——— River
○ Station (Roanoke River Mile)

Reach 1 — River Mile 131.0 to 129.7
Reach 2 — River Mile 129.7 to 128.7
Reach 3 — River Mile 128.7 to 128.0
Reach 4 — River Mile 128.0 to 126.9
Reach 5 — River Mile 126.9 to 125.6
Reach 6 — River Mile 125.6 to 124.8
Reach 7 — River Mile 124.8 to 117.6

161

depth, cross-sectional area, and velocity were measured at various locations. During the 1995 survey, VDEQ conducted a time-of-travel study through a portion of the river. Dye was released from the outfall of the Burlington Industries Plant and tracked by collecting dye samples, which were then analyzed with a fluorometer. The hydraulic geometry conditions associated with these two surveys are presented in Table 6-4. Figure 6-10 shows a time-of-travel plot for the flows associated with the two surveys. Also shown in Figure 6-10 is the time of travel for the 7-day 10-year low flow (7Q10) condition (225 cfs).

Samples taken from the river (see sampling stations in Figure 6-9) were analyzed for $CBOD_u$, organic nitrogen, ammonium, nitrite + nitrate, total phosphorus, orthophosphate, alkalinity, total solids, and total dissolved solids. *In situ* measurements of DO, pH, temperature, and conductivity were conducted using a Surveyor II Hydrolab.

The same water quality parameters were analyzed for the wastewater treatment effluent samples. Treatment plant flows were obtained from the continuous monitoring records. Table 6-5 lists the point source flows and CBOD and NBOD loads discharged to the Roanoke River during the surveys.

Long-term CBOD tests of the wastewater effluent samples were conducted for a period of 50 days to determine the $CBOD_u$ concentrations and the decay rate of the CBOD. Figure 6-11 shows the $CBOD_u$ concentrations in effluents of the Alta Vista and Burlington Plants. The data were analyzed with the least square method to

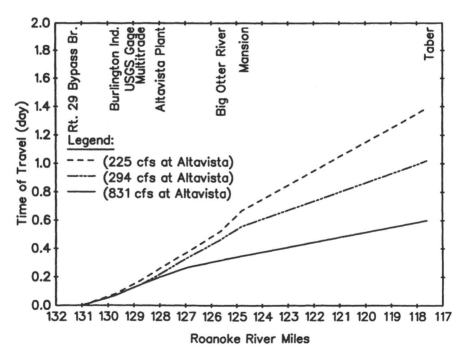

Figure 6-10 Time of travel in Roanoke River from Alta Vista to Tabor, Virginia.

TABLE 6-5 Point Source Loads to the Roanoke River

Point Source	Flow (mgd)	CBOD$_u$ (lb/day)	NBOD (lb/day)	DO Deficit (mg/L)
September 1995 Survey				
Burlington Industries	3.88	639	105	4.5
Alta Vista Plant	1.86	109	4.1	2.6
Big Otter River	78.9	4,644	26	1.0
October 1993 Survey				
Burlington Industries	3.14	368	112	3.6
Alta Vista Plant	1.56	323	4.1	0.0
Big Otter River	54.0	902	451	1.0
Existing Permits[a]				
Burlington Industries	4.50	2,332	112	4.0
Multitrade	0.31	62	0	4.0
Alta Vista Plant	3.60	675	4.1	4.0
Doubled Alta Vista Permit Load[a]				
Burlington Industries	4.50	2,332	112	4.0
Multitrade	0.31	62	0	4.0
Alta Vista Plant	3.6	1,300	4.1	4.0

[a]Permit loads in CBOD$_5$ (not CBOD$_u$).

determine the first-order CBOD decay rate at 0.07 day^{-1}. Such a low rate suggests well-stabilized effluents. Using the following equation:

$$\frac{CBOD_u}{CBOD_5} = \frac{1}{1 - e^{-5K_1}} \tag{6-10}$$

a CBOD$_u$ to CBOD$_5$ ratio of 3.39 was derived for the Roanoke River near Alta Vista using 1993 and 1995 data. Such a high ratio clearly indicates that the effluents are stabilized. Its residue is in a refractory form, which biodegrades very slowly, resulting in a low DO consumption rate (in-stream deoxygenation). This ratio is also consistent with literature values for well-treated effluents (Lung, 1998).

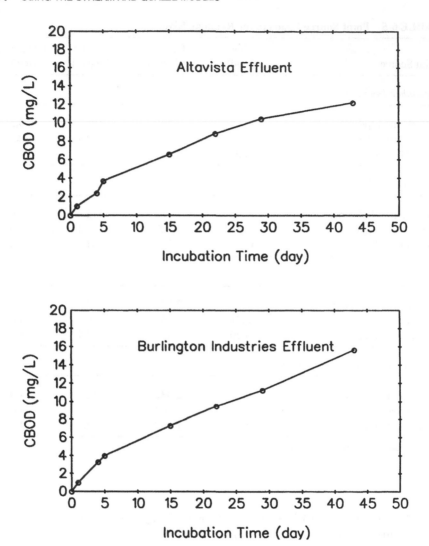

Figure 6-11 Long-term CBOD analysis results of final effluents of Alta Vista Plant and Burlington Industries.

6.5.4 Model Calibration and Verification Analyses

Data from the October 1995 survey was used for model calibration as the river was near the 7Q10 condition. The 1993 data were then used to verify the model using the calibrated model coefficients. Results of the analysis for these two data sets are presented in Figure 6-12; comparing model results of $CBOD_u$, NBOD, and DO with measured data. In general, model results match the data quite well. The low in-stream deoxygenation rate contributes to the relatively constant $CBOD_u$ concentrations in

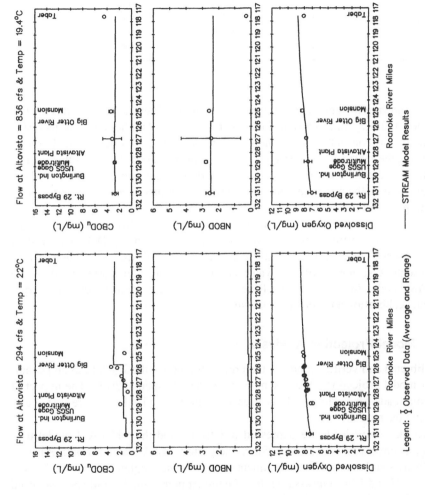

Figure 6-12 STREAM model calibration and verification results for the Roanoke River.

165

TABLE 6-6   Kinetic Coefficients (at 20°C) for Model Calibration

Reach Mile Point	K_r (day^{-1})	K_d (day^{-1})	K_n (day^{-1})	K_a (day^{-1})
131.0–129.7	0.0	0.0	0.0	2.107
129.7–128.7	0.07	0.07	0.0	0.987
128.8–128.0	0.07	0.07	0.0	0.815
128.0–126.9	0.07	0.07	0.0	0.994
126.9–125.6	0.07	0.07	0.0	3.082
125.6–124.8	0.07	0.07	0.0	3.061
124.8–117.6	0.07	0.07	0.0	4.790

the river, showing insignificant decay of well-treated effluents. Also noted is the relatively low CBOD$_u$ concentrations in the ambient water. Another point of interest is that ammonium concentrations in the study area are low, even below the detection limit during the 1995 survey. Perhaps the most significant observation is that the DO concentration profile does not show the classic depression, sag, and recovery along the river. Instead, the minimum DO concentration is located near the discharge point of Burlington Industries. DO concentrations increase progressively in the downstream direction, eventually approaching the saturated level. Note that the first stage of the classic DO profile is missing, a common feature of many DO concentration profiles observed in small rivers and streams receiving well-treated effluents. Table 6-6 presents a summary of the water column kinetic coefficients used in the STREAM model for the Roanoke River, a unique set of values developed from the model calibration and verification analyses.

6.5.5   Model Projection Analysis

The calibrated and verified model is then reconfigured for the 7Q10 (225 cfs) condition for model projections. Water temperature was assumed at 30°C. The impact of the permit loads for Burlington Industries and Alta Vista (Table 6-5) are first evaluated using the model.

Since the permit loads are expressed in CBOD$_5$ and the STREAM model uses CBOD$_u$, a conversion is necessary. Since the permit CBOD loads for both Burlington Industries and the Alta Vista treatment plant are much higher than the loads measured in 1993 and 1995, one assumes that the higher permit loads would be associated with this less stable waste, thereby yielding a higher deoxygenation rate, K_d. A K_d value of 0.2 day^{-1} is considered appropriate for wastewater effluent following secondary treatment (Thomann and Mueller, 1987), which is associated with a CBOD$_u$ to CBOD$_5$ ratio of 1.58. This ratio is used to convert the CBOD$_5$ loads in Table 6-5 to CBOD$_u$ loads for the model projection analysis. (Using a CBOD$_u$ to CBOD$_5$ ratio of 1.58 associated with a K_d rate of 0.2 day^{-1} would result in a greater oxygen demand in the river than the ratio of 3.39 and K_d of 0.07 day^{-1} as used in the model calibration and verification analyses.)

The top portion of Figure 6-13 shows the model results with the current permit loads under the 7Q10 condition. The DO concentrations are consistently above the 5.4 mg/L minimum and gradually approach the saturation concentration of 7.61 mg/L (30°C). Different DO concentrations in the effluents show a slight difference in the predicted DO concentrations in the river (Figure 6-13, top). In general, a significant assimilative capacity exists in the Roanoke River near Alta Vista under the 7Q10 condition.

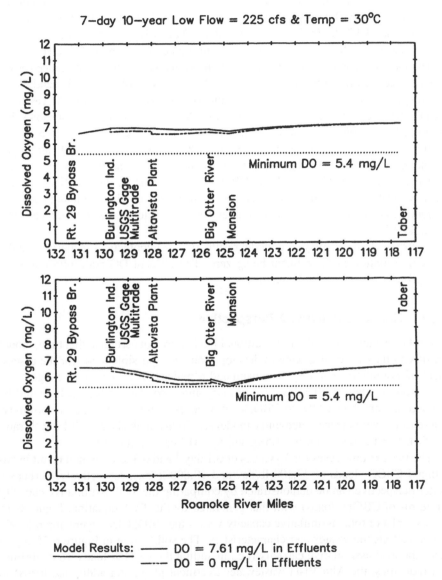

Figure 6-13   STREAM model prediction results under 7-day 10-year low flow condition in Roanoke River.

The model is then used to quantify the assimilative capacity of the Roanoke River from River Mile 131 (Rt. 29 bypass) to River Mile 125 (Mansion) under the 7Q10 condition, where it is assumed that any additional CBOD loads will enter at River Mile 131. A K_d rate of 0.2 day^{-1} for CBOD deoxygenation in the river is selected for secondary treatment of future wasteloads. The additional load would reduce the DO concentration but would not be sufficient to depress the DO level to below 5.4 mg/L anywhere in the river. Results of the model analysis are shown in the lower graph of Figure 6-13 for variable DO concentrations in the effluents of assumed treatment levels. A minimum DO of 5.4 mg/L is reached at River Mile 124.8 with an additional CBOD$_u$ load of 17,500 lb/day entering at River Mile 131. Again, varying DO concentrations in the effluents has an insignificant impact on in-stream DO levels.

The calibrated and verified model is also used to quantify the significance (or insignificance) of the current point source loads to the Roanoke River. Essentially, a unit response analysis is conducted by turning off the loads one at a time to calculate the DO deficit contributed by individual loads. The top part of Figure 6-14 shows that the total DO deficit at the upstream, that is, nonpoint loads, contributes the largest portion of DO deficit, decreasing from 1 mg/L to less than 0.2 mg/L along the Roanoke River. The next largest contributor is the CBOD load from Burlington Industries, followed by the Alta Vista CBOD load. In general, these deficits are extremely small and insignificant, much less than 0.1 mg/L. The result suggests that current CBOD loads from the point sources are not a factor in the DO balance of the Roanoke River.

Figure 6-14 also displays the model results for the hypothetical scenario with the DO concentration in the effluent being zero. The effluent DO concentration has a negligible effect on the model results.

6.5.6 Model Sensitivity: A Perspective

Since the characteristics of these additional loads were not certain at that time, model sensitivity runs were conducted to develop a range of possible loads for the assessment of assimilative capacity. Thomann and Mueller (1987) reported a range of K_d rates from 0.1 day^{-1} to 0.3 day^{-1} for primary and secondary effluents. It should be pointed out that the CBOD$_u$ to CBOD$_5$ ratio varies with these K_d rates accordingly. These ratios were incorporated into a model sensitivity analysis: $K_d = 0.1$ day^{-1} (ratio = 2.54), $K_d = 0.2$ day^{-1} (ratio = 1.58), and $K_d = 0.3$ day^{-1} (ratio = 1.29).

The two extreme cases for K_d values of 0.1 day^{-1} and 0.3 day^{-1} would result in additional CBOD$_u$ loads of 37,500 lb/day and 10,800 lb/day, respectively. Perhaps a better perspective can be demonstrated by comparing the total assimilative capacity (in terms of CBOD$_5$ loads) with that generated by the TVA equation. Figure 6-15 shows a plot of total assimilative capacity versus the CBOD deoxygenation rate, K_d, which reflects the wastewater characteristics. The solid curve in Figure 6-15 represents the total assimilative capacity (i.e., including the CBOD$_5$ loads from Burlington Industries, the Alta Vista wastewater treatment plant, and additional loads) in the Roanoke River near Alta Vista under the 7Q10 condition while maintaining a minimum DO concentration of 5.4 mg/L. This solid curve applies to the K_d values

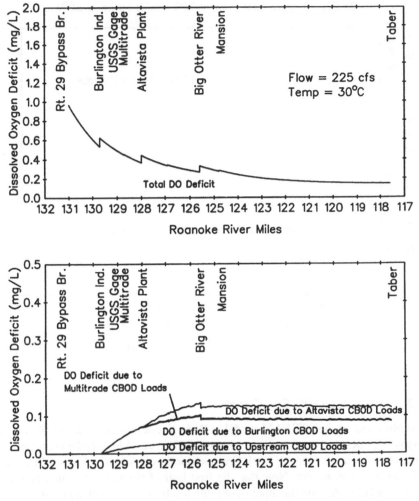

Figure 6-14 Unit response analysis results of Roanoke River using STREAM model.

ranging from 0.1 day^{-1} to 0.3 day^{-1}, characterizing treatment levels from advanced secondary to advanced primary. The dashed-dot line represents the total assimilative capacity predicted by the TVA equation. The model results yield a much higher assimilative capacity than the TVA equation does by including the receiving water BOD kinetics, which in turn depends on the wastewater characteristics. Without the BOD kinetics, the TVA equation assumes that the effluent is marginally treated. In fact, Figure 6-15 suggests that the TVA equation is associated with very high K_d rates, resulting in conservative predictions. Measured K_d rates from long-term CBOD tests of the Alta Vista effluents in this study are close to 0.07 day^{-1}, indicating a well-treated waste. A significant assimilative capacity is therefore expected at such a low CBOD deoxygenation rate.

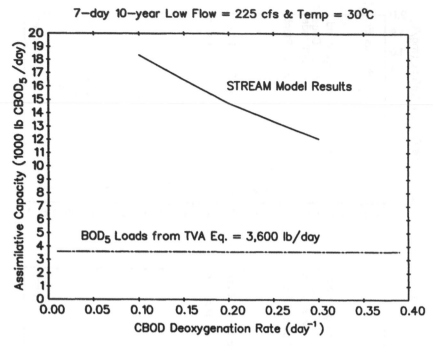

Figure 6-15   Assimilative capacity analysis of the Roanoke River near Alta Vista.

6.5.7   The Classic DO Sag Curve

Applications of the STREAM model to Rock Creek and the Roanoke River share many similarities, but there is one significant difference. While the DO concentration profiles (see Figure 6-5) along Rock Creek show the classic sag curve, profiles along the Roanoke River do not. Note that the first half of the sag curve, that is, the depression of DO following the input of organic loads, is missing in the Roanoke River (see Figure 6-12). Instead, only the second half of the sag curve, that is, the recovery portion, is present. This result is found for many wastewater treatment plants that are producing excellent effluent quality these days following wastewater treatment plant expansion and upgrade. As mentioned before, such a result is even more pronounced for small streams whose flows consist primarily of wastewater flow. A good example is the Shirtee Creek in Alabama. Figure 6-16 shows the results of the STREAM model for Shirtee Creek under three survey conditions. Note the overall low flows in the stream during the surveys and the significance of the wastewater flows from the two point sources: Avondale Plant (textile effluent from Russell Industry) and Earl Ham Plant (municipal wastewater). The missing DO depression is due mainly to the weakness of the wastewater flow. With the majority of the stream flow coming from the wastewater flows, the shallow water column experiences significant aeration, resulting in a rise of the DO concentrations in all three field surveys. Similarly, the STREAM model DO results also display this rise, matching the measured DO levels

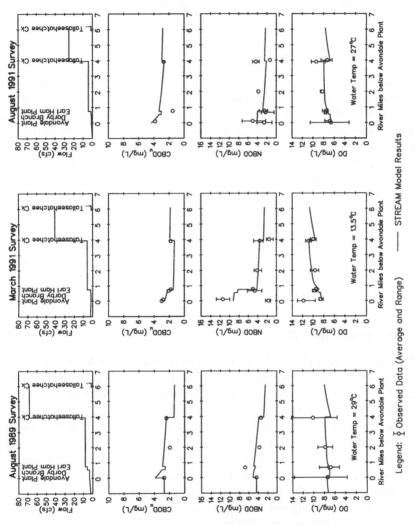

Figure 6-16 STREAM model results of Shirtee Creek in Alabama.

along the stream quite well. Thus this DO concentration profile looks different from the classic DO sag curve following the input from point sources in many small streams.

6.6 APPLICATION OF THE QUAL2E MODEL TO THE BLACKSTONE RIVER

6.6.1 The QUAL2E Model

The QUAL2E model is presently the most widely used water quality model for one-dimensional streams and rivers (Chapra, 1997). Its basic structure is very similar to that of the STREAM model, that is, it shares the same mass balance equations. While the STREAM model can simulate three water quality constituents, the QUAL2E model is capable of accommodating 15 water quality constituents. In addition to simulating nutrients and algae levels in the water column, the QUAL2E model calculates temperature and conservative substances simultaneously.

The development, refinement, and enhancement of the QUAL2E model can be found in Chapra (1997) and Brown and Barnwell (1987) and will not be repeated here. It should be pointed out that the QUAL2E model is capable of addressing steady-state water quality problems (e.g., BOD/DO and nutrient/eutrophication) in one-dimensional streams and rivers and is usually run until a steady state is reached. The only time-variable water quality problem that the model can address presently is diurnal fluctuation of DO in the water column.

6.6.2 The Blackstone River

The Blackstone River is 48 miles long, flowing from south central Massachusetts into northeastern Rhode Island (Figure 6-17). The Blackstone River has its headwaters in Worcester, Massachusetts, from whence it then flows south into Rhode Island where it discharges into the Seekonk River at Pawtucket. The river was the site of the region's first textile mill, Slater's Mill, in 1793. Through the 1800s, the river became the hardest working river in the United States, with one dam for every mile of water (Wright et al., 1998). It is an important natural, recreational, and cultural resource to both Massachusetts and Rhode Island. The river is a major source of fresh water to Rhode Island's most valuable resource, Narragansett Bay, a productive and diverse estuary important for fishing, shellfishing, tourism, and recreation.

In 1987, the U.S. Environmental Protection Agency (EPA) created the Narragansett Bay Project (NBP) through the National Estuary Program. The NBP produced a Comprehensive Conservation and Management Plan (CCMP), which detailed present and future long-term management actions. These actions are carried forward by governmental and local agencies. One of the primary recommendations of the CCMP was a coordinated assessment and sampling project for the Blackstone River in both Massachusetts and Rhode Island. To help support and implement some of the CCMP recommendations, the EPA established the Blackstone River Initiative (BRI) in 1991.

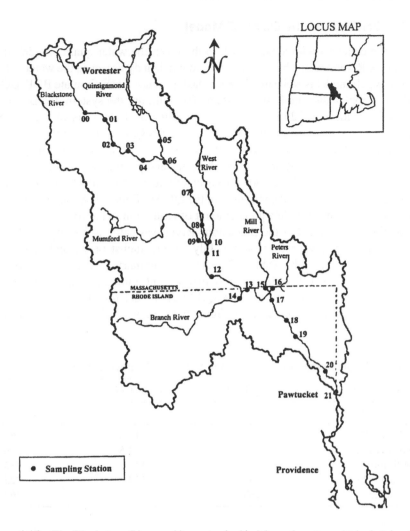

Figure 6-17 The Blackstone River and its watershed in Massachusetts and Rhode Island (including sampling stations).

The Massachusetts Department of Environmental Protection (MADEP) later selected the Blackstone River as one of four rivers for the development of a TMDL management plan. Phase I of the TMDL was designed for dry weather, steady-state conditions. Three comprehensive water quality surveys were completed in the summer and fall of 1991 to support the water quality modeling effort for TMDL. Figure 6-17 shows the locations of the water quality monitoring stations. For the dry weather surveys, there were 15 stations along the Blackstone River and 6 near the mouth of the major tributaries. In addition, two point source discharges were sampled in conjunction with the stream monitoring program: the Upper Blackstone Water Pollution Abatement District (UBWPAD) and Woonsocket WWTF.

6.6.3 Configuring the QUAL2E Model

Due to the concerns with algal growth in the river, the water quality model for the Blackstone River had to address the algal and nutrient dynamics in the water column. The QUAL2E model is well suited for this task because it can simulate BOD/DO and nutrient/eutrophication in a given system. To configure the system, the Blackstone River was divided into 25 reaches, with computational elements of 0.20 miles in length for a total of 229 elements (Wright et al., 1998). Reach divisions were developed based on the following considerations: spatial variation of hydraulic geometry along the study area, locations of the tributaries and point sources, and locations of dams. A unique feature of the study area is that there are many low head dams. A total of 19 dams are included and configured in the QUAL2E model. As shown later in the model results, dams provide a good amount of reaeration to the river.

The QUAL2E model of the Blackstone River was calibrated using the data collected during three dry weather surveys: July 10–11, August 14–15, and October 2–3, 1991 (Wright et al., 1998). Table 6-7 presents a summary of measured flows in the Blackstone River, its tributaries, and five major point sources for these three surveys. The river flows incorporated into the QUAL2E model for model calibration and verification analyses are presented in Figure 6-18. Flows during the July and August

TABLE 6-7   Measured Flows (cfs) for the 1991 Dry Weather Surveys

Gaging Station	July 10–11	August 14–15	October 2–3
Blackstone River			
U.S. Steel[a]	13.5	14.0	69.1
Northbridge[a]	77.4	84.5	236
Millville[a]	98.7	118	483
Woonsocket[a]	137	152	725
Lonsdale[a]	189	200	760
Tributaries			
Quinsigamond River[b]	7.3	8.6	60.5
Branch River[b]	26.0	30.5	122
Point Sources			
UBWPAD Plant	38.4	44.6	64.7
Millbury Plant	0.6	0.8	1.3
Grafton Plant	1.6	1.6	1.6
Northbridge Plant	1.8	1.2	1.8
Woonsocket Plant	8.3	11.5	13.4

Source: From Wright et al. (1998).

[a]USGS temporary gaging station.

[b]USGS permanent gaging station.

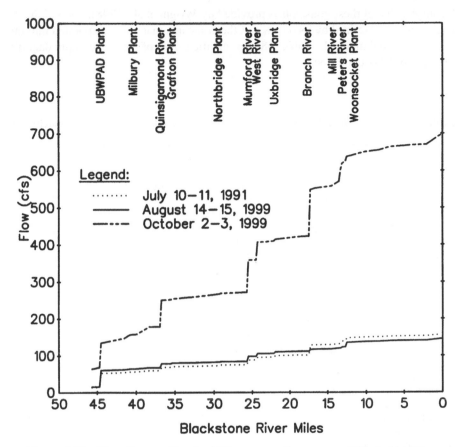

Figure 6-18   River flow profiles for 1991 water quality surveys of Blackstone River.

surveys are very close and represent a low flow river condition with the UBWPAD plant contributing a significant portion of the river flow. The October flow is much higher, showing the diminishing significance of the UBWPAD plant flow.

The QUAL2E model allows for variations of river velocity and depth with the river flow rate in the following manner:

$$U = aQ^b \qquad (6\text{-}11)$$
$$D = cQ^d \qquad (6\text{-}12)$$

where

$$U = \text{stream velocity (ft/sec)}$$
$$Q = \text{stream flow (cfs)}$$
$$D = \text{stream depth (ft)}$$
$$a, b, c, \text{ and } d = \text{empirical constants}$$

A complete list of these constants is provided by Wright et al. (1998). Each of the 25 reaches is characterized by its own set of the constants that are incorporated into the QUAL2E model input file. Table 6-8 presents the listing of the model input data file for the July 1991 condition.

Table 6-8   Input Data File for QUAL2E Model of Blackstone River (Massachusetts Portion)

```
TITLE01               STREAM QUALITY MODEL—QUAL2E; BLACKSTONE RIVER, MA
TITLE02               SURVEY
TITLE03   YES         CONSERVATIVE MINERAL   I    CHLORI  MG/L
TITLE04   NO          CONSERVATIVE MINERAL   II
TITLE05   NO          CONSERVATIVE MINERAL III
TITLE06   NO          TEMPERATURE
TITLE07   YES         5-DAY BIOCHEMICAL OXYGEN DEMAND
TITLE08   YES         ALGAE AS CHL-A IN UG/L
TITLE09   YES         PHOSPHORUS CYCLE AS P IN MG/L
TITLE10                  (ORGANIC-P; DISSOLVED-P)
TITLE11   YES         NITROGEN CYCLE AS N IN MG/L
TITLE12                  (ORGANIC-N; AMMONIA-N; NITRITE-N;' NITRATE-N)
TITLE13   YES         DISSOLVED OXYGEN IN MG/L
TITLE14   NO          FECAL COLIFORM IN NO./100 ML
TITLE15   NO          ARBITRARY NON-CONSERVATIVE
ENDTITLE
LIST DATA INPUT
WRITE OPTIONAL SUMMARY
NO FLOW AUGMENTATION
NO STEADY STATE
NO TRAP CHANNELS
PRINT LCD/SOLAR DATA
NO PLOT DO AND BOD
FIXED DNSTM CONC (YES=1) =        0.        5D-ULT BOD CONV K COEF =       0.25
INPUT METRIC             =        0.        OUTPUT METRIC            =       0.
NUMBER OF REACHES        =       15.        NUMBER OF JUNCTIONS      =       0.
NUM OF HEADWATERS        =        1.        NUMBER OF POINT LOADS    =       8.
TIME STEP (HOURS)        =     1.00         LNTH. COMP. ELEMENT (MI) =      0.2
MAXIMUM ROUTE TIME (HRS) =      196         TIME INC. FOR RPT2 (HRS) =      6.0
LATITUDE OF BASIN (DEG)  =     42.5         LONGITUDE OF BASIN (DEG) =     83.3
STANDARD MERIDIAN (DEG)  =      75.         DAY OF YEAR START TIME   =     196.
EVAP. COEF.,(AE)         =   .00068         EVAP. COEF.,(BE)         =  0.00027
ELEV. OF BASIN (METERS)  =     150.         DUST ATTENUATION COEF.   =     0.13
ENDATA1
O UPTAKE BY NH3 OXID(MG O/MG N)=    3.5  O UPTAKE BY NO2 OXID(MG O/MG N)=   1.07
O PROD BY ALGAE (MG O/MG A)    =    1.6  O UPTAKE BY ALGAE (MG O/MG A)  =   2.00
N CONTENT OF ALGAE (MG N/MG A) =   .100  P CONTENT OF ALGAE (MG P/MG A) =  0.050
ALG MAX SPEC GROWTH RATE(1/DAY)=   2.50  ALGAE RESPIRATION RATE (1/DAY) =   0.20
N HALF SATURATION CONST (MG/L) =   0.15  P HALF SATURATION CONST (MG/L) =  0.025
LIN ALG SHADE CO (1/H-UGCHA/L) =  0.011  NLIN SHADE (1/H-(UGCHA/L)**2/3)=  0.017
LIGHT FUNCTION OPTION (LFNOPT) =      1  LIGHT SATURATION COEF (INT/MIN)=    .06
DAILY AVERAGING OPTION (LAVOPT)=      2  LIGHT AVERAGING FACTOR (AFACT) =   0.92
NUMBER OF DAYLIGHT HOURS (DLH) =     14  TOTAL DAILY SOLAR RADTN (INT)  =   1400
ALGY GROWTH CALC OPTION(LGROPT)=      1  ALGAL PREF FOR NH3-N (PREFN)   =    0.0
```

Table 6-8 (*Continued*)

```
ALG/TEMP SOLR RAD FACTOR(TFACT)=    0.45   NITRIFICATION INHIBITION COEF  =     0.6
ENDATA1A
ENDATA1B
STREAM REACH    1. RCH= MILBURY ST.       FROM      45.8   TO        44.0
STREAM REACH    2. RCH= McCRAKEN RD.      FROM      44.0   TO        41.4
STREAM REACH    3. RCH= RIVERLIN ST.      FROM      41.4   TO        40.8
STREAM REACH    4. RCH= MILBURY WWTP      FROM      40.8   TO        39.8
STREAM REACH    5. RCH= SINGING DAM       FROM      39.8   TO        38.2
STREAM REACH    6. RCH= PLEASANT ST.      FROM      38.2   TO        36.8
STREAM REACH    7. RCH= FISHERVILLE DAM   FROM      36.8   TO        35.4
STREAM REACH    8. RCH= GRAFTON WWTP      FROM      35.4   TO        32.0
STREAM REACH    9. RCH= RIVERDALE ST.     FROM      32.0   TO        29.2
STREAM REACH   10. RCH= NORTHBRIDGE WWTP  FROM      29.2   TO        27.8
STREAM REACH   11. RCH= RICE CITY POND DA FROM      27.8   TO        26.0
STREAM REACH   12. RCH= RT 16 UXBRIDGE    FROM      26.0   TO        25.6
STREAM REACH   13. RCH= MUMFORD RIVER     FROM      25.6   TO        23.2
STREAM REACH   14. RCH= WEST RIVER        FROM      23.2   TO        19.2
STREAM REACH   15. RCH= USGS MILLVILLE    FROM      19.2   TO        18.2
ENDATA2
ENDATA3
FLAG FIELD RCH=  1.         9.         1.2.2.2.2.2.6.2.2.
FLAG FIELD RCH=  2.        13.         2.2.2.2.2.2.2.2.2.2.2.2.2.
FLAG FIELD RCH=  3.         3.         2.2.2.
FLAG FIELD RCH=  4.         5.         6.2.2.2.2.
FLAG FIELD RCH=  5.         8.         2.2.2.2.2.2.2.2.
FLAG FIELD RCH=  6.         7.         2.2.2.2.2.2.2.
FLAG FIELD RCH=  7.         7.         6.2.2.2.2.2.2.
FLAG FIELD RCH=  8.        17.         6.2.2.2.2.2.2.2.2.2.2.2.2.2.2.2.
FLAG FIELD RCH=  9.        14.         2.2.2.2.2.2.2.2.2.2.2.2.2.2.
FLAG FIELD RCH= 10.         7.         6.2.2.2.2.2.
FLAG FIELD RCH= 11.         9.         2.2.2.2.2.2.2.2.2.
FLAG FIELD RCH= 12.         2.         2.2.
FLAG FIELD RCH= 13.        12.         6.2.2.2.2.2.6.2.2.2.2.2.
FLAG FIELD RCH= 14.        20.         2.2.2.2.2.2.6.2.2.2.2.2.2.2.2.2.2.2.2.2.
FLAG FIELD RCH= 15.         5.         2.2.2.2.5.
ENDATA4
HYDRAULICS RCH=  1.    300.0 .    .073    .494    .530    .221    .040
HYDRAULICS RCH=  2.    300.0     .250    .320    .822    .109    .040
HYDRAULICS RCH=  3.    300.0     .082    .308    .790    .280    .040
HYDRAULICS RCH=  4.    300.0     .072    .334   4.000    .000    .040
HYDRAULICS RCH=  5.    300.0     .161    .356    .280    .448    .040
HYDRAULICS RCH=  6.    300.0     .011    .827   3.370    .080    .040
HYDRAULICS RCH=  7.    300.0     .063    .447    .030    .963    .040
HYDRAULICS RCH=  8.    300.0     .058    .574    .733    .295    .040
HYDRAULICS RCH=  9.    300.0     .009    .736    .537    .329    .040
HYDRAULICS RCH= 10.    300.0     .010    .713    .411    .413    .040
HYDRAULICS RCH= 11.    300.0     .074    .335    .509    .331    .040
HYDRAULICS RCH= 12.    300.0     .625    .155    .064    .761    .040
HYDRAULICS RCH= 13.    300.0     .771    .108    .179    .528    .040
HYDRAULICS RCH= 14.    300.0     .537    .127    .448    .379    .040
HYDRAULICS RCH= 15.    300.0     .012    .581   1.452    .310    .040
```

(*continues*)

Table 6-8   (*Continued*)

```
ENDATA5
ENDATA5A
REACT COEF RCH=    1.   0.100     0.00    0.150   3.    0.00     0.00     0.00
REACT COEF RCH=    2.   0.100     0.00    0.550   3.    0.00     0.00     0.00
REACT COEF RCH=    3.   0.100     0.00    0.550   3.    0.00     0.00     0.00
REACT COEF RCH=    4.   0.100     0.00    0.550   3.    0.00     0.00     0.00
REACT COEF RCH=    5.   0.100     0.00    0.550   3.    0.00     0.00     0.00
REACT COEF RCH=    6.   0.100     0.00    0.550   3.    0.00     0.00     0.00
REACT COEF RCH=    7.   0.100     0.00    0.370   3.    0.00     0.00     0.00
REACT COEF RCH=    8.   0.100     0.00    0.370   3.    0.00     0.00     0.00
REACT COEF RCH=    9.   0.100     0.00    0.230   3.    0.00     0.00     0.00
REACT COEF RCH=   10.   0.100     0.00    0.230   3.    0.00     0.00     0.00
REACT COEF RCH=   11.   0.100     0.00    0.150   3.    0.00     0.00     0.00
REACT COEF RCH=   12.   0.100     0.00    0.150   3.    0.00     0.00     0.00
REACT COEF RCH=   13.   0.100     0.00    0.150   3.    0.00     0.00     0.00
REACT COEF RCH=   14.   0.100     0.00    0.150   3.    0.00     0.00     0.00
REACT COEF RCH=   15.   0.100     0.00    0.150   3.    0.00     0.00     0.00
ENDATA6
N AND P COEF   RCH=    1.   0.200   0.05   0.100    5.0   0.200   0.35   0.05   0.50
N AND P COEF   RCH=    2.   0.200   0.05   1.000    5.0   2.000   0.35   0.05   0.50
N AND P COEF   RCH=    3.   0.200   0.05   0.100    5.0   0.200   0.35   0.05   0.50
N AND P COEF   RCH=    4.   0.200   0.05   0.100    5.0   0.200   0.35   0.05   0.50
N AND P COEF   RCH=    5.   0.200   0.05   1.000    5.0   2.000   0.35   0.05   0.50
N AND P COEF   RCH=    6.   0.200   0.05   1.000    5.0   2.000   0.35   0.05   0.50
N AND P COEF   RCH=    7.   0.200   0.05   1.000    5.0   2.000   0.35   0.05   0.50
N AND P COEF   RCH=    8.   0.200   0.05   1.000    5.0   2.000   0.35   0.05   0.50
N AND P COEF   RCH=    9.   0.200   0.05   1.000    5.0   2.000   0.35   0.05   0.50
N AND P COEF   RCH=   10.   0.200   0.05   1.000    5.0   2.000   0.35   0.05   0.50
N AND P COEF   RCH=   11.   0.200   0.05   0.880    5.0   1.760   0.35   0.05   0.50
N AND P COEF   RCH=   12.   0.200   0.05   0.880    5.0   1.760   0.35   0.05   0.50
N AND P COEF   RCH=   13.   0.200   0.05   0.880    5.0   1.760   0.35   0.05   0.50
N AND P COEF   RCH=   14.   0.200   0.05   0.600    5.0   1.200   0.35   0.05   0.50
N AND P COEF   RCH=   15.   0.200   0.05   0.100    5.0   0.200   0.35   0.05   0.50
ENDATA6A
ALG/OTHER COEF RCH=    1.   3.00   00.00   0.01   0.00
ALG/OTHER COEF RCH=    2.   3.00   00.00   0.01   0.00
ALG/OTHER COEF RCH=    3.   3.00   00.00   0.01   0.00
ALG/OTHER COEF RCH=    4.   3.00   00.00   0.01   0.00
ALG/OTHER COEF RCH=    5.   3.00   00.00   0.01   0.00
ALG/OTHER COEF RCH=    6.   3.00   00.00   0.01   0.00
ALG/OTHER COEF RCH=    7.   3.00   00.00   0.01   0.00
ALG/OTHER COEF RCH=    8.   3.00   00.00   0.01   0.00
ALG/OTHER COEF RCH=    9.   3.00   00.00   0.01   0.00
ALG/OTHER COEF RCH=   10.   3.00   00.00   0.01   0.00
ALG/OTHER COEF RCH=   11.   3.00   01.00   0.01   0.00
ALG/OTHER COEF RCH=   12.   3.00   01.00   0.01   0.00
ALG/OTHER COEF RCH=   13.   3.00   01.00   0.01   0.00
ALG/OTHER COEF RCH=   14.   3.00   01.00   0.01   0.00
ALG/OTHER COEF RCH=   15.   3.00   01.00   0.01   0.00
ENDATA6B
INITIAL COND-1 RCH=    1.   72.10   6.40   1.00 109.80   0.00   0.00   0.000   0.0
INITIAL COND-1 RCH=    2.   72.10   6.30   2.60  94.20   0.00   0.00   0.000   0.0
```

Table 6-8 (*Continued*)

INITIAL COND-1 RCH=	3.	72.10	7.50	2.40	92.30	0.00	0.00	0.000	0.0
INITIAL COND-1 RCH=	4.	72.10	8.10	2.50	93.10	0.00	0.00	0.000	0.0
INITIAL COND-1 RCH=	5.	72.10	8.10	2.40	90.00	0.00	0.00	0.000	0.0
INITIAL COND-1 RCH=	6.	72.10	7.10	2.20	88.40	0.00	0.00	0.000	0.0
INITIAL COND-1 RCH=	7.	72.10	7.10	2.00	85.60	0.00	0.00	0.000	0.0
INITIAL COND-1 RCH=	8.	72.10	7.10	2.10	85.50	0.00	0.00	0.000	0.0
INITIAL COND-1 RCH=	9.	72.10	7.30	1.95	84.70	0.00	0.00	0.000	0.0
INITIAL COND-1 RCH=	10.	72.10	7.30	1.90	83.20	0.00	0.00	0.000	0.0
INITIAL COND-1 RCH=	11.	72.10	6.10	1.80	83.00	0.00	0.00	0.000	0.0
INITIAL COND-1 RCH=	12.	72.10	6.10	1.70	75.00	0.00	0.00	0.000	0.0
INITIAL COND-1 RCH=	13.	72.10	6.10	1.70	75.00	0.00	0.00	0.000	0.0
INITIAL COND-1 RCH=	14.	72.10	6.60	1.40	68.00	0.00	0.00	0.000	0.0
INITIAL COND-1 RCH=	15.	72.10	6.60	1.40	68.00	0.00	0.00	0.000	0.0
ENDATA7									
INITIAL COND-2 RCH=	1.	1.85	0.39	0.190	0.00	2.764	0.00	0.048	
INITIAL COND-2 RCH=	2.	1.95	0.15	0.170	0.02	4.782	0.00	1.013	
INITIAL COND-2 RCH=	3.	2.20	0.15	0.190	0.03	4.670	0.00	0.937	
INITIAL COND-2 RCH=	4.	1.50	0.15	0.180	0.04	4.590	0.00	0.813	
INITIAL COND-2 RCH=	5.	1.50	0.14	0.170	0.04	4.450	0.00	0.813	
INITIAL COND-2 RCH=	6.	1.50	0.14	0.160	0.04	4.380	0.00	0.813	
INITIAL COND-2 RCH=	7.	3.45	0.12	0.180	0.04	3.900	0.00	0.602	
INITIAL COND-2 RCH=	8.	3.45	0.12	0.220	0.05	3.930	0.00	0.602	
INITIAL COND-2 RCH=	9.	14.10	0.11	0.160	0.07	3.920	0.00	0.565	
INITIAL COND-2 RCH=	10.	14.10	0.11	0.180	0.08	4.120	0.00	0.565	
INITIAL COND-2 RCH=	11.	18.80	0.09	0.140	0.07	3.680	0.00	0.464	
INITIAL COND-2 RCH=	12.	18.80	0.08	0.120	0.06	2.910	0.00	0.464	
INITIAL COND-2 RCH=	13.	18.80	0.08	0.120	0.06	2.910	0.00	0.464	
INITIAL COND-2 RCH=	14.	15.85	0.07	0.090	0.05	2.570	0.00	0.230	
INITIAL COND-2 RCH=	15.	15.85	0.07	0.090	0.05	2.570	0.00	0.230	
ENDATA7A									
INCR INFLOW-1 RCH=	1.	1.190	72.10	6.6	0.0	17.4	0.0	0.0	0.0
INCR INFLOW-1 RCH=	2.	1.709	72.10	6.6	0.0	17.4	0.0	0.0	0.0
INCR INFLOW-1 RCH=	3.	1.229	72.10	6.6	0.0	17.4	0.0	0.0	0.0
INCR INFLOW-1 RCH=	4.	0.269	72.10	6.6	0.0	17.4	0.0	0.0	0.0
INCR INFLOW-1 RCH=	5.	3.014	72.10	6.6	0.0	17.4	0.0	0.0	0.0
INCR INFLOW-1 RCH=	6.	0.077	72.10	6.6	0.0	17.4	0.0	0.0	0.0
INCR INFLOW-1 RCH=	7.	0.269	72.10	6.6	0.0	17.4	0.0	0.0	0.0
INCR INFLOW-1 RCH=	8.	0.960	72.10	6.6	0.0	17.4	0.0	0.0	0.0
INCR INFLOW-1 RCH=	9.	1.037	72.10	6.6	0.0	17.4	0.0	0.0	0.0
INCR INFLOW-1 RCH=	10.	0.134	72.10	6.6	0.0	17.4	0.0	0.0	0.0
INCR INFLOW-1 RCH=	11.	0.230	72.10	6.6	0.0	17.4	0.0	0.0	0.0
INCR INFLOW-1 RCH=	12.	0.096	72.10	6.6	0.0	17.4	0.0	0.0	0.0
INCR INFLOW-1 RCH=	13.	0.115	72.10	6.6	0.0	17.4	0.0	0.0	0.0
INCR INFLOW-1 RCH=	14.	1.382	72.10	6.6	0.0	17.4	0.0	0.0	0.0
INCR INFLOW-1 RCH=	15.	0.338	72.10	6.6	0.0	17.4	0.0	0.0	0.0
ENDATA8									
INCR INFLOW-2 RCH=	1.	0.000	0.00	0.05	0.0	0.180	0.0	0.06	
INCR INFLOW-2 RCH=	2.	0.000	0.00	0.05	0.0	0.180	0.0	0.06	
INCR INFLOW-2 RCH=	3.	0.000	0.00	0.05	0.0	0.180	0.0	0.06	
INCR INFLOW-2 RCH=	4.	0.000	0.00	0.05	0.0	0.180	0.0	0.06	

(*continues*)

Table 6-8 *(Continued)*

```
INCR INFLOW-2   RCH=    5.  0.000   0.00   0.05    0.0  0.180    0.0   0.06
INCR INFLOW-2   RCH=    6.  0.000   0.00   0.05    0.0  0.180    0.0   0.06
INCR INFLOW-2   RCH=    7.  0.000   0.00   0.05    0.0  0.180    0.0   0.06
INCR INFLOW-2   RCH=    8.  0.000   0.00   0.05    0.0  0.180    0.0   0.06
INCR INFLOW-2   RCH=    9.  0.000   0.00   0.05    0.0  0.180    0.0   0.06
INCR INFLOW-2   RCH=   10.  0.000   0.00   0.05    0.0  0.180    0.0   0.06
INCR INFLOW-2   RCH=   11.  0.000   0.00   0.05    0.0  0.180    0.0   0.06
INCR INFLOW-2   RCH=   12.  0.000   0.00   0.05    0.0  0.180    0.0   0.06
INCR INFLOW-2   RCH=   13.  0.000   0.00   0.05    0.0  0.180    0.0   0.06
INCR INFLOW-2   RCH=   14.  0.000   0.00   0.05    0.0  0.180    0.0   0.06
INCR INFLOW-2   RCH=   15.  0.000   0.00   0.05    0.0  0.180    0.0   0.06
ENDATA8A
ENDATA9
HEADWTR-1 HDW=   1.MILBURY ST.          13.70  73.4   7.08  0.89 114.0   0.0   0.0
ENDATA10
HEADWTR-2  HDW=  1.  0.00   0.00   2.80  0.64  0.20   0.0  0.65   0.0  0.01
ENDATA10A
POINTLD-1 PTL=   1.UBWPAD        0.00     38.4  73.4  6.0   2.0  91.2   0.0   0.0
POINTLD-1 PTL=   2.MILBURY WWTP  0.00    0.598  73.4  6.0  44.0 282.0   0.0   0.0
POINTLD-1 PTL=   3.QUINSIGAMOND  0.00    8.950  73.4  6.33 0.50 68.13   0.0   0.0
POINTLD-1 PTL=   4.GRAFTON WWTP  0.00    1.640  73.4  6.0   9.0 102.0   0.0   0.0
POINTLD-1 PTL=   5.NORTHBR WWTP  0.00    1.780  73.4  6.0   6.2  45.0   0.0   0.0
POINTLD-1 PTL=   6.MUMFORD RIVR  0.00    13.15  73.4  8.08 0.63  20.0   0.0   0.0
POINTLD-1 PTL=   7.WEST RIVER    0.00     7.18  73.4  6.59 0.63  43.0   0.0   0.0
POINTLD-1 PTL=   8.UXBRIDG WWTP  0.00    3.875  73.4  6.0   5.0 102.0   0.0   0.0
ENDATA11
POINTLD-2  PTL=  1.  0.0  0.00    0.0  0.75  0.44   0.0   5.40  0.00  0.90
POINTLD-2  PTL=  2.  0.0  0.00    0.0  0.00 21.00   0.0   3.0   0.00  4.13
POINTLD-2  PTL=  3.  0.0  0.00   1.50 0.514  0.07   0.0   0.14  0.00  0.07
POINTLD-2  PTL=  4.  0.0  0.00   0.00 0.00   2.00   0.0   3.0   0.00  4.3
POINTLD-2  PTL=  5.  0.0  0.00    0.0 0.00   5.97   0.0   3.0   0.00  2.3
POINTLD-2  PTL=  6.  0.0  0.00   1.20 0.436  0.05   0.0   0.15  0.00  0.16
POINTLD-2  PTL=  7.  0.0  0.00   1.45 0.466  0.04   0.0   0.10  0.00  0.01
POINTLD-2  PTL=  8.  0.0  0.00   0.00 0.000  0.23   0.0   3.00  0.00  3.67
ENDATA11A
DAM DATA        DAM=    1    2    2  1.60  0.70  1.0    4.
DAM DATA        DAM=    2    3    3  1.60  0.70  1.0    4.
DAM DATA        DAM=    3    5    2  1.60  0.70  1.0   14.
DAM DATA        DAM=    4    5    4  1.60  0.70  1.0    4.
DAM DATA        DAM=    5    5    7  1.60  0.70  1.0    4.
DAM DATA        DAM=    6    7    3  1.60  0.70  1.0    4.
DAM DATA        DAM=    7    7    7  1.60  0.70  1.0    4.
DAM DATA        DAM=    8    9    2  1.60  1.05  1.0   10.
DAM DATA        DAM=    9   11    2  1.60  0.70  1.0   10.
ENDATA12
ENDATA13
ENDATA13A
LOCAL CLIMATOLOGY                       .25    25.       20.  980.       2.5
BEGIN   RCH   1
PLOT    RCH   1   0   0   0   0   6
BEGIN   RCH   2
PLOT    RCH   0   2   0   4   5   0
```

6.6.4 Mass Transport Calibration

The QUAL2E model is used to simulate chloride concentrations in the Blackstone River using the conservative constituent slot. Based on the flow profiles shown in Figure 6-18, a mass transport model of the Blackstone River is formed. Boundary conditions from point sources and tributaries are incorporated to drive the model. Model results for the three water quality surveys are presented in Figure 6-19 for

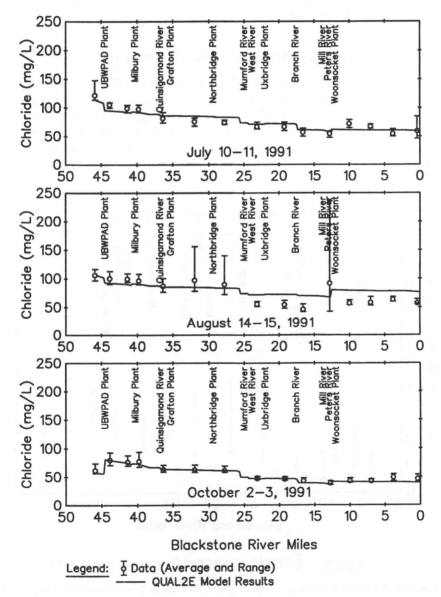

Figure 6-19   Mass transport model results of Blackstone River using the QUAL2E model.

comparison with the measured chloride concentrations. In general, the model results match the data quite well, except the lower 25 miles in the August survey when the chloride concentrations calculated by the model are higher than the data. The mass transport modeling analysis of a one-dimensional river system simply serves as a mass balance check for all the inputs and outputs of a conservative constituent, in this case, chloride. This exercise also checks the point source chloride loads from wastewater treatment plants and tributaries. Figure 6-19 also shows that chloride concentrations are progressively decreasing in the downstream direction, with the Mumford River and Branch River providing much dilution.

6.6.5 Deriving the CBOD Deoxygenation Coefficient

The procedure described in Section 4.1.3 to quantify the CBOD deoxygenation coefficient, K_d, is not suitable for the Blackstone River study. A plot of $CBOD_5$ concentration versus distance is shown in Figure 6-20, suggesting an increasing $CBOD_5$ load in the downstream direction instead of a downward trend as normally expected. This upward trend occurs because the unfiltered CBOD samples include algal biomass. (This problem clearly indicates that samples from waters known to have significant

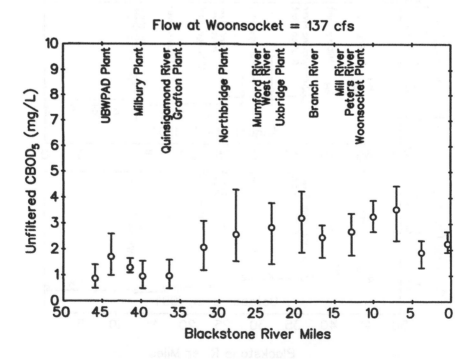

Figure 6-20 Unfiltered $CBOD_5$ concentration profiles in Blackstone River, July 10–11, 1991.

algae should be filtered so that the filtered CBOD data can be used in developing the K_d rate.) Without an independent derivation of the K_d rate, one has to use information such as Figure 4-9 to obtain a preliminary estimate of this rate. Subsequent model calibration and sensitivity analysis would fine-tune this coefficient to be 0.1 day^{-1}. A $CBOD_u$ to $CBOD_5$ ratio of 4 is used for the model to yield results in $CBOD_5$.

6.6.6   Reaeration Coefficient

The O'Connor–Dobbins equation for reaeration is chosen based on the river velocity and depth (see Figure 4-14) in the water column. Subsequent model calibration and sensitivity analysis substantiate this selection. Another factor in choosing the O'Connor–Dobbins equation is its ability to simulate the diurnal variation of DO in the river (Wright et al., 1998).

The 19 dams included in the model provide additional reaeration to the water column, simulated with the relationship developed by Butts and Evans (1973). As seen in the results, this reaeration due to the dams has a significant impact on the river's DO budget.

6.6.7   Other Kinetic Coefficients Related to BOD/DO Modeling

As stated in Wright et al. (1998), nitrification rates, β_1 and β_2, are developed from the plot of ammonia loadings versus time of travel. The SOD values are developed using field measurement data obtained in September 1992. In general, the SOD values vary from 1.61 gm O_2/m^2/day to 5.92 gm O_2/m^2/day, consistent with the values reported in the literature. Also, the highest values are found behind the impoundment. While the SOD has a significant impact on the DO balance in the impoundments, dam reaeration is an important oxygen input, which allows the DO concentration to recover.

6.6.8   Model Calibration Procedure

A brief summary of steps to fine-tune the model is outlined below to further improve model calibrations.

1. Perform independent estimates of the exogenous variables such as river flows, mass transport patterns, boundary conditions, and environmental conditions. These variables should first be derived from the data to minimize uncertainties (see Sections 6.6.1 to 6.6.7).

2. Independently develop and derive as many kinetic coefficients as possible using the available data. One example is quantifying the deoxygenation coefficient, K_d, from the CBOD data. Unfortunately, data deficiency prevented such an exercise for the Blackstone River.

3. Consult literature for coefficients that cannot be derived from the available data. Examples are algal growth rate, respiration rate, nutrient half-saturation constants, and so on. In this situation, experience in water quality modeling is required to assign reasonable coefficient values.

4. Conduct model sensitivity runs and component analyses to quantify the relative importance of the coefficients to the dissolved oxygen budget as well as the algal-nutrient dynamics in the water column.

Note that model calibrations are not intended to "curve fit" the data. Rather, the calibrated coefficients are obtained through a series of model sensitivity runs with reasonable and narrow ranges of their values derived from literature and other modeling studies. In model calibrations, adjusting the kinetic coefficients and constants (within narrow ranges) to improve the calibration of a certain water quality constituent(s) often results in an adverse outcome of matching other water quality constituents. These are the constraints in the model calibration process that will eventually lead to the development of a unique set of model coefficients.

6.6.9 Model Results

The model results for the three surveys are presented in Figures 6-21 to 6-23. In each figure, the river flow profile that produced the mass transport model calibration in Figure 6-19 is displayed. The water quality parameters shown in Figures 6-21 to 6-23 are unfiltered $CBOD_5$, TKN, ammonium, nitrite and nitrate, orthophosphate, chlorophyll a, and dissolved oxygen. The QUAL2E model produced results for dissolved $CBOD_5$ and cannot be directly compared with the unfiltered measured CBOD. Decomposition of dead algae consumes DO, so the addition of this amount to the calculated CBOD value allows the sum to be compared with the unfiltered CBOD data. The solid curve represents this result in the CBOD plot. Also shown is a dotted curve, representing the dissolved $CBOD_5$. It is obvious that once the algal biomass is added to the dissolved CBOD, the results match the data quite well. The dotted curve shows a progressive decline of the dissolved CBOD concentrations along the river due to biological oxidation in the water column.

Model results of other water quality constituents, such as nutrient components, match the data closely. The sharp rise in total Kjeldahl nitrogen (TKN) and ammonium concentrations near Mile Point 13 reflects the input of the Woonsocket Plant, which has secondary treatment but no nitrification. Its nitrate input is small and thereby not reflected in the receiving water concentration. The calculated chlorophyll a levels start to rise near river mile 32 and match the measured concentrations quite well. The drop and rise in the dissolved oxygen concentration profile are the effect of SOD behind the dam and the dam reaeration, respectively. To conclude, the DO standard of 5 mg/L is maintained during these three surveys.

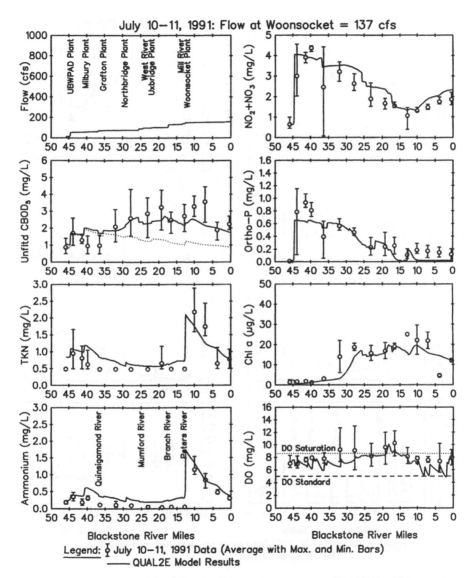

Figure 6-21 QUAL2E modeling analysis of Blackstone River using July 1991 data.

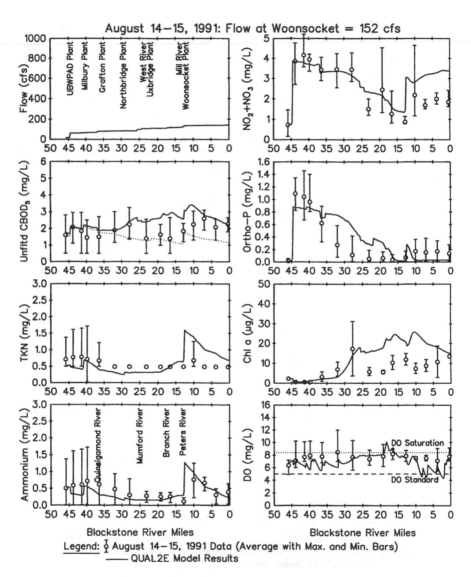

Figure 6-22 QUAL2E modeling analysis of Blackstone River using August 1991 data.

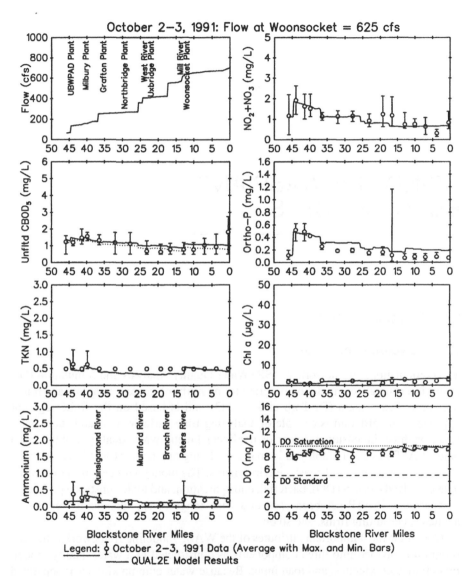

Figure 6-23   QUAL2E modeling analysis of Blackstone River using October 1991 data.

CHAPTER 7

USING THE WASP AND EUTRO MODELS

7.1 THE WASP MODEL

7.1.1 Evolution of WASP

The computational framework WASP (Water Quality Analysis and Simulation Program) is a generalized modeling program for quantifying contaminant fate and transport in surface waters (Di Toro et al., 1983) and is based on the finite-segment approach. The program is capable of analyzing time-variable or steady state, one-, two-, or three-dimensional, linear or nonlinear kinetic water quality problems. To date, WASP has been employed in many modeling applications that have included river, lake, estuarine, and coastal environments. The model has been used to investigate dissolved oxygen (DO), bacterial, eutrophication, and toxic substance problems. Furthermore, WASP has been used in a wide range of applications by regulatory agencies, consulting firms, and so on.

One of the most important attributes of the WASP modeling framework is that the program is divided into three components as constructed in a water quality model: mass transport, kinetics, and load input. Because water column kinetics is separated from mass transport, the user can work on developing mass transport and kinetics at the same time. Perhaps the most significant feature is the flexibility for the user to develop and specify the water column kinetics for any given water quality problem(s) in the original version of WASP (Di Toro et al., 1983). While specifying kinetics is a valuable feature, the difficulty lies in linking up the mass transport and kinetics modules. The user is required to possess both water quality modeling skills and programming capability, a rare combination. When the U.S. Environmental Protection Agency's (EPA's) Center for Exposure Assessment Modeling (CEAM) adopted this modeling framework in the late 1980s, this difficulty was alleviated by hardwiring eutrophication kinetics in the WASP model, thus resulting in the EUTRO model. WASP5 is the EPA's latest version of the WASP and it is currently supported and distributed by CEAM in Athens, Georgia (Ambrose et al., 1993a,b). EUTRO5 is the eutrophication component of WASP5, modeling eight water quality constituents in the water column and sediment bed. Because the user is no longer required to write the kinetics when using the EPA's model, the flexibility of addressing more complex eutrophication problems is lost while user ease is gained.

7.1.2 Contents of the Chapter

This discussion of the WASP/EUTRO model looks at examples that demonstrate the efficient utilization of code in this highly versatile program. Many of the model configurations presented in this chapter are not specifically stated in the model manual but are of use to the reader. Options of using the program are explored to provide hints or shortcuts to specific applications through easily followed case studies. This presentation is designed to assist the reader in learning how to configure and apply the code to a variety of water quality problems, either in steady-state or time-variable computations.

7.2 KEY ATTRIBUTES

7.2.1 General Features

The WASP model structure is based on the finite-segment approach, which employs a unique way of solving a three-dimensional problem by performing three one-dimensional calculations, thereby significantly reducing computational effort. In fact, this structure is identical to that found in the HAR03 model. While the HAR03 model provides only the steady-state solutions, the WASP model performs time-variable calculations, using an explicit time integration scheme. In addition, the WASP model uses a one-step Euler solution technique for numerical integration while the DYN-HYD module uses a predictor-corrector numerical scheme. By running the WASP model time variably to steady state, one can analyze steady-state behavior of a natural water system. Varying integration steps during the time-variable model simulations helps to maximize the size of the integration steps to achieve the best computation efficiency.

The modeling framework allows the user to configure the model to suit the application. For example, the maximum number of segments, the maximum number of boundary conditions and loads, and the maximum number of advective flows and dispersive flows can be specified for each application. While the user could potentially optimize the computer resources by recompiling WASP to the dimensionality of each application, this step is not required if the project dimensions are less than those provided in the distributed executable image (up to 100 segments, 5 boundaries, and

5 loadings). Although applications with dimensions exceeding that will require a re-configuration of the model, the source code readily available from the EPA makes this reconfiguration effort easy. Running the model however, is often limited by the computer system's random access memory (RAM) and its ability to handle the needed size of configuration.

The WASP model user manual lists four levels of applications/configurations to address different complexities of water quality problems (Ambrose et al., 1993b). Recent experience has shown that another level of configuration is valuable in many modeling application. For instance, the modeling framework can be used to calculate concentrations of conservative substances (salinity, chloride, specific conductivity, and total dissolved solids) in mass transport modeling. Further, nonconservative substances can be modeled if they follow simple linear reaction kinetics. Examples of this type of application include total phosphorus or nitrogen (first-order decay from the water column) and total suspended solids (with a settling velocity).

This mass transport modeling feature proves to be valuable, despite the fact that the hydrodynamic module, DYNHYD is limited only to one-dimensional calculations and does not support two- or three-dimensional hydrodynamic calculations. The mass transport calculation can therefore be used to calibrate mass transport coefficients using conservative tracers, a much more practical approach toward quantifying two- or three-dimensional mass transport.

In many wasteload allocation studies, independently quantifying the change of sediment oxygen demand (SOD) over time (probably decades) is critical to assessing the long-term recovery of the natural water system following nutrient load reductions. The WASP has the capability to include a sediment system in the model configuration; thus SOD and nutrient release flux rates can be calculated instead of being assigned as external parameters. In the SOD calculations, negative DO concentrations in the sediment are often encountered and allowed by the model.

7.2.2   Segmentation and Segment Volume

Applying the WASP/EUTRO modeling framework (using a finite difference box type configuration) requires that the water system be segmented into a number of completely mixed water volume elements. Several points are considered in determining the length and configuration of model segments. In general, where receiving water concentrations show or are expected to show a rapid rate of change in concentration, small segment lengths are selected to reproduce these gradients in the water column. A second consideration in selecting segment sizes is numerical accuracy. In the box type models, accuracy of the calculated solution is a function of segment size. For example, segmentation of a river into 100 equal size segments will produce more accurate results than segmenting the system into 2 equal size segments. The term associated with accuracy based on segment size is called numerical error, and it can be thought of as an additional mixing or dispersion term. To minimize errors, this term should be on the order of or less than the actual dispersion coefficients.

In many water bodies such as lakes and estuaries, vertical segmentation is needed to account for stratification of the water quality constituents such as salinity, nutrients, and DO. The WASP/EUTRO modeling framework has several nice features in its input data group C to treat vertical segmentations, a feature that the QUAL2E model does not offer (Brown and Barnwell, 1987). For example, each segment is identified as a surface or nonsurface segment, but reaeration is allowed only in the surface segments. Further, the segments above and below a given segment are also identified for several reasons: to properly quantify light penetration through the water column, to account for settling of solids from one segment to the segment below, and to account for reception of solids from the segment above the given segment. Additional details of settling of solids can be found in applications later in this chapter.

7.2.3 Specifying the Transport

Total transport can be decomposed into several individual flow fields for:

1. Surface water transport carrying both dissolved and particulate matter
2. Pore water transport carrying only dissolved matter
3. Solids type 1 carrying only particulate matter, for example, organics
4. Solids type 2 carrying only particulate matter, for example, algae
5. Solids type 3 carrying only particulate matter, for example, inorganics
6. Evaporation/precipitation adding or taking away water only

Users must specify these flows via direct measurements, either calibrated to tracer studies or quantified by a hydrodynamic model. Individual flow patterns are specified by inflow and continuity balance as a fraction of the inflow. Flows for solids are interpreted as a settling, deposition, scour, or sedimentation velocity in m/s (Ambrose et al., 1993b). The fraction of flow design is unique in that addition or removal of flows from a point source or a group of sources can be accomplished easily without reconfiguring any other sources. Another key element in configuring the flows is establishing the time series of the flows; this provides time-variable flow fields.

Another option for implementing the transport pattern in WASP is incorporating the bulk flow instead of the net flow across the segment interface. For example, if the flow from segment 1 to segment 2 is 10 m³/s and the flow from segment 2 to segment 1 is 6 m³/s, the WASP model will use a net advective flow of 4 m³/s and a dispersive flow of 6 m³/s, thereby creating some mixing across the interface.

7.2.4 Model Configuration

Some applications of the WASP/EUTRO program require configuring the model to suit the site-specific situation. From the computational standpoint, it is important to run the configured site-specific model in the most efficient manner, that is, to minimize the storage requirement on the computer and streamline the output for post processing and interpreting the model results. All the configuration attributes, such as the maximum number of flows, boundary conditions, and so on, are contained in the WASP.INC file, which users can easily edit to suit their needs without changing the main program code. Once this file is edited, the entire code must be recompiled and linked to form the executable file.

It should be pointed out that not every application requires recompilation. While recompilation optimizes computer resources, modern personal computers (PCs) have no problem with the executable file that is provided by the EPA. Recompilation is required only when the application dimensions exceed those provided. When that case arises, one option would be to use the special, large version of the program that is available on request from the EPA. This large version might take up excessive computer memory, however, and becomes wasteful if the user only needs 150 segments and has to make room in the computer memory for 1000 segments.

7.3 MASS TRANSPORT MODELING APPLICATIONS

The WASP/EUTRO model has been applied to the Upper Mississippi River and Lake Pepin (Figure 7-1) in a wasteload allocation study (Lung and Larson, 1995). Concerns about accelerated eutrophication in Lake Pepin have been raised in recent years, particularly following algal blooms and fish kills in 1988. One of the control alternatives to reverse eutrophication is the reduction of point source phosphorus

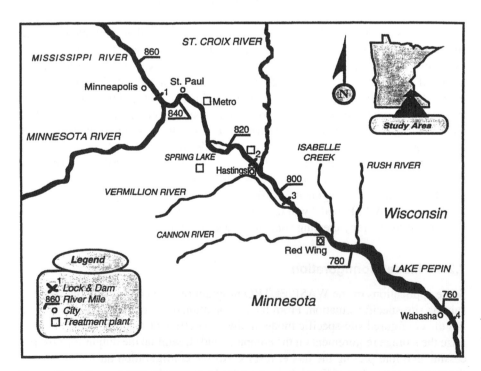

Figure 7-1 Upper Mississippi River and Lake Pepin from Lock & Dam No. 1 to Lock & Dam No. 4.

loads to the lake. The Metro Plant, located approximately 50 miles upstream of the lake entrance, is the largest (250 mgd) point source of phosphorus. The first step of the modeling study is to develop the mass transport for the EUTRO model. Mass transport patterns are quantified for 1988, 1990, and 1991 on a steady-state and time-variable basis. Mass transport information is included in WASP/EUTRO input data groups A–D. The derivation and compilation of the data for these four groups are presented in the following sections.

7.3.1   Model Configuration

The EUTRO model is configured to perform mass transport calculations, using a conservative tracer such as chloride, specific conductivity, or total dissolved solids. Only the first system variable slot, ammonia, is used. Input for the other seven system variables is not required, thereby significantly reducing the effort to prepare the input data file and improving the computational efficiency. In the input data file, the number of system variables, NSYS, is set to one (1). However, the total number of systems, SY in the WASP.CMN file is still set to eight (8). For conservative substances, the kinetic coefficients (for the first system variable, ammonia) must be set to zero. The nonzero first-order kinetic coefficient can be set to characterize the decay of a nonconservative substance such as fecal coliform bacteria, total phosphorus, or any uncoupled system variable with a first-order decay. The model can also be reconfigured for modeling total suspended solids by incorporating an average settling velocity for the solids. Thus there are a variety of configurations associated with the mass transport model.

Another way of using the model to simulate conservative substances would be to use the WASP model alone, without linking the EUTRO module. The user would simply comment out two calls to the WASPB subroutine in the WASP code. (The WASPB subroutine is the main block in the EUTRO code that has the formulations of the water column kinetics.) Remember, however, that changing this code requires a recompilation of the program.

7.3.2   Model Segmentation and Hydraulic Geometry

In an earlier wasteload allocation study for the Metro Plant's biochemical oxygen demand (BOD) loads, a water quality model was developed by Hydroscience, Inc. (1979). The river from Lock & Dam No. 1 to Lock & Dam No. 2 was divided into 81 segments, following the Hydroscience segmentation. That segmentation scheme is adopted for this study. Additional segments are added (see Lung and Larson, 1995) to include the portion from Lock & Dam No. 2 to the lake outlet. Beginning at UM834.1, the water column is divided into two layers (Figure 7-2). The average depth of the surface segments in this portion of the river is 4 ft. In addition, lateral segmentation is introduced in the Spring Lake area, generating a three-dimensional configuration (Figure 7-2). In the portion of the study area from Lock & Dam No. 2 to the lake outlet, another 80 segments are added in a two-layer fashion, following a close examination of the data and information on river hydraulic geometry and

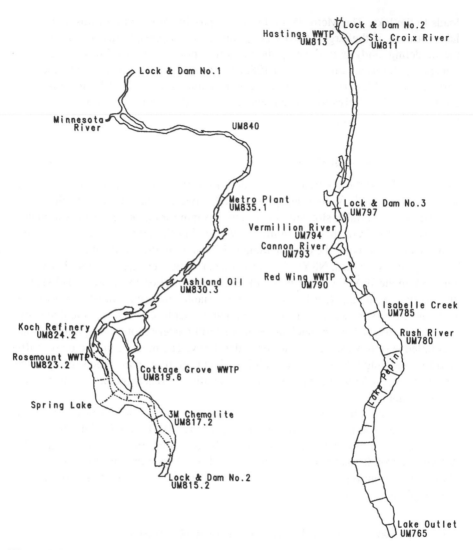

Figure 7-2 WASP/EUTRO model segmentation for the Upper Mississippi River and Lake Pepin.

hydrographics. Thus a total of 174 segments are used for this mass transport analysis. The depth of the surface segments from Lock & Dam No. 2 to the lake outlet is 4.5 ft.

The next step in developing the input data file is calculating the segment volumes. In general, hydrographic information is needed for this purpose. The Army Corps of Engineers keeps complete records of the hydrographic data as they have been dredging the river channel from time to time. They also regulate the locks and dams and maintain records on the flows over the dams as well as the pool and tailwater elevations. Data from 1988, 1990, and 1991, retrieved from the database, are summarized in Figure 7-3 for Locks & Dams Nos. 1 to 4. Although the flows vary greatly from

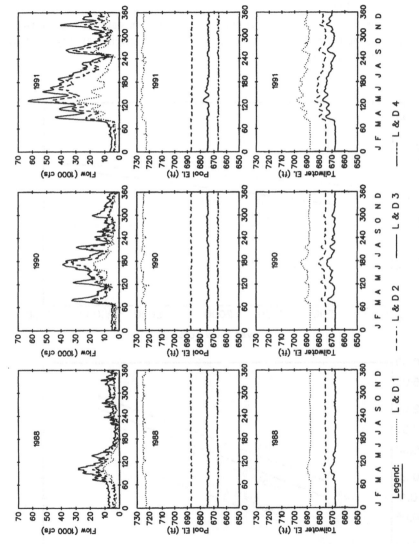

Figure 7-3 Pool (tailwater) elevations and river flows in the Upper Mississippi River and Lake Pepin.

one season to another and from one year to another, the pool elevations remain relatively constant throughout the year and from one to another due to regulations. Such a unique feature suggests that the segment volumes remain constant. Variations of the flow would only affect the retention time in the system. It should be pointed out that constant volumes do not exist in free-flowing rivers where the river stage strongly depends on the river flow.

The next step is to determine the empirical hydraulic constants in the following equations, using the information such as the relationship between velocity and flow and between depth and flow (Figure 7-4):

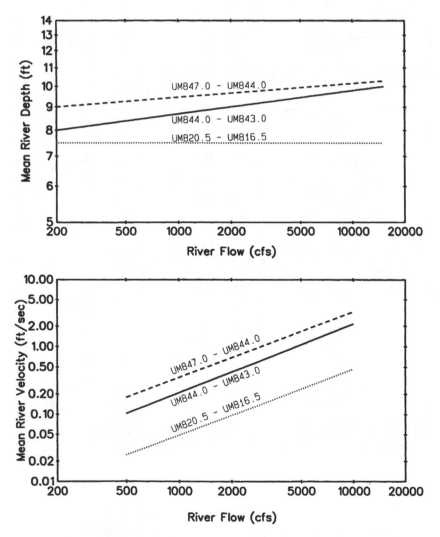

Figure 7-4   Relationships of velocity versus flow and depth versus flow in the Upper Mississippi River and Lake Pepin.

$$u = aQ^b \qquad (7\text{-}1)$$

$$h = cQ^d \qquad (7\text{-}2)$$

where

u = average channel velocity (ft/s)
Q = flow rate (cfs)
h = average channel depth (ft)
$a, b, c,$ and d = empirical constants

The hydraulic constants are summarized in Table 7-1. In the area where the water column is divided into two layers, the values of c in Table 7-1 represent the total water depths. In the Upper Mississippi River, the coefficients a and b do not vary significantly in the longitudinal direction. Further, the b values are very close to 1, indicating that the average velocity is more or less proportional to the river flow in the channel, further suggesting that the segment volumes do not change with the flow. The river portion from UM839.2 to UM793.7 is characterized with $d = 0.0$, reflecting that

TABLE 7-1 Hydraulic Constants for Upper Mississippi River EUTRO Model

River Miles	Segments	a	b	c	d
847.0–844.0	1–7	0.00039	0.97976	6.886	0.04486
844.0–843.0	8	0.00021	1.0	5.9155	0.0546
843.0–841.5	9–11	0.00020	1.0	10.505	0.01209
841.5–839.2	12–15	0.00020	1.0	7.6658	0.01726
839.2–836.5	16–19	0.00010	0.99326	13.0	0.0
836.5–825.8	20–35, 61–72	0.00014	0.97095	10.0	0.0
825.8–824.5	36–37, 73–74	0.00008	1.00824	12.3	0.0
824.5–823.5	38, 75	0.00008	1.00824	12.05	0.0
823.5–820.5	39–42, 76–79	0.00012	0.92438	10.0	0.0
820.5–816.5	43–48, 80–85	0.00008	0.91760	6.40	0.0
816.5–815.2	49–50, 86–87	0.00003	1.05644	15.7	0.0
815.0–813.5	95, 135	0.00011	0.94094	14.0	0.0
813.5–811.4	96–99, 136–139	0.00012	1.0	11.2	0.0
811.4–809.5	100–101, 140–141	0.00008	1.0	11.0	0.0
809.5–808.0	102, 142	0.00009	1.0	9.8	0.0
808.0–801.5	103–109, 143–149	0.00009	1.0	11.0	0.0
801.5–797.0	110–114, 150–154	0.00006	1.02745	10.3	0.0
797.0–793.7	115–116, 155–156	0.00014	0.96674	15.5	0.0
793.7–788.0	117–121, 157–161	0.00014	0.96674	8.5771	0.02897
788.0–786.0	122, 162	0.00007	1.01363	12.751	0.01764
786.0–783.0	123–124, 163–164	0.00022	0.91903	7.8911	0.01842
783.0–781.0	125, 165	0.00004	1.02783	10.866	0.0203
781.0–776.4	126–128, 166–168	0.00003	0.97328	14.13	0.01128
776.4–764.9	129–134, 169–174	0.00001	0.96379	20.0	0.0

Source: Developed from a report by Federal Water Pollution Control Administration (1965).

the channel depth does not change with the flow. In other sections of the river, the d values are very small and water depths vary slightly with the flow. Note that values for a and c must be converted from the English units to SI units so that they are compatible with the EUTRO model input requirements. Figure 7-4 shows the relationships in Eq. 7-1 and Eq. 7-2 for several locations in the Upper Mississippi River in log-log plots. The horizontal line in the depth versus flow plot indicates that the water surface is constant with the river flow. In addition, river velocities decrease from the tailwater of Lock & Dam No. 1 to the pool behind Lock & Dam No. 2. Note that the three parallel lines in the velocity versus flow plot suggest that the b values in Eq. 7-1 are constant from one location to another. Independent data and information are crucial to the development of Table 7-1 and Figure 7-4 for the study.

The hydraulic geometry constants are primarily used to calculate stream reaeration coefficients in the model. Another use of these constants is to calculate the time of travel as a consistency check of the hydraulic geometry. For example, the calculated time of travel along the river from Lock & Dam No. 1 to Lock & Dam No. 2 for four different flow conditions is shown in Figure 7-5. Note that the river flow at St. Paul during summer 1988 averages 2,024 cfs. The Metro Plant adds 240 mgd (371 cfs) flow to the river at UM835.1, resulting in a total flow of 2,395 cfs, which is routed downstream toward Lock & Dam No. 2.

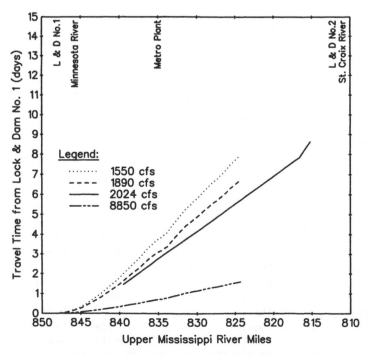

Figure 7-5   Times of travel in the Upper Mississippi River under various river flow conditions.

7.3.3 Advective and Dispersive Transport

The main advective flow for the model is the flow at Lock & Dam No. 1 at the upstream boundary of the model. The U.S. Geological Survey (USGS) flow data provide daily average flows at Anoka (approximately 17 miles above Lock & Dam No. 1) and at St. Paul (about 8 miles below Lock & Dam No. 1). Flows at Anoka are not used as the gaging station is located too far upstream of Lock & Dam No. 1. To maintain reasonable flow balances and accurate time of travel in the system, daily average flows from the Minnesota River (entering the Mississippi River at 4.5 miles upstream of the St. Paul gaging station) are subtracted from the flows measured at St. Paul to derive the flows at Lock & Dam No. 1. This approach yields accurate advective flows at the upstream boundary of the model, particularly under low flow conditions in 1988. For the same reason of maintaining flow balances and providing accurate flows to Lake Pepin, flows at Prescott (UM811.4) are used in the mass transport analysis. That is, flows at St. Paul are added to the Metro Plant flows to form the flows below the plant, which are in turn subtracted from the river flows at Prescott to derive the flows from the St. Croix River. The derived flows for the Minnesota and St. Croix Rivers are slightly different (averaging within 10 to 12%) from the flows estimated by the USGS. Table 7-2 lists the

TABLE 7-2   Major Advective Flows (cfs) for the Upper Mississippi River EUTRO Model in 1988

Date	Julian Day	Lock & Dam No. 1	Minnesota River	St. Croix River
1/11	11	3,390	540	2,421
1/19	19	3,065	505	2,611
2/10	41	3,520	490	2,209
2/16	47	2,900	500	2,479
2/29	60	3,350	750	1,859
3/14	74	5,890	4,610	4,684
4/4	95	8,665	5,135	8,786
4/19	110	9,055	3,045	6,296
5/2	123	3,822	5,418	3,227
5/31	152	3,542	2,048	3,447
6/13	165	1,380	1,130	2,606
7/5	187	991	509	1,599
7/18	200	1,091	419	1,479
8/1	214	736	3,324	1,382
8/15	228	1,487	373	2,282
9/6	250	2,725	355	2,080
9/19	263	2,583	297	1,710
10/3	277	5,091	329	3,242
10/21	295	2,847	333	2,502
10/31	305	2,803	317	2,982
11/14	319	3,195	405	2,572
11/28	333	4,567	443	3,192
12/12	347	2,200	350	3,909

dates and flows in 1988 at Lock & Dam No. 1, the Minnesota River, and the St. Croix River. Information from this table is incorporated into the data file.

There is only one (advective) flow field in the mass transport calculation of dissolved substances such as conductivity, chloride, or total dissolved solids. The variable, IQOPT in data group D, should be set at 1, thereby neglecting all other five advective flow fields. The number of flow functions are five: the flow at Lock & Dam No. 1, the Minnesota River, the Metro Plant, the St. Croix River, and the Cannon River. Each of these five flow functions consists of different numbers of flows. Since the flow at Lock & Dam No. 1 is routed from the upstream boundary to the downstream boundary, it has the most number of flows (202). Table 7-3 shows how these 202 flows (in fractions) are being routed through the system following WASP model convention. The default maximum number of flows (S2) in the WASP.CMN file must be set to accommodate this value of 202. Table 7-3 shows a constant flow of 1,670 cfs at Lock & Dam No. 1. For time-variable flows, each flow function is then entered in time breaks for the entire year.

TABLE 7-3   Advective Flow for Data Group B

1	4	*	+	*	+	*	+	*	+	*	+	D: FLOWS	
7		1.0		.0283		cubic	feet/second	to	cubic	meters/second			
202													
	1.0	0	1		1.0	1	2		1.0	2	3	1.0	3	4
	1.0	4	5		1.0	5	6		1.0	6	7	1.0	7	8
	1.0	8	9		1.0	9	10		1.0	10	11	1.0	11	12
	1.0	12	13		1.0	13	14		1.0	14	15	1.0	15	16
	1.0	16	17		1.0	17	18		1.0	18	19	1.0	19	20
	1.0	20	21		1.0	21	22		1.0	22	23	0.327	23	24
0.327	24	25		0.327	25	26		0.327	26	27	0.327	27	28
0.327	28	29		0.327	29	30		0.327	30	31	0.327	31	32
0.327	32	33		0.327	33	34		0.327	34	35	0.327	35	36
0.327	36	37		0.311	37	38		0.295	38	39	0.295	39	40
0.311	40	41		0.329	41	42		0.202	42	43	0.166	43	44
0.166	44	45		0.249	45	46		0.200	46	47	0.366	47	48
0.440	48	49		0.440	49	50		0.673	23	61	0.673	61	62
0.673	62	63		0.673	63	64		0.673	64	65	0.673	65	66
0.673	66	67		0.673	67	68		0.673	68	69	0.673	69	70
0.673	70	71		0.673	71	72		0.673	72	73	0.673	73	74
0.640	74	75		0.607	75	76		0.607	76	77	0.640	77	78
0.313	78	79		0.276	79	80		0.249	80	81	0.285	81	82
0.284	82	83		0.213	83	84		0.340	84	85	0.393	85	86
0.393	86	87		0.016	37	51		0.016	51	40	0.016	38	52
0.049	52	53		0.113	53	54		0.149	54	56	0.156	56	46
0.182	41	55		0.163	55	44		0.055	55	57	0.073	45	57
0.333	57	58		0.167	58	48		0.093	48	59	0.093	59	60
0.093	60	50		0.027	53	88		0.054	88	90	0.073	90	83

TABLE 7-3 (*Continued*)

0.127	78	89	0.127	89	81	0.037	89	91	0.073	82	91
0.254	91	92	0.127	92	85	0.074	85	93	0.074	93	94
0.074	94	87	0.467	87	50	0.033	74	51	0.033	75	52
0.033	51	77	0.205	46	57	0.144	83	91	0.200	78	41
0.327	50	95	0.163	44	56	0.036	42	55	0.091	42	53
0.036	43	54	0.156	56	45	0.037	79	89	0.027	80	88
0.091	81	90	0.072	90	82	0.166	58	47	0.127	92	84
0.673	50	135	0.327	95	96	0.327	96	97	0.327	97	98
0.327	98	99	0.327	99	100	0.327	100	101	0.327	101	102
0.327	102	103	0.327	103	104	0.327	104	105	0.327	105	106
0.327	106	107	0.327	107	108	0.327	108	109	0.327	109	110
0.327	110	112	0.327	112	114	0.673	135	136	0.673	136	137
0.673	137	138	0.673	138	139	0.673	139	140	0.673	140	141
0.673	141	142	0.673	142	143	0.673	143	144	0.673	144	145
0.673	145	146	0.673	146	147	0.673	147	148	0.673	148	149
0.673	149	150	0.673	150	152	0.673	152	154	0.673	154	114
0.327	114	115	0.327	115	116	0.327	116	117	0.327	117	118
0.327	118	119	0.327	119	120	0.327	120	121	0.327	121	122
0.327	122	123	0.327	123	124	0.327	124	125	0.327	125	126
0.327	126	127	0.327	127	128	0.327	128	129	0.327	129	130
0.327	130	131	0.327	131	132	0.327	132	133	0.327	133	134
0.327	134	0	0.673	114	155	0.673	155	156	0.673	156	157
0.673	157	158	0.673	158	159	0.673	159	160	0.673	160	161
0.673	161	162	0.673	162	163	0.673	163	164	0.673	164	165
0.673	165	166	0.673	166	167	0.673	167	168	0.673	168	169
0.673	169	170	0.673	170	171	0.673	171	172	0.673	172	173
0.673	173	174	0.673	174	0						

2   Flows at Lock & Dam No.1

1670.   0.   1670.   360.

The longitudinal and lateral dispersion coefficients used in the modeling study by Hydroscience, Inc. (1979) are used as preliminary estimates of dispersion coefficients for the first 94 segments of the WASP/EUTRO model. They are the results of several model calibration runs. Dispersion coefficients for the segments in Pools 3 and 4 of the model are first derived from literature values (Fischer et al., 1979) for rivers (see Chapter 3). Information from sources such as Stefan and Anderson (1977), Stefan and Demetracopoulos (1979), and Stefan and Wood (1976) supplements the derivation of the dispersion coefficients. Subsequent model calibration and sensitivity analyses further fine-tune these coefficients.

Since only the water column is modeled, the number of exchange fields (NRFLD) in data group B should be set at 1 (i.e., not modeling pore water exchanges in the sediment system). The dispersion coefficients are then presented in seven groups (i.e., number of exchanges for field No. 1 is 7) for the entire spatial domain of the model. Time-variable values of exchange coefficients for each of these seven groups follow.

7.3.4 Boundary Conditions and Point Source Loads

Since there is only one system variable to be modeled, data groups E and F are significantly reduced in size. It is a common practice to enter the tributary inputs as concentration (in mg/L) in group E and the point source discharges (municipal and industrial) as loads (in lb/day) in group F. In group E, a boundary condition must be entered for every boundary segment. Both boundary conditions and loads can be entered in a number of breaks to characterize the seasonal variation of these quantities. Note that if the user provides loads in lb/day, the scale factor or units conversion factor should be set to 0.454 to convert the input to the model internal units of kg/day.

7.3.5 Parameters, Constants, Miscellaneous Time Functions, and Initial Conditions

There is no parameter needed for the mass transport calculation, setting the number of parameters in data group G to zero. Since the mass transport model calculates concentrations of a conservative substance, constants (in group H) for that system (i.e., No. 1) must also be set to zero. Consequently, the number of global constants and the number of system 1 constants are set to zero. Initial conditions (on Julian day 11) for all 174 segments are entered. Because there is no miscellaneous time function associated with the mass transport model runs, the number of time functions should be zero. Since the conservative constituent (total dissolved solids or specific conductivity) considered is dissolved, the dissolved fraction should be set at 100% (i.e., 1.0) in group J for initial concentrations. In this time-variable model run covering the entire year, accurate initial concentrations are critical to the success of the calculation. Usually, though, accurate initial concentrations in a time-variable model to a steady-state condition are not essential.

7.3.6 Integration Steps

In a time-variable model run, the starting time must be specified. For the Upper Mississippi River and Lake Pepin in 1988, the first available data points are the measurements collected on January 11. Thus the model run starts on that day and the first day of simulation should be set at 12.0 in group A. To minimize the computational effort and achieve the maximum run efficiency, time-variable integration steps should be used. However, it is difficult to determine the appropriate integration step sizes prior to the runs. An easy way to determine the time-variable integration time steps is to examine, from a preliminary run, the *.TRN file for all segments for the entire year. In the *.TRN file, parameter 10 is retention time, an estimate of hydraulic retention time in each segment, calculated by dividing the segment volume by flow. Parameter 11 is exchange time, an estimate of the retention time in a segment due to dispersive exchange, calculated by dividing the segment volume by dispersive flow. Parameter 12 is the total retention time, an estimate of a constituent's time spent in a segment, calculated by 0.9 times segment volume divided by the sum of advective and dispersive flows and kinetic loss rate (volume times loss rate constant). Using a factor of 0.9 serves as a safety factor. For conservative substances, this last parameter is an estimate of the transport retention time. For a given simulation period, the shortest time among all segments should be used as the integration step size to ensure numerical stability. In the time-variable calculation, varying the flow and dispersion coefficient will change the integration step sizes accordingly.

7.3.7   Model Shakedown

One of the key requirements in the model calculation is the balance of advective flows. Although the WASP/EUTRO program conducts an internal check of flow balance, it is always prudent to perform an independent check. Simply set the initial and boundary conditions of 1.0 mg/L for system 1, turn off all kinetic coefficients for the ammonia system, run the simulation to a steady state, and check for any deviation of predicted ammonia concentrations from 1.0 mg/L. If the concentrations remain to be 1.0 mg/L, a flow balance is achieved. If the concentrations deviate from 1.0 mg/L, mass balance is violated.

Note that WASP sometimes issues erroneous flow balance warnings, especially for complicated flow patterns using multiple flow functions. These warnings cause the program execution to pause. The user should simply hit return (perhaps more than once) to proceed with the simulation.

7.3.8   Model Results and Output

The complete model output is routed to the *.EDF file at any print interval(s). A one-year, time-variable run will generate a sizable output file. For the Upper Mississippi River and Lake Pepin modeling study, a separate, concise output file is generated, storing only the time-variable results of the first system variable for post processing. A simple change in the subroutine EUTRODMP (i.e., to choose output parameters in any given segments) is made to generate a specific *.EDF output file. This procedure, which requires a reconfiguration and recompilation of the WASP/EUTRO source code, has proven highly efficient in many model applications.

Figure 7-6 shows the model results compared with measured total dissolved solids concentrations at five locations: UM839.1, UM831.0, UM826.7, UM815.6, and UM796.9 in the Upper Mississippi River upstream of Lake Pepin in 1988. Data collected at UM847.7 are used as the upstream boundary condition for the model runs (Figure 7-6). In general, the model results mimic the temporal trends of total dissolved solids in the riverine portion of the study area. For this conservative substance, the effluent concentrations at the wastewater treatment plants are estimated to 300 mg/L as no direct measurements of total dissolved solids are made at the plants. This total dissolved solids concentration is derived using the measured chloride concentrations in the Metro Plant effluent and based on a regression between total dissolved solids and chloride concentrations. Due to its large flows, the Metro Plant provides the most total dissolved solids load among the municipal wastewater treatment plants. Figure 7-6 shows that the total dissolved solids concentrations at UM

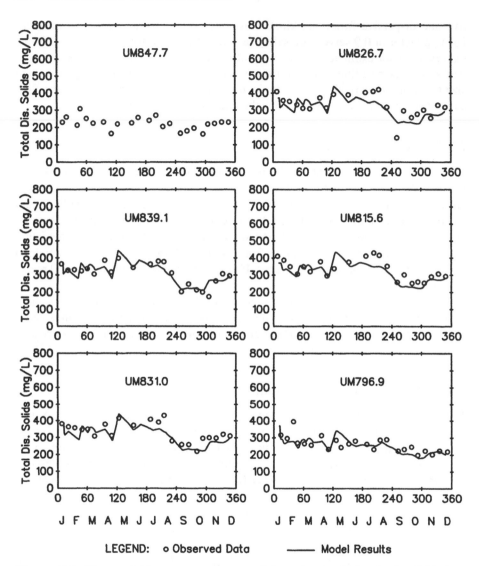

Figure 7-6 Time-variable mass transport model results (total dissolved solids) from WASP/EUTRO versus data for the Upper Mississippi River and Lake Pepin in 1988.

831.0 increase slightly over those at UM839.1, reflecting the Metro Plant input (Lung, 1996a; Lung and Larson, 1995).

Although no total dissolved solids data are available for Lake Pepin, partial year monitoring of specific conductivity levels in Lake Pepin has been made since 1990 and 1991. Figure 7-7 shows the model calculated conductivity levels and measured values in the Upper Mississippi River and Lake Pepin (at 12 locations) in 1990. Model results match the data very well, reproducing the seasonal variations. Since the

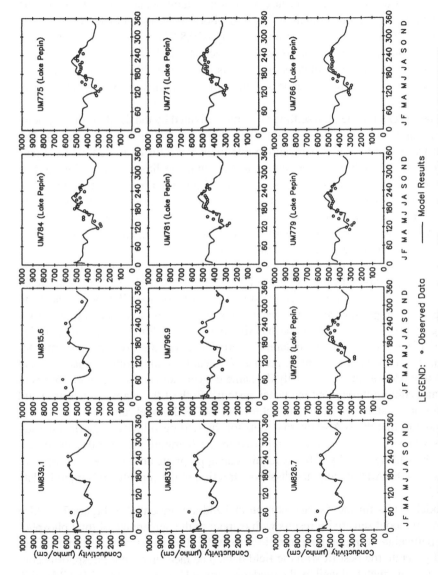

Figure 7-7 Time-variable mass transport model results (specific conductivity) from WASP/EUTRO versus data for the Upper Mississippi River and Lake Pepin in 1990.

river flows in 1990 are high, that is, higher than average, spatial attenuation of conductivity throughout the study area from Lock & Dam No. 1 to the outlet of Lake Pepin does not take place.

The mass transport models are then used to calculate the total phosphorus concentrations in the study area. Because a portion of total phosphorus settles into the sediment, the key parameter in modeling phosphorus is the assignment of the settling velocity. For this calculation, a settling velocity of 0.8 ft/day is selected. Further, the portion of total phosphorus that is in the particulate settleable form never reaches 100%. The estimated particulate phosphorus is about 20 to 30% of total phosphorus in this study area, yielding an approximation of 25% in this example (EnviroTech, 1993).

In the EUTRO model, the settling velocity is divided by the depth of the segment, resulting in a first-order kinetic coefficient, referred to as settling rate (day^{-1}). The deeper water will have a slower settling rate, thereby requiring more time for a particle to travel through the water column and reach the sediment. Such a kinetic formulation results in lower settling rates in Lake Pepin than in the upstream riverine portion. Other factors such as velocity and river flow also affect the overall amount of phosphorus removed from the water column by this process. For example, lower flows would allow more time for particles to settle, thus compensating for the lower settling rate in the water column of Lake Pepin.

There are two ways to implement a settling velocity along with the percent of particulate phosphorus in the EUTRO model. First, the settling velocity may be set at 0.8 ft/day and the fraction of dissolved phosphorus may be set to 0.75 (corresponding to 25% particulate phosphorus in the water column) in the initial concentrations block (data group J). The alternate approach is simply to use a settling velocity of 0.2 ft/day (= 25% of 0.8 ft/day) while maintaining a zero fraction of dissolved phosphorus. Although the first option has a more precise physical meaning, both approaches yield identical model results (i.e., exactly the same concentrations for total phosphorus in the water column throughout the water system). Both of these approaches are valid only for one-system configuration, not for full EUTRO model runs.

Also note that the WASP internal units for the settling velocity of particulates are in m/s while the units commonly reported in the literature are in ft/day or m/day. Using the settling velocity in ft/day requires an appropriate units conversion factor of 0.3048/86,400. Additionally, the dissolved fraction of the constituent is input in data group J.

Model results for the riverine portion in 1988 are presented in Figure 7-8. Also shown are model sensitivity results with respect to the percentage of total phosphorus in particulate form, ranging from 14% to 100%. It is seen that the estimated 25% would generate the best fit with the field data. The measured total phosphorus concentrations are matched well by the model results at UM839.1, UM831.0, UM826.7, UM815.6, and UM796.9. The significant increases in total phosphorus concentrations at UM831.0 are due to the Metro Plant input. Total phosphorus concentrations are measured on a routine basis in the study area from Lock & Dam No. 1 through Lake Pepin in 1990 and 1991. Figure 7-9 presents the model results and measured total phosphorus concentrations at 12 locations in the study area in 1990. Each of the

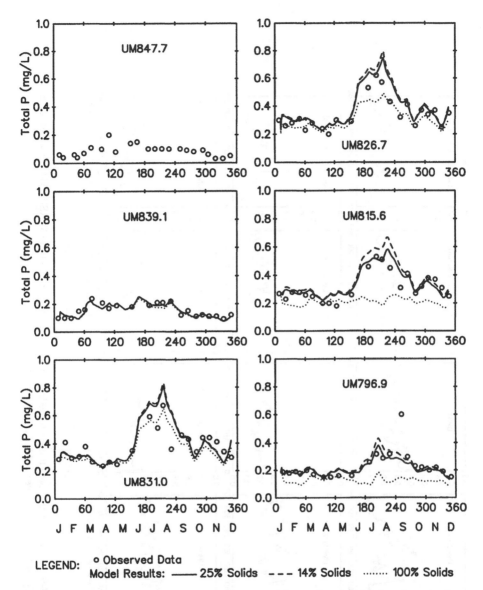

LEGEND: o Observed Data
Model Results: —— 25% Solids – – – 14% Solids ········ 100% Solids

Figure 7-8 Time-variable mass transport model results (total phosphorus) from WASP/EUTRO versus data for the Upper Mississippi River in 1988, showing effect of varying settling velocity.

time-series plots shows the seasonal trend of total phosphorus for the entire year. Model results in general match the data for the riverine portion (from UM839.1 to UM796.9) very well. In the lower portion of Lake Pepin, the model calculated phosphorus concentrations are slightly below the data (see UM771 and UM766) during the later part of the year, possibly due to sediment releases of phosphorus, a

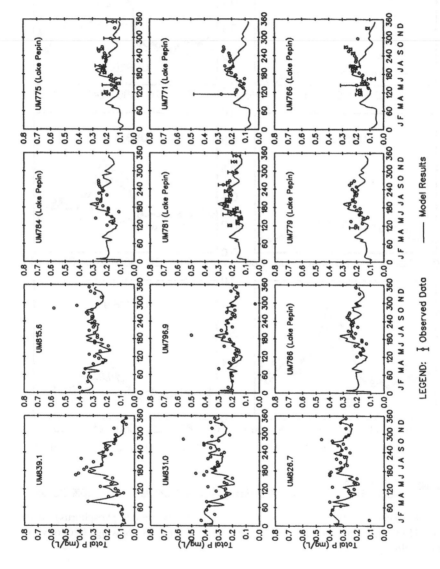

Figure 7-9 Time-variable mass transport model results (total phosphorus) from WASP/EUTRO versus data for the Upper Mississippi River and Lake Pepin in 1990.

mechanism that is not yet incorporated in this modeling analysis. Note that at Lock & Dam No. 2 (UM815.6) the model results match the data closely—without sediment release.

In analyzing the 174-segment Upper Mississippi River and Lake Pepin model, over 90 of the segments have particulate matter settling from the water column into the bottom of the system. In the WASP/EUTRO input structure, all 90 of these segments would be considered as boundary segments, thereby requiring that boundary conditions be specified for each of the 90 segments in the model's Boundary Condition Block E. To minimize the number of boundary conditions, it is wise to add an extra segment below all of the 90 segments, essentially serving as a catch-all pan. In this way, all 90 of the segments become internal segments, eliminating the need to specify boundary conditions for them. This extra 175th segment is referred to only in data group C and need not be actually incorporated into the input. In other words, listing of segment attributes in data group C still stops at segment 174.

7.3.9   Statistical Analysis of Model Results

A qualitative evaluation of the success of the model calibration in this study can be made by inspection of the agreement between the calculated temporal distributions and the data. Statistical techniques have been proposed as quantitative measures of comparison (Thomann, 1980), although the interpretation of these results may be ambiguous and should, therefore, be used with caution. In the Upper Mississippi River and Lake Pepin modeling study, difficulty arises as depth averaged water quality constituent concentrations calculated by the model are compared to grab samples at a single depth in the water column at the sampling location. Assuming the data from all locations are characterized by the same bias throughout the year, the results of statistical analyses may provide a basis for relative comparisons between various locations for time-variable model simulations.

A variety of statistical comparisons may be appropriate to quantify model comparison with field data (Thomann, 1980):

1. Regression analyses
2. Relative error
3. Comparison of means
4. Root mean square (rms) error

Each of these quantitative measures displays model credibility from different statistical viewpoints. Some are apparently useful for diagnostic purposes, while others appear to be directly of value in succinctly describing model verification and calibration status.

For the Upper Mississippi River mass transport models, the rms error measure is selected to quantify the goodness of fit between the model results and field data. The rms error is evaluated as follows:

$$r = \left[\frac{\Sigma(x_i - C_i)^2}{N} \right]^{0.5} \tag{7-3}$$

where

r = rms error
x_i = model results
C_i = field data
N = number of data points

These rms errors are statistically well-behaved and provide a direct measure of model errors. For this study, the rms errors are calculated over time (one year) at single locations with field data. The analysis is conducted for total dissolved solids, conductivity, and total phosphorus. To evaluate the goodness of fit, relative rms errors must be determined for the model results. The analysis results (Table 7-4) indicate that most rms errors of total dissolved solids and conductivity are approximately 3 to 15% of the annual averaged concentrations, which is considered a very good fit. The phosphorus model results yield rms errors about 20 to 40% of the annual averaged concentrations. The increased variations for total phosphorus are not surprising as an additional mechanism, settling, is involved and its kinetic formulation is an approximation at best. Nevertheless, the temporal distributions of total phosphorus at various locations in the study area during 1988, 1990, and 1991 are mimicked closely by the model, using the settling velocity from the recent AESOP model post audit study (EnviroTech, 1993). In fact, relative rms errors for the total phosphorus model results would reduce significantly if some of the apparently biased data points were removed (see Figures 7-8 and 7-9).

7.3.10 Data to Support the Mass Transport Modeling Calculations

In the Upper Mississippi River and Lake Pepin modeling study, no additional field-sampling program was conducted. The available data are considered adequate to support the modeling analysis. Table 7-5 lists the sources of the data and information for the model input data groups.

The vast majority of data come from four sources: Metropolitan Waste Control Commission (MWCC), Minnesota Pollution Control Agency (MPCA), Wisconsin Department of Natural Resources (WDNR), and the USGS. In the model input files, the MPCA and WDNR data play minor roles compared to the MWCC data, which describe 122 of 174 model segments of the Mississippi River, the two major tributaries (Minnesota and St. Croix), the largest point source (Metro Plant), and more.

7.3.11 Other Applications

The previous example of the Upper Mississippi River and Lake Pepin demonstrates a one-dimensional configuration of the EUTRO modeling framework (except in

TABLE 7-4 Relative rms Errors (%) of Mass Transport Model Results

Monitoring Location	Model Segment	Total Dissolved Solids			Conductivity			Total Phosphorus		
		1988	1990	1991	1988	1990	1991	1988	1990	1991
Mississippi River:										
UM839.1	15	10.7	6.0	8.6	11.5	6.6	8.1	14.1	29.3	38.7
UM831.0	28	12.6	9.8	9.2	11.3	9.1	7.6	22.4	22.3	41.1
UM826.7	35	14.4	10.8	9.2	10.3	8.6	8.0	19.6	21.9	26.6
UM815.6	50	12.7	10.5	8.7	10.2	8.0	7.0	31.0	22.1	26.6
UM796.9	115	15.1	9.7	9.4	9.8	12.2	14.3	32.8	25.6	26.0
Lake Pepin:										
UM786	122					11.7	7.1	25.1	34.5	
UM784	123					10	9.4	25.0	34.3	
UM781	125					11.1	3.3	22.8	38.2	
UM779	126					11.1	5.0	23.6	33.7	
UM775	129					11.5	5.2	27.5	38.5	
UM771	132					11.8	3.5	33.3	39.0	
UM766	134					13.2	4.5	38.2	41.7	

TABLE 7-5   Data Used in Mass Transport Modeling of Upper Mississippi River

EUTRO Data Group	Sources of Data and Information
B. Exchange flows	Reports from previous studies: Hydroscience (1979) Stefan and Anderson (1977) Stefan and Wood (1976) Falch et al. (1979)
C. Segment volumes Hydraulic geometry Depth vs. flow Velocity vs. flow	Army Corps of Engineers Reservoir Regulation Manuals Hydrographic information from Federal Water Pollution Control Administration
D. Advective flows River and tributaries Treatment plants	USGS surface water records Point source discharge records
E. Boundary conditions	Receiving water quality data collected by: MWCC MPCA [with the Minnesota Department of Natural Resources (MDNR)] WDNR
F. Loads	Wastewater treatment plant records
J. Initial conditions	MWCC data[a]

[a]Also used to assemble the receiving water data for comparison with model results.

some portions of the study area that are configured in a two-dimensional fashion) for a time-variable modeling analysis of conservative substances such as total dissolved solids and conductivity. A more comprehensive view to mass transport modeling is gained, however, by presenting different spatial configurations in other applications. Two other mass transport modeling studies are presented in this section: the Maryland Coastal Bay (horizontal two-dimensional segmentation) and Patuxent Estuary (longitudinal-vertical two-dimensional segmentation).

The Maryland Coastal Bay is a shallow coastal water system and has been an area of water quality concern due largely to rapid development in recent years. The shallow depths of the water and relatively small freshwater inputs result in poor flushing rates. Slow flushing rates make these bodies of water, and especially the tributaries, susceptible to a variety of water quality stresses, such as nutrient enrichment. To investigate the linkage between land use and the eutrophic conditions of the bay, a modeling study was conducted using the WASP/EUTRO model (Lung, 1994).

The water column is divided into 30 horizontal segments (Figure 7-10) for a water quality model to assess the impact of land use changes in the watershed. Figure 7-11 shows the advective transport pattern during the summer months of 1993, resulting from a small freshwater inflow from the St. Martin River. The advective flows in the

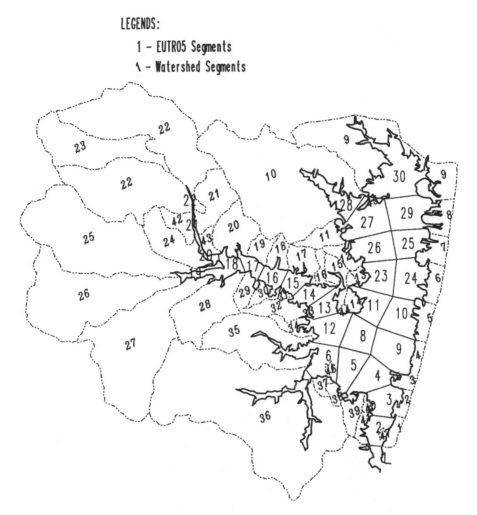

Figure 7-10   Watershed and water segmentation of Maryland Upper Coastal Bay for the WASP/EUTRO model.

open bay area are derived from a hydrodynamic model with a fine grid system (Lung, 1994). Such a complicated advective flow pattern requires routing 29 flows in Block D of the input data file, a very involved effort. Dispersion coefficients between the segment interfaces are determined by reproducing the measured salinity concentrations in the system. The calibrated dispersion coefficients are shown in Figure 7-11. Tidal actions are the primary force responsible for the horizontal mixing in the open water of the study area. The dispersion coefficients that characterize the horizontal mixing decrease in the inland direction. Under the summer 1993 conditions, they range from 3.85 mi^2/day at the mouth of the Isle of Wight Bay (near Ocean City) to about 0.03 mi^2/day in the St. Martin River at Bishopville (Figure 7-11). Like the advective flows, the spatially variable dispersion coefficients require an elaborate effort to enter them in the

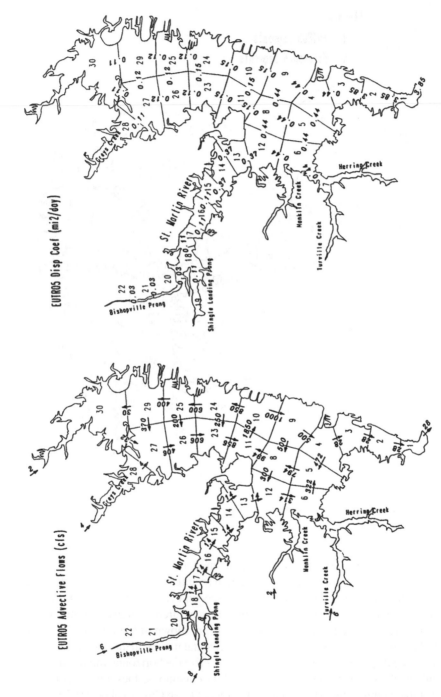

Figure 7-11 Steady-state advective and dispersive transport patterns in Maryland Upper Coastal Bay, summer 1993.

input data file for the WASP/EUTRO model. Figure 7-12 shows the EUTRO model results for salinity distribution under the summer 1993 condition. Also shown for comparison are the tidally averaged salinity concentrations derived from measured values in July, August, and September 1993. In general, the model results mimic the measured spatial distribution in the open waters. Figure 7-12 also shows the longitudinal profiles of salinity in the St. Martin River in the summer months of 1993. The spatial trend of progressively decreasing salinity levels in the upstream direction along the St. Martin River is mimicked by the mass transport model. The small salinity gradients in the open bay make tuning the dispersion coefficients somewhat difficult because the model results are insensitive to changes in small dispersion coefficients.

Another example of mass transport modeling is the Patuxent Estuary, which is a partially mixed estuarine system located in Maryland (Figure 7-13). Persistent, low DO problems are observed in the lower estuary, and the lowest DO levels exist in the bottom waters near Broomes Island at Station XDE5339 (Figure 7-13). These waters, which are partly derived from the Chesapeake Bay mainstem, are depleted in oxygen before they enter the Patuxent Estuary (Domotor et al., 1989). One hypothesis to explain these unacceptable levels is that the low DO concentrations observed at Broomes Island are caused primarily by the intrusion of these low oxygen Bay waters. Another hypothesis suggests that the DO problem originates primarily within the Patuxent Estuary itself, and is aggravated by the intrusion of oxygen-deficient Bay waters. Furthermore, repeated algal blooms have been observed over the years in the upper estuary near Nottingham at Station PXT0402 (Figure 7-13). Subsequent decomposition of algal biomass (being transported downstream) has caused significantly low DO levels in the Broomes Island area. A two-layer circulation in the Patuxent Estuary may transport nutrient released from the downstream sediments upstream to support the algal growth in the upper estuary. Clearly, a two-layer model is required to properly characterize the mass transport patterns of this estuarine system (Lung, 1992).

An analytical method to quantify the longitudinal and vertical velocities in a partially mixed estuary has been developed by Lung and O'Connor (1984) and is presented in Chapter 3. That same methodology is used here to develop the advective flows and vertical dispersion coefficients in this two-layer mass transport model. Figure 7-14 shows the two-layer mass transport pattern in the Patuxent Estuary under a freshwater flow of 100 cfs at the fall line. At the mouth of the river, the tidally averaged surface and bottom layer flows reach 12,281 cfs and 12,025 cfs, respectively. (Their difference is 256 cfs, representing a progressive increase in the freshwater flow in the downstream direction.) Such a significant (over a hundredfold) increase in longitudinal flows results from the density-driven circulation pattern characterized in partially mixed estuaries in the coastal plain region. In addition to the longitudinal advective flows, there are vertical flows and vertical dispersion coefficients between the two layers in a spatially variable fashion along the estuary. Incorporating such a transport pattern in the EUTRO input structure is a time-consuming effort.

Applying this methodology to a number of flow conditions at different times of the year and running the EUTRO model time-variably for the one-year period results in a calibration of the time-variable, tidally averaged mass transport model for the Patuxent Estuary. This exercise takes full advantage of the WASP/EUTRO modeling

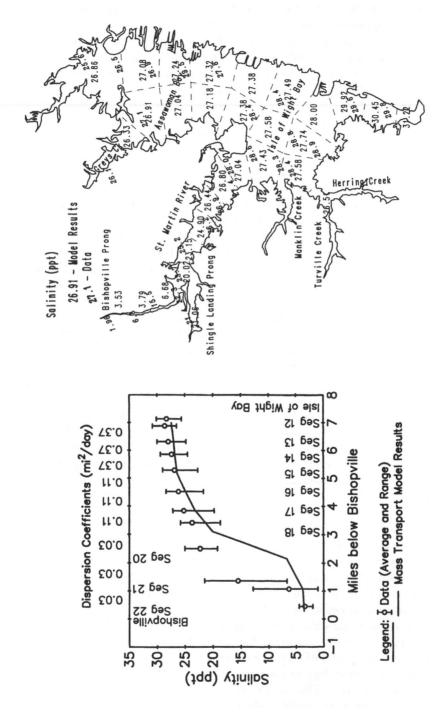

Figure 7-12 Mass transport model results (salinity) from WASP/EUTRO versus data for Maryland Upper Coastal Bay, summer 1993.

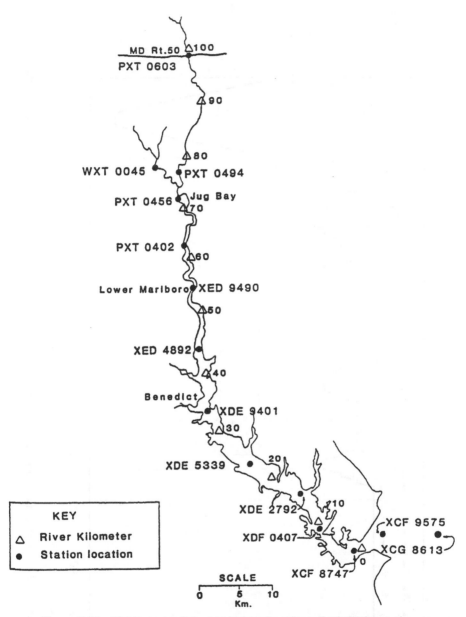

Figure 7-13   The Patuxent Estuary, showing locations of monitoring stations.

framework for its time-variable computation capability. Note that time-variable and spatial-variable advective flows and vertical dispersion coefficients must be incorporated into the input data file. Model calculated salinity and 1983 observed data are summarized in Figure 7-15, showing temporal plots for eight locations. The temporal trend of salinity at these locations along the estuary is reproduced, and significant spring runoff starting on day 75 decreases the salinity levels considerably. As the

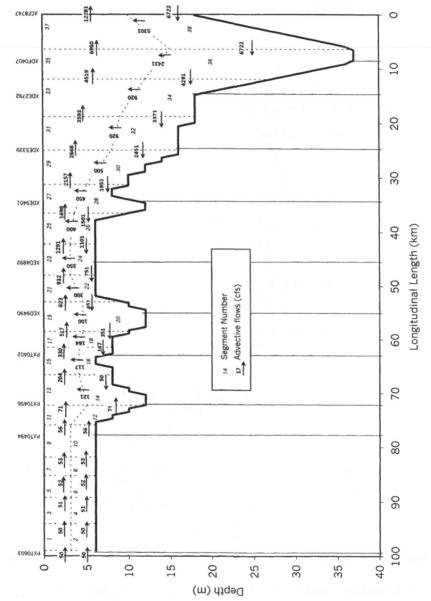

Figure 7-14 Tidally averaged two-layer mass transport patterns in the Patuxent Estuary under freshwater flow of 2,180 cfs.

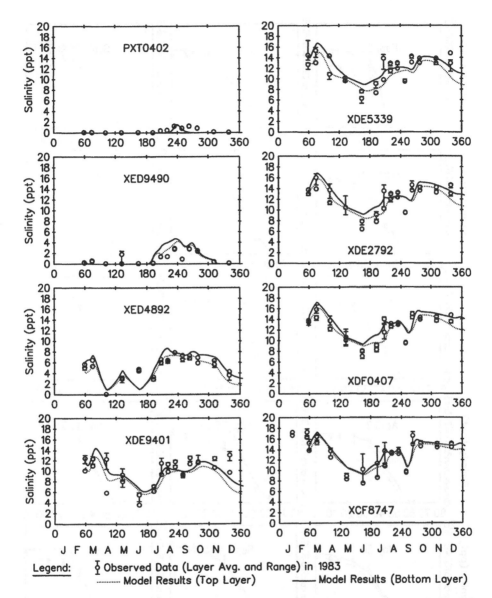

Figure 7-15   Time-variable mass transport model results (salinity) from WASP/EUTRO5 versus data for the Patuxent Estuary in 1983.

summer low flow condition approaches, the salinity levels increase, a pattern that continues throughout the summer months and into early fall. Vertical salinity gradients exist and are more pronounced in the spring high flow months, as expected. Longitudinal plots comparing model calculations and measured salinity during 1983 are presented in Figure 7-16. In general, the two-layer mass transport model mimics the

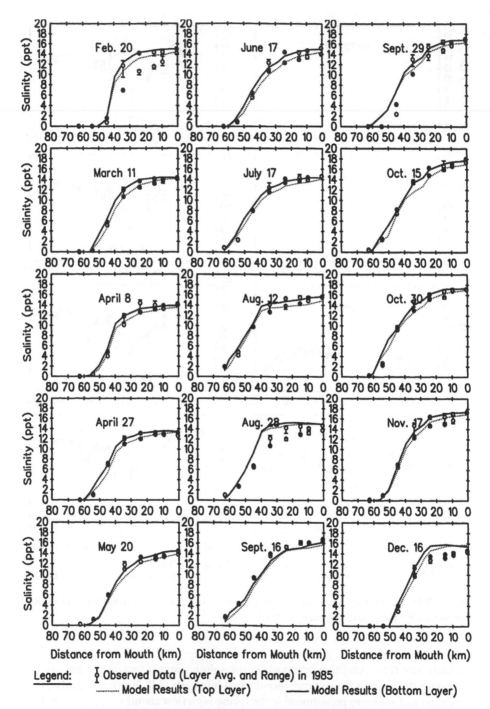

Legend: ⚲ Observed Data (Layer Avg. and Range) in 1985
.... Model Results (Top Layer) ———— Model Results (Bottom Layer)

Figure 7-16 Longitudinal salinity profiles—model results versus data at different times of 1985 along the Patuxent Estuary.

salinity distributions in the Patuxent Estuary well. Seasonal changes in salinity are reproduced by this simple and straightforward methodology.

In all previous examples, the combined WASP/EUTRO model is used for the mass transport modeling of conservative substances. However, running the WASP model alone, that is, without the EUTRO model, offers another method of modeling mass transport. This procedure involves making two changes in the WASP source code. Search for all the call WASPB statements and comment them out. The revised WASP source code must be recompiled to generate a new executable image. This action decouples the EUTRO model from the WASP model since the WASPB subroutine is the code for all water column kinetics. By not linking the EUTRO module, the user runs the WASP code alone for mass transport calculations, resulting in a more efficient operation. There should be no changes in the WASP input data file for this procedure.

7.4 WATER QUALITY MODELING APPLICATION

7.4.1 Reconfiguring the Model

The calibrated transport model of the Upper Mississippi River and Lake Pepin is now expanded to form the water quality model using the full features of the WASP/EUTRO modeling framework. In the water quality model, information such as settling velocity for phytoplankton and particulate organics must be included. In addition, kinetic coefficients for all eight system variables should be incorporated, as well as parameters for each segment and miscellaneous forcing functions. Point source and tributary loads, boundary conditions, and environmental conditions are also incorporated into the model on a time-variable basis for the entire year. SOD and sediment nutrient release fluxes are included in the model, adjusted according to the water column temperatures.

The EUTRO model accepts spatial variations of water column temperatures and light extinction coefficients. For the Upper Mississippi River and Lake Pepin, three temperature and five light extinction coefficient time-variable functions are employed to characterize the spatial and temporal variations of these two parameters.

Nutrient loads in 1988 originate from the four main sources: the Upper Mississippi River at Lock & Dam No. 1, the Minnesota River, the Metro Plant, and the St. Croix River. Time series plots of total phosphorus, orthophosphate, ammonia, and nitrite/nitrate loads are shown in Figure 7-17. Also shown are the associated flows from these sources. Of the four sources, the Upper Mississippi River carries a significant nitrate load, primarily due to nonpoint inputs in that watershed. Furthermore, the Metro Plant is a significant contributor of phosphorus in 1988. Its steady input of phosphorus is particularly important during the summer months when the river phosphorus loads are much depressed.

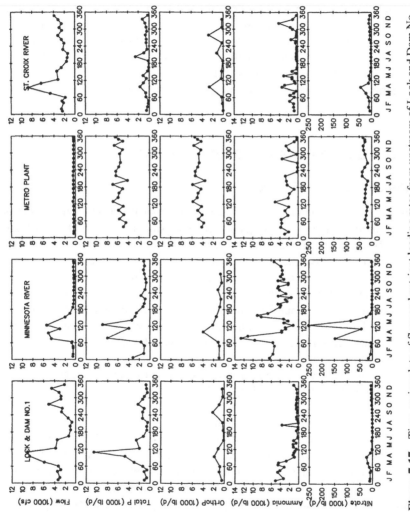

Figure 7-17 Time-series plots of flow and nutrient loading rates from upstream of Lock and Dam No. 1, the Minnesota River, Metro Plant, and the St. Croix River in 1988 as input to the water quality model of the Upper Mississippi River and Lake Pepin using the WASP/EUTRO model.

7.4.2 Ultimate Carbonaceous Biochemical Oxygen Demand (CBOD$_u$) Concentrations

The EUTRO model simulates ultimate, not 5-day, CBOD, so the modeler must input either ultimate values or 5-day values with a conversion factor. In the Upper Mississippi River and Lake Pepin modeling study, only a limited number of ultimate measures are available for the study years, while unfiltered 5-day CBOD (CBOD$_5$) is routinely monitored in the receiving water and most point sources (Larson, 1993). Therefore, unfiltered CBOD$_5$ is used as the base for developing model inputs to this system. The following methodology is used to develop the filtered CBOD$_u$ concentrations.

Because phytoplankton are modeled separately from CBOD in EUTRO, the algal component of CBOD must be subtracted from unfiltered CBOD$_5$ samples according to the following formulas:

$$L_u = \text{Chl } a \ (0.033)(2.667)$$

$$y = L_u\left[1 - e^{-0.15(5.0)}\right]$$

$$\text{filtered CBOD}_5 = \text{unfiltered CBOD}_5 - y$$

where

$$
\begin{aligned}
L_u &= \text{oxygen (mg/L) required to oxidize the algal biomass} \\
\text{Chl } a &= \text{viable chlorophyll } a \ (\mu\text{g/L}) \\
0.033 &= \text{carbon to chlorophyll } a \text{ ratio in algal biomass (mg C/}\mu\text{g Chl } a) \\
2.667 &= \text{oxygen to carbon ratio (mg O}_2\text{/mg C)} \\
y &= \text{5-day oxygen equivalent due to algal respiration} \\
0.15 &= \text{algal endogenous respiration rate (day}^{-1})
\end{aligned}
$$

The formulas are not applied where algae are negligible or absent. Note that the carbon to chlorophyll a ratio, oxygen to carbon ratio, and algal endogenous respiration rate are site specific for the Upper Mississippi River and Lake Pepin. Values for other study areas may be different. Ultimate to 5-day ratios are then determined for each of the model inputs based on the best available data. The filtered CBOD$_5$, as measured or calculated, is multiplied by the ultimate to 5-day ratio to derive model inputs for the CBOD$_u$ system. The ratios vary from 1.8 to 3.0 for the study (Larson, 1993).

7.4.3 Nonliving Organic Nitrogen and Phosphorus

As with CBOD, the algal components of organic nitrogen and organic phosphorus must be subtracted from the inputs to these two systems, because these components are modeled separately in the phytoplankton system. Note that only the viable algal components are subtracted, not the detritus ("nonliving"). The formulas used to derive the model inputs are described below:

$$\text{algal N} = \text{Chl } a \ (0.033)(0.21)$$

$$\text{algal P} = \text{Chl } a \ (0.033)(0.03)$$

$$\text{NLON} = \text{TKN} - \text{NH}_4 - \text{algal N}$$

$$\text{NLOP} = \text{TP} - \text{PO}_4 - \text{algal P}$$

where

Chl a	=	viable chlorophyll a (μg/L)
0.033	=	carbon to chlorophyll a ratio in algal biomass (mg C/μg Chl a)
0.21	=	nitrogen to carbon ratio in algal biomass (mg N/mg C)
0.03	=	phosphorus to carbon ratio in algal biomass (mg P/mg C)
TKN	=	total Kjeldahl nitrogen
NH$_4$	=	ammonia nitrogen
TP	=	total phosphorus
PO$_4$	=	orthophosphate

Algal stoichiometric ratios are taken from Hydroscience (1979) and have been substantiated by the model calibration analysis.

7.4.4 Boundary Conditions

As stated earlier, boundary conditions in the EUTRO model are determined by how flows are structured in the transport system (data group D). In a river system like the Upper Mississippi, one may choose to describe a single flow from the headwaters to the end of the model, or one can add branches for major tributaries and point sources and describe and route their flows separately. In other words, tributary and point source loads can be defined in two ways in the EUTRO model:

1. Flows in data group D coupled with water quality concentrations in data group E (boundary conditions)
2. Loadings in data group E (forcing functions)

The end result is the same with either the modeler or the model calculating the loads. The advantage lies in the first option, where one can adjust the flows and water quality concentrations independently.

For each flow defined in data group D, concentrations for the eight water quality systems must be described in data group E. In the Upper Mississippi River model, only the headwaters (Lock & Dam No. 1), Minnesota River, and St. Croix River are ascribed actual values in data group E. Concentrations for the other four flows, including the Metro Plant, are set to zero, and the actual loads are given in data group E.

In addition to the seven boundary conditions mentioned above, the model has boundaries after the two final segments in Lake Pepin: No. 134 (in the top layer) and

No. 174 (in the bottom layer). Because the model calculations use the backward difference in spatial differentiation to quantify advective mass transport and since there is no longitudinal dispersion, water quality conditions at the downstream end of the model do not affect the computation. Therefore, the concentrations for these two boundaries are all set to zero (they could even be set to negative values). A total of nine boundary conditions exist in the phosphorus study model: six of them act simply as placeholders, while the boundaries representing the three major rivers are fully defined.

7.4.5 Forcing Functions

A total of 15 tributaries and point sources are included in the model of the Upper Mississippi River and Lake Pepin. The six tributaries are: Minnesota River, St. Croix River, Vermillion River, Cannon River, Isabelle Creek, and Rush River. The Minnesota and St. Croix Rivers are treated as boundary conditions in the model, while the other four tributaries are handled as forcing functions. Eight point sources with process or wastewater discharges averaging over 1 mgd output directly to the Mississippi River study area: Metro Wastewater Treatment Plant (WWTP), Ashland Oil, Koch Refinery, Rosemont WWTP, Cottage Grove WWTP, 3M Chemolite, Hastings WWTP, and Red Wing WWTP. Loadings from the eight point sources are also treated as forcing functions in the model. A comprehensive derivation of these loads can be found in Larson (1993).

7.4.6 Environmental Parameters

Data group G in the WASP model describes environmental conditions that vary spatially along segments in the model and includes factors, such as light and temperature, that influence the kinetics among the eight water quality systems. In the time-variable model calculations, the modeler can declare up to four time functions for water temperature and up to five time functions for the light extinction coefficient (K_e). While the values for temperature and K_e are given in data group I (see Section 7.4.8), each segment in data group G must point to a specific set of temperatures and K_e, the ones most appropriate for the segment. A close examination of spatial and temporal differences in the field data is required to set up the time-variable parameters for the model input. Figure 7-18 shows the time-variable functions of water column temperature, light extinction coefficient, and solar radiation measured in 1988 in the study area. They are incorporated into the model input groups G and I.

Sediment nutrient fluxes and oxygen demand rates are also incorporated as a function of water temperature. Finally, two other environmental parameters, canopy shading and zooplankton grazing, are used strictly to simulate the suppression of algae from the confluence of the Minnesota River to river mile 831, a phenomenon that has been noted in previous modeling studies. Canopy shading (ITOTLIM) is selected because it functions as a multiplication factor to light and, as a consequence, to the algal growth rate. Zooplankton grazing (ZOOP) is selected because its value is added to the algal death rate.

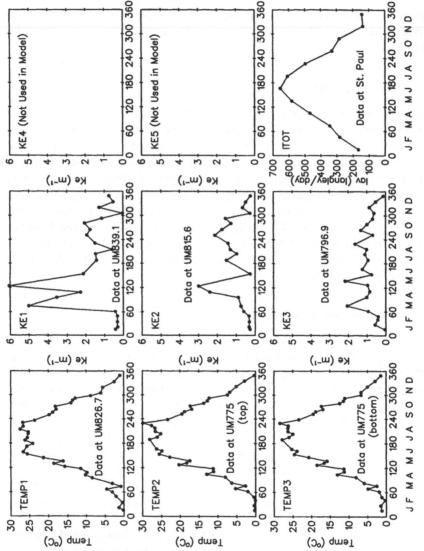

Figure 7-18 Time-series plots of water temperature and light extinction coefficients at various locations along the Upper Mississippi River and Lake Pepin in 1988.

7.4.7 Constants

Data group H describes environmental conditions that are held steady, or constant, during the entire model run, remaining the same for every segment and time step. In EUTRO the constants represent kinetic coefficients for reactions among the eight water quality systems. Rather than use values from the literature as the starting points in model calibration, the best choice would be to select constants that have proven successful in recent modeling of BOD/DO, nutrients, and phytoplankton in the Upper Mississippi River (MPCA, 1981; Lung, 1996a).

Only two constants are changed between the wasteload allocation study (MPCA, 1981) and post audit (Lung, 1996a). The CBOD deoxygenation rate is decreased from 0.25 day^{-1} to 0.073 day^{-1} at 20°C, based on the field and lab data (see Chapter 4). This lower value has been confirmed in a model post audit by Lung (1996a), and is consistent with literature values reported for advanced secondary treatment (U.S. EPA, 1995). A range of algal growth rates (1.3 to 2.5 day^{-1}) have been used in previous modeling analyses (Hydroscience, 1979; MPCA, 1981) with the speculation that the algal group dominating the river shifts seasonally and with changing flows. The post audit study (Lung, 1996a) found that a value of 1.9 day^{-1} provided the best calibration of DO and chlorophyll a concentrations. The final values for the constants following comprehensive model calibration analyses are listed in Table 7-6. Group H in the input data file is presented in Table 7-7. Note the efficient way of configuring this input block, that is, grouping all kinetic coefficients as global constants and setting all system-specific constants to zero entries.

TABLE 7-6   Upper Mississippi River and Lake Pepin Kinetic Coefficients

Constant	Code	Value
Nitrification rate	K12C	0.07 day^{-1} at 20°C
Temperature coefficient	K12T	1.04
Half-saturation constant	KN1T	0.00 mg O$_2$/L
Denitrification rate	K20C	0.00 day^{-1} at 20°C
Temperature coefficient	K20T	0.00
Half-saturation constant	KNO3	0.00 mg O$_2$/L
Saturated growth rate of phytoplankton	K1C	1.90 day^{-1} at 20°C
Temperature coefficient	K1T	1.066
Endogenous respiration rate of phytoplankton	K1RC	0.075 day^{-1} at 20°C
Temperature coefficient	K1RT	1.08
Nonpredatory phytoplankton death rate	K1D	0.11 day^{-1} at 20°C
Zooplankton grazing rate	K1G	1.0 L/cell-day

(continues)

TABLE 7-6 (*Continued*)

Phytoplankton stoichiometry		
Oxygen to carbon ratio	OCRB	2.67 mg O_2/mg C
Carbon to chlorophyll *a* ratio	CCHL	33 mg C/mg Chl *a*
Phosphorus to carbon ratio	PCRB	0.03 mg P/mg C
Nitrogen to carbon ratio	NCRB	0.21 mg N/mg C
Half-saturation constants for phytoplankton growth		
Carbon	KMPHY	0.0 mg C/L
Nitrogen	KMNG1	0.005 mg N/L
Phosphorus	KMPG1	0.001 mg P/L
Nutrient limitation option	NUTLM	1 = multiplicative
Light formulation switch	LGHTS	1 = Di Toro
Saturation light intensity for phytoplankton	IS1	350 langley/day
BOD deoxygentation rate	KDC	0.073 day^{-1} at 20°C
Temperature coefficient	KDT	1.04
Half-saturation constant	KBOD	0.0 day^{-1}
Mineralization rate of dissolved organic N	K71C	0.03 day^{-1} at 20°C
Temperature coefficient	K71T	1.045
Mineralization rate of dissolved organic P	K83C	0.03 day^{-1} at 20°C
Temperature coefficient	K83T	1.045
Fraction of dead and respired phytoplankton		
Recycled to organic nitrogen	FON	0.30
Recycled to organic phosphorus	FOP	0.30

**TABLE 7-7 Group H Configuration of Upper Mississippi River and Lake Pepin
EUTRO Model**

```
   *    +    *    +    *    +    *    +    *    +    *    +    H: CONSTANTS
GLOBALS        1
               31
K12C          11    0.07 K12T          12    1.04
KNIT          13    0.00 K20C          21    0.00
K20T          22    0.00 KNO3          23    0.00
K1C           41    1.90 K1T           42    1.066
LGHTS         43    1    CCHL          46    33.
IS1           47    350. KMNG1         48    0.005
KMPG1         49    0.001K1RC          50    0.075
K1RT          51    1.08 K1D           52    0.11
K1G           53    1.0  NUTLM         54    1
PCRB          57    0.03 NCRB          58    0.21
KMPHY         59    0    KDC           71    0.073
```

TABLE 7-7 (*Continued*)

KBOD	75	0.0	KDT	72	1.04
OCRB	81	2.67	K71C	91	0.03
K71T	92	1.045	FON	95	0.30
K83C	100	0.03	K83T	101	1.045
FOP	104	0.30			
NH3	0				
NO3	0				
PO4	0				
PHYT	0				
CBOD	0				
DO	0				
ON	0				
OP	0				

7.4.8 Time Functions

Data group I describes environmental conditions that vary over time. Inputs to this data group are paired with Julian dates, which sequentially number the days in the year. In time-variable calculations, the values change over the annual cycle according to the available data and desired resolution. The following time functions are used: total daily solar radiation (ITOT); fraction of daylight, or photoperiod (F); wind speed (WIND); air temperature (AIRTMP); three functions for water temperature (TEMP1 to TEMP3); five functions for the light extinction coefficients (KE1 to KE5); normalized sediment ammonium flux (TFNH4); and normalized sediment phosphate flux (TFPO4). They are included in Table 7-8.

TABLE 7-8   Group I Configuration of Upper Mississippi River and Lake Pepin EUTRO Model

	14	* + * + * + * + * + * + * + * + * + * + * + * + * +							I: TIME FUNCTION
TEMP1	39	1	Riverine = UM 826.7							
	0.7	15.	0.6	23.	1.7	36.	1.5	51.	
	1.2	63.	2.2	73.	4.8	78.	5.7	85.	
	4.8	92.	11.8	99.	6.8	106.	10.0	113.	
	13.4	120.	9.0	127.	18.9	134.	17.7	141.	
	22.2	155.	23.0	162.	24.3	169.	22.5	176.	
	25.3	183.	23.7	191.	24.7	197.	25.9	204.	
	21.4	219.	22.7	225.	21.3	232.	25.2	239.	
	23.0	247.	20.6	255.	18.9	260.	13.3	268.	
	13.9	275.	11.3	281.	10.0	288.	10.3	295.	
	3.7	323.	0.3	337.	1.0	349.			
TEMP2	27	2	Lake Pepin, top layer = UM 775 top							
	0.7	15.	0.7	51.	2.8	80.	9.6	113.	

(*continues*)

TABLE 7-8 (*Continued*)

9.8	128.	18.7	134.	18.1	141.	22.9	150.
25.6	163.	25.5	171.	23.4	176.	25.8	184.
25.1	190.	26.3	199.	27.1	204.	24.1	211.
22.8	218.	25.3	225.	23.9	234.	26.4	241.
23.2	248.	22.0	255.	17.3	262.	14.5	267.
10.9	296.	3.0	325.	0.0	349.		
TEMP3 27	3	Lake Pepin, bottom layer = UM 775 bottom					
1.5	15.	2.3	51.	3.8	80.	9.8	113.
9.9	128.	15.3	134.	16.9	141.	22.3	150.
23.6	163.	24.8	171.	22.9	176.	25.9	184.
25.0	190.	25.9	199.	27.0	204.	22.4	211.
22.7	218.	22.7	225.	22.9	234.	25.4	241.
23.2	248.	21.9	255.	17.2	262.	14.5	267.
10.0	296.	3.0	325.	0.0	349.		
ITOT 12	5	Solar Radiation (ly/day)					
152.	15.	269.	46.	298.	74.	395.	105.
372.	135.	551.	166.	535.	196.	432.	227.
305.	258.	216.	288.	146.	319.	124.	349.
F 12	6	Photoperiod (fraction of day)					
.383	15.	.435	46.	.495	74.	.562	105.
.618	135.	.649	166.	.636	196.	.588	227.
.523	258.	.459	288.	.399	319.	.367	349.
WIND 12	7	Wind Speed (m/sec)					
5.0	15.	4.2	46.	5.3	74.	5.1	105.
4.6	135.	4.3	166.	3.8	196.	4.6	227.
5.2	258.	5.6	288.	5.1	319.	3.9	349.
KE1 23	8	Light Extinction (/m)		UM 839.1			
0.85	15.	0.28	23.	0.45	36.	0.26	51.
0.40	63.	0.50	78.	7.67	92.	12.26	106.
3.37	120.	3.07	134.	7.32	155.	6.18	169.
4.55	183.	10.71	197.	5.07	219.	0.97	247.
19.18	260.	1.84	275.	1.04	288.	1.01	310.
1.06	323.	0.56	337.	0.45	349.		
KE2 24	9	Light Extinction (/m)		UM 815.6			
0.10	15.	0.10	23.	0.23	36.	0.12	51.
0.54	63.	0.78	78.	2.62	92.	2.32	106.
1.53	120.	2.95	134.	4.88	155.	3.54	169.
4.65	183.	3.61	197.	2.92	219.	6.42	232.
3.40	247.	9.27	260.	1.82	275.	2.24	288.
0.99	310.	0.71	323.	0.43	337.	0.42	349.
KE3 25	10	Light Extinction (/m)		UM 781			
0.09	15.	0.05	51.	0.29	80.	0.94	113.
0.42	128.	0.44	134.	0.26	141.	0.70	150.
1.22	163.	1.26	171.	2.10	184.	1.48	190.
1.42	199.	1.32	211.	1.71	218.	0.65	225.

TABLE 7-8 (*Continued*)

	0.72	234.	0.90	241.	1.91	248.	1.03	255.	
	2.58	262.	1.48	267.	1.31	296.	0.50	325.	
	0.28	349.							
KE4	25	11		Light Extinction (/m)		UM 775			
	0.09	15.	0.07	51.	0.10	80.	0.60	113.	
	0.58	128.	0.36	134.	0.31	141.	0.21	150.	
	1.22	163.	1.10	171.	1.43	184.	1.72	190.	
	1.80	199.	1.96	204.	1.71	211.	2.27	218.	
	2.54	225.	1.90	234.	1.27	241.	0.96	248.	
	1.01	255.	1.71	262.	1.19	267.	0.62	296.	
	0.32	349.							
KE5	25	12		Light Extinction (/m)		UM 766			
	0.07	15.	0.07	51.	0.17	80.	0.35	113.	
	0.98	128.	1.27	134.	1.18	141.	1.11	150.	
	1.33	163.	1.27	171.	1.99	184.	1.33	190.	
	1.21	199.	1.28	204.	1.41	211.	1.28	218.	
	1.26	225.	1.15	234.	1.10	241.	1.44	248.	
	0.93	255.	0.87	262.	0.92	267.	0.20	325.	
	0.16	349.							
TFNH4	12	13		Ammonia Flux (temp. adj.)					
	1.0	15.	1.0	46.	1.0	74.	1.0	105.	
	1.0	135.	1.0	166.	1.0	196.	1.0	227.	
	1.0	258.	1.0	288.	1.0	319.	1.0	349.	
TFPO4	12	14		Phosphate Flux (temp. & DO adj.)					
	1.0	15.	1.0	46.	1.0	74.	1.0	105.	
	1.0	135.	1.0	166.	1.0	196.	1.0	227.	
	1.0	258.	1.0	288.	1.0	319.	1.0	349.	
AIRTM	12	21		Air Temperature (degrees C)					
	-10.8	15.	-4.2	46.	1.3	74.	9.5	105.	
	16.6	135.	22.7	166.	22.4	196.	21.7	227.	
	15.0	258.	8.4	288.	-4.2	319.	-6.0	349.	

7.4.9 Model Results

Results from the 1988 time-variable model are summarized in Figures 7-19, 7-20, and 7-21 for five locations in the Upper Mississippi River (UM839.1, UM831.0, UM826.7, UM815.6, and UM796.9) and the outlet of Lake Pepin (UM764.3). Temporal plots of water quality constituents show the comparison between the model results and field data for CBOD, TKN, ammonium, nitrite/nitrate nitrogen, total phosphorus, orthophosphate, viable chlorophyll *a*, and dissolved oxygen. The model results follow the temporal trends of the water quality constituents closely, mimicking seasonal variations of the water quality at each location. Comparing the plots between UM839.1 and UM831.0, Figure 7-19 illustrates the impact of the Metro Plant

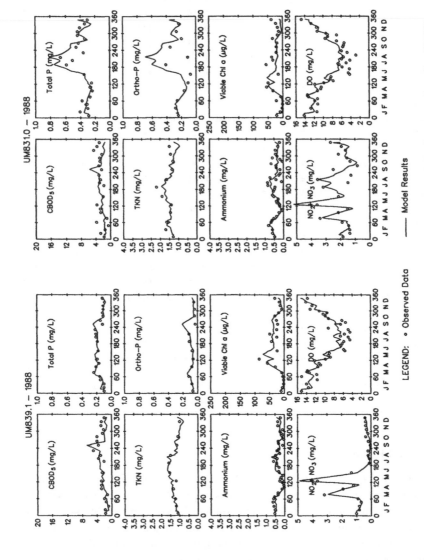

Figure 7-19 Time-variable water quality model results from WASP/EUTRO versus data for CBOD$_5$, TKN, ammonium, nitrite/nitrate, total phosphorus, orthophosphate, viable chlorophyll a, and DO at UM839.1 and UM831.0 in the Upper Mississippi River and Lake Pepin, 1988.

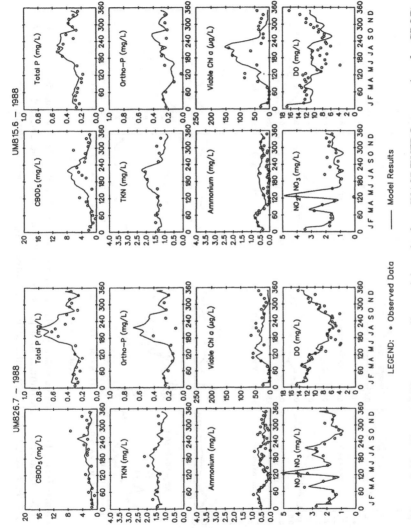

Figure 7-20 Time-variable water quality model results from WASP/EUTRO versus data for CBOD₅, TKN, ammonium, nitrite/nitrate, total phosphorus, orthophosphate, viable chlorophyll *a*, and DO at UM826.7 and UM815.6 in the Upper Mississippi River and Lake Pepin, 1988.

233

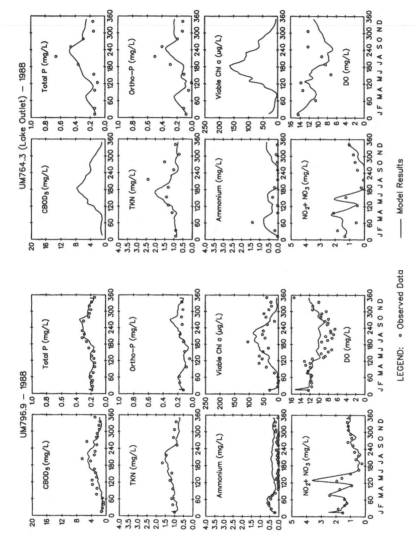

Figure 7-21 Time-variable water quality model results from WASP/EUTRO versus data for CBOD₅, TKN, ammonium, nitrite/nitrate, total phosphorus, orthophosphate, viable chlorophyll *a*, and DO at UM796.9 and UM764.3 (lake outlet) in the Upper Mississippi River and Lake Pepin, 1988.

on the nutrient concentrations in the river; that is, total phosphorus and orthophosphate concentrations are increased following the inflow from the Metro Plant. The increases are particularly pronounced during the summer months in 1988, as also indicated in the steady-state modeling analysis. At Lock & Dam No. 2 (UM815.6), the chlorophyll *a* levels reach 150 µg/L in August (Figure 7-20). Subsequent inflow from the St. Croix River reduces the nutrient concentrations somewhat, resulting in lower chlorophyll *a* concentrations at UM796.9 (Figure 7-21).

Figure 7-21 also shows the same temporal plots for the outlet of Lake Pepin in 1988. Again, the model results match the data well at the lake outlet for 1988, but due to data limitation, no comparison between the model results and field data can made for temporal trends in Lake Pepin. Note the depression of ammonium and nitrate concentrations during the summer months, suggesting potential nitrogen limitation for algal growth in Lake Pepin if all other conditions are optimal. While the inorganic nitrogen concentrations are lowered significantly during the summer months of 1988, the phosphorus concentrations in the water column are strongly supported by the sediment fluxes, resulting in sufficient phosphorus in Lake Pepin.

Additional results from model calibration, verification, and projection can be found in Larson (1993). The EUTRO model is again configured for a steady-state modeling analysis of the summer conditions in 1988, 1990, and 1991 (Lung and Larson, 1995). Since the mass transport, boundary conditions, loads, and environmental parameters (input data groups B through I) do not vary with time, model input files for steady-state analyses are greatly simplified and reduced. Note that initial conditions are not critical in the EUTRO model, as it runs to steady state. They only affect the time for the model calculations to reach a steady state. In general, a few trial runs are made to determine the time to reach steady state. In the Upper Mississippi River and Lake Pepin case, the model calculations reach a steady state in about 120 days for the summer 1988 condition. Therefore, constant values for groups B through I are used for time at 0.0 and time 360.0, a period sufficiently long enough to ensure constant input of all these values. In addition, a single integration time step is used for the simulations. While a low flow, such as the one in the summer of 1988, would allow a greater integration step than a high flow period (e.g., summer 1990 or summer 1991), the time to reach a steady-state condition would also increase. For example, the summer 1990 and 1991 model runs take only 30 days to reach steady state. Figure 7-22 shows the steady-state model results versus data in summer (June–August) 1988.

7.4.10 Statistical Analysis of Model Results

For the upper Mississippi River and Lake Pepin water quality model, a regression analysis is conducted to quantify the goodness of fit for the model results. A perspective on the adequacy of a model can be obtained by regressing the calculated values with the observed values (Thomann, 1980). Let the testing equation be:

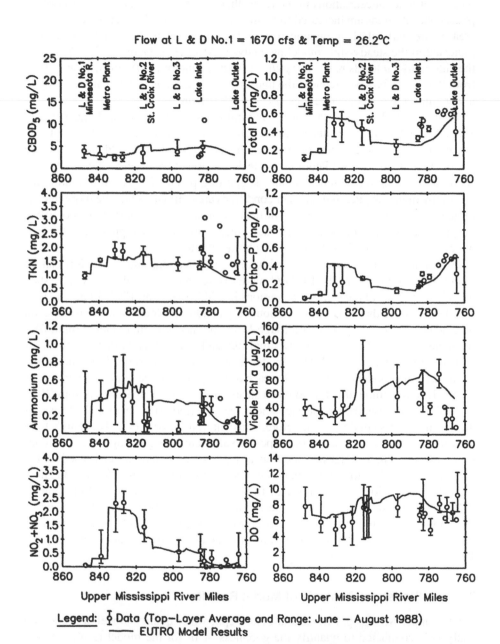

Figure 7-22 Steady-state model results versus data for CBOD$_5$, TKN, ammonium, nitrite/nitrate, total phosphorus, orthophosphate, viable chlorophyll a, and DO under summer 1988 flow condition in the Upper Mississippi River and Lake Pepin.

$$C_{obs} = \alpha + \beta C_{cal} + \varepsilon \tag{7-4}$$

where α and β are the true intercept and slope, respectively, between the calculated concentrations, C_{cal}, and the observed values, C_{obs}, and ε is the error of C_{cal}. With Eq. 7-4, standard linear regression statistics can be computed, including:

- The square of the correlation coefficient, r^2, (the percentage of variance accounted for), between calculated and observed,
- Standard error of estimate, representing the residual error between model and data,
- Slope estimate, b, of β and intercept estimate, a, of α, and
- Test of significance on the slope and intercept.

The null hypothesis on the slope and intercept is given by $\beta = 1.0$ and $\alpha = 0.0$. Therefore, the test statistics $(b - 1)/s_b$ and α/s_a are distributed as Student's t and $n - 2$ degrees of freedom. The variance of the slope and intercept, s^2_b and s^2_a, are computed according to standard formulas. A two-tailed "t" test is conducted on b and a, separately, with a 5% probability in each tail, that is, a critical value of t of about 2 provides the rejection limit of the null hypothesis (Thomann, 1980).

Figure 7-23 shows the regression model results for total phosphorus, orthophosphate, chlorophyll a, and DO in 1988. The solid lines represent the actual regression between the calculated and observed concentrations, while the dashed lines are the theoretical slope of 1:1. Also shown in Figure 7-23 are the calculated slope, intercept, and correlation coefficients for the regression. Note that all data points and associated model results for the study system throughout 1988 are presented.

The results of the two-tailed t test for the period from 1985 to 1996 are summarized in Table 7-9. For 1988, α/s_a and $(b - 1)/s_b$ values for DO are greater than 2.0, indicating that the intercept is not significantly different from 0 and the slope is not significantly different from 1. Similar results are obtained for chlorophyll a in 1988 [i.e., with both α/s_a and $(b - 1)/s_b$ less than 2.0]. Results for orthophosphate in 1988 show that the intercept and slope are significantly different from 0 and 1, respectively. For total phosphorus in 1988, the slope of the regression is significantly different from 1.

Over this 12-year period, DO results match the data quite well, with exceptions only in 1986 and 1990. This outcome is not surprising as DO levels in the water column are closely regulated by water temperature throughout the year. Also, model calibration results for orthophosphate improve in the last five years of model simulations, showing both α/s_a and $(b - 1)/s_b$ values less than 2.

1988 REGRESSIONS

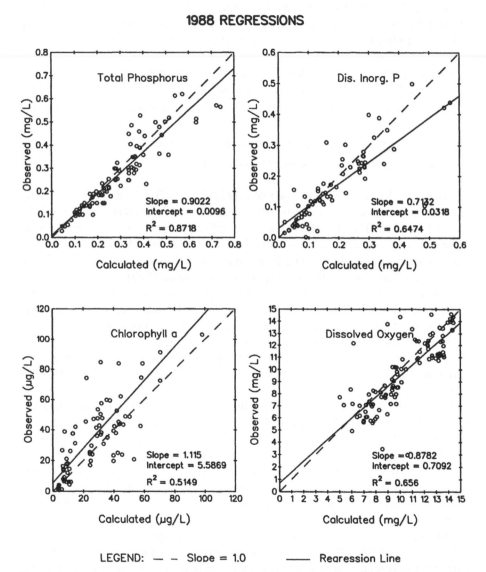

LEGEND: − − Slope = 1.0 ——— Rearession Line

Figure 7-23 Quantitative analysis of the goodness of fit between calculated concentrations and data for total phosphorus, orthophosphate, chlorophyll *a*, and DO in 1988 for the Upper Mississippi River and Lake Pepin.

TABLE 7-9   Results of a Two-Tail t Test for Intercept and Slope

Year	Total Phosphorus		Ortho-P		Chlorophyll a		Dissolved Oxygen	
	α/s_a	$(b-1)/s_b$	α/s_a	$(b-1)/s_b$	α/s_a	$(b-1)/s_b$	α/s_a	$(b-1)/s_b$
1985	1.627	**2.229**[a]	**2.724**	**3.095**	1.416	0.244	1.681	1.969
1986	**2.623**	**2.335**	0.363	0.465	1.062	0.001	1.635	**2.785**
1987	1.107	**2.385**	1.334	**2.309**	**2.858**	1.861	1.471	1.223
1988	0.781	**2.589**	**2.247**	**4.624**	1.331	0.926	1.062	1.945
1989	**2.752**	**6.298**	1.926	**4.886**	**2.431**	**2.584**	1.571	0.794
1990	**3.614**	**5.762**	0.366	**2.643**	**5.010**	**2.261**	**2.996**	**3.286**
1991	**3.736**	**5.647**	**2.895**	**6.659**	**4.548**	1.229	0.668	1.004
1992	**2.130**	1.937	1.872	1.893	**4.133**	**3.361**	0.027	0.186
1993	1.780	1.246	0.968	1.034	**2.908**	1.714	1.304	1.230
1994	1.661	**2.928**	1.363	0.934	**3.582**	0.997	0.711	0.542
1995	**4.622**	**6.444**	1.526	1.063	**6.113**	**3.416**	0.398	0.122
1996	**3.370**	**3.689**	**2.093**	1.673	**3.835**	**2.965**	0.421	0.610

[a] Bold numbers indicate values greater than 2.0.

7.4.11 Data to Support Eutrophication Modeling

A significant amount of data is used for modeling the Upper Mississippi River and Lake Pepin. Table 7-10 lists the sources and extent of the data used for this study. Data sources for mass transport (groups B, C, and D) are the same as those used for the mass transport model as presented in Table 7-5. MPCA data are used to describe initial conditions in the Lake Pepin segments. MPCA data are also used to calculate loads from two smaller tributaries (Vermillion and Cannon Rivers) and four point sources that discharge outside the MWCC collection system. WDNR data describe the loads from Isabelle Creek and Rush River, which are small but direct tributaries to Lake Pepin. And the MPCA and WDNR data become important in the field data files against which the models are calibrated. MPCA data for Lake Pepin and WDNR data for the lake's outlet describe the water quality of the focal point area in the modeling study.

In most modeling studies, existing data collected from routine sampling programs are the most commonly used. It is not usual to develop a sampling program to collect specific data exclusively to support the modeling work. Yet in using existing data, there is often a mismatch between what is available and what is required by the WASP/EUTRO model. Table 7-11 summarizes such a situation facing the total maximum daily load (TMDL) modeling team at Maryland Department of the Environment (MDE). As shown in Table 7-11, some compromises and adjustments have been made to accommodate the model input requirements in the modeling effort.

TABLE 7-10   Data Used in Eutrophication Modeling of the Upper Mississippi River

EUTRO Data Group	Sources of Data and Information

B. Exchange flows	Reports from previous studies:
	    Hydroscience (1979)
	    Stefan and Anderson (1977)
	    Stefan and Wood (1976)
	    Falch et al. (1979)

C. Segment volumes
    Hydraulic geometry
    Depth vs. flow
    Velocity vs. flow
	Army Corps of Engineers Reservoir Regulation Manuals
	Hydrographic information from Federal Water
	    Pollution Control Administration

D. Advective flows
    River and tributaries
    Treatment plants
	
	USGS surface water records
	Point source discharge records

E. Boundary conditions	Receiving water quality data collected by:
	    MWCC
	    MPCA [with the Minnesota Department of Natural
	        Resources (MDNR)]
	    WDNR

F. Loads	Wastewater treatment plant records
G. Environmental parameters[a]	Data from MWCC, MPCA, MDNR, and WNDR
H. Constants	From literature values followed by model calibration

I. Time functions
    Total daily solar radiation
    Fraction of daylight
    Windspeed and air
        temperature
	
	University of Minnesota Soil Science Department
	Minnesota State Climatological Office
	National Oceanic and Atmospheric Administration

J. Initial conditions	MWCC data[b]

[a]Sediment nutrient fluxes were measured by James et al. (1992) and James and Barko (1992).
[b]Also used to assemble the receiving water data for comparison with model results.

TABLE 7-11   Matching WASP/EUTRO Data Requirements with Commonly Available Water Quality Data

System	WASP/EUTRO Required Input	Laboratory Analysis	Estimated As	Model Output
1.NH_3	Total NH_3	Dis. NH_3	Dis. only	Dis. NH_3
2.NO_{23}	Total NO_{23}	Dis. NO_{23}	Dis. only	Dis. NO_{23}

TABLE 7-11   (*Continued*)

System	WASP/EUTRO Required Input	Laboratory Analysis	Estimated As	Model Output
3.PO$_4$	Total PO$_4$	Dis. PO$_4$		
4.Phytoplankton carbon	Viable Chl *a*	Viable Chl *a*		Viable Chl *a*
5.BOD	CBOD$_u$	BOD$_5$	CBOD$_u$	BOD$_5$
6.DO	As measured	DO		Average DO
7.Organic nitrogen	NLON[a]	TPN[b], TDN[c], NH$_3$ NO$_{23}$, Viable Chl *a*	NLON[d]	TON[e]
8.Organic phosphorus	NLOP[f]	TPP[g], TDP[h], PO$_4$ Viable Chl *a*	NLOP[i]	TOP[j]

[a]NLON (nonliving organic nitrogen) = dis. + part. – algal N.
[b]Total particulate nitrogen.
[c]Total dissolved nitrogen.
[d]NLON = TDN + TPN – NH$_3$ – NO$_{23}$ – algal N.
[e]Total organic nitrogen.
[f]NLOP (nonliving organic phosphorus) = dis. + part. – algal P.
[g]Total particulate phosphorus.
[h]Total dissolved phosphorus.
[i]NLOP = TDP + TPP – PO$_4$ – algal P.
[j]Total organic phosphorus.

7.5   NUMERICAL TAGGING

While the model results show that phosphorus load reductions at the Metro Plant would have a minimal effect on reducing the algal biomass in Lake Pepin (Lung, 1995), an issue left unresolved by the modeling analysis is the fate of phosphorus from the Metro Plant. That is, to what extent is phosphorus from the Metro Plant transported to Lake Pepin under both existing and potential reduced loading conditions? For example, some phosphorus from the Metro Plant could be incorporated into the phytoplankton biomass in Lake Pepin, some could be deposited into the river sediments behind the dams, and some could be accumulated in the sediment of Lake Pepin. More specifically, we would like to know the percentage of phosphorus from the Metro Plant in the algal biomass in Lake Pepin.

To address this question, one could think of the component analysis routinely performed in BOD/DO modeling to quantify the contribution of individual BOD sources to DO deficits. The procedure to perform such an analysis is presented in Chapter 6. Results of a component analysis for the Upper Mississippi River from Lock and Dam No. 1 and Lock and Dam No. 2 are presented in Figure 7-24. The sum of the DO deficits due to individual oxygen consumption components is equal to the total DO

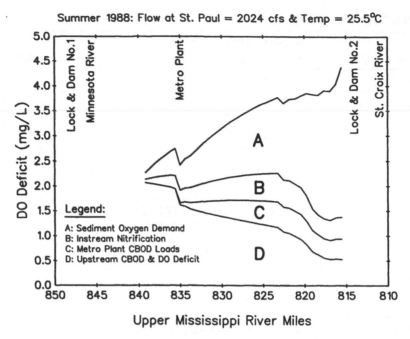

Figure 7-24   A component analysis of the BOD/DO model of the Upper Mississippi River from Lock and Dam No. 1 to Lock and Dam No. 2.

deficit calculated from a model run with all four components being turned off at the same time. Superimposing one deficit over the other is a key feature of many BOD/DO models, which formulate the BOD and DO deficit in a linear relationship. Unfortunately, algal growth and nutrient dynamics in eutrophication models such as WASP/EUTRO are formulated in a nonlinear fashion. When the algal biomass calculated from individual model runs associated with a given phosphorus source are added up, the sum is much different than the algal biomass calculated from a model run associated with all phosphorus sources. Therefore, the simple component analysis for BOD/DO modeling is not appropriate to address this question in eutrophication modeling. A different strategy is needed.

The numerical tagging technique is quite similar to the $^{32}PO_4$ technique that limnologists use in tracking phosphorus in natural water systems by measuring the amount of $^{32}PO_4$ in various phosphorus compartments. Instead of a radioactive tracer, a numerical tracer is injected into one of the nutrient sources in the eutrophication model. The numerical tagging analysis requires that the EUTRO model be modified to enable phosphorus loads from individual sources such as the Metro Plant to be numerically labeled and tracked separately from other phosphorus sources (Lung, 1996b). To accomplish this, additional system variables have been incorporated to represent the labeled phosphorus compartments: labeled orthophosphate (variable 9), labeled nonliving organic phosphorus (variable 10), and labeled phytoplankton (variable 11). The revised model kinetics diagram is shown in Figure 7-25.

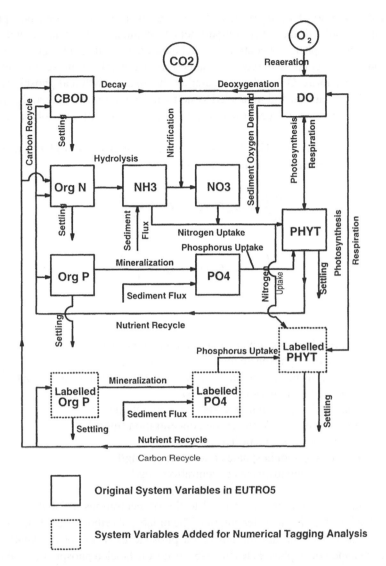

Figure 7-25 Modified EUTRO kinetics relationship for the numerical tagging model by incorporating labeled organic phosphorus, labeled orthophosphate, and labeled phytoplankton.

Again, mass balance is the key to this numerical tagging effort. Special care is required to preserve the nonlinear relationship between algal growth rate and phosphorus concentrations. While the kinetic interrelationships among these labeled compartments are separate (Figure 7-25) but the same as those for the unlabeled compartments, algal growth rates must be calculated based on the total concentration of labeled and unlabeled orthophosphate simply because the phytoplankton should not discriminate between these two phosphorus sources. When either labeled or unlabeled orthophosphate is exhausted, algal growth and associated phosphorus uptake should shift to the other compartment to maintain mass balance and avoid negative orthophosphate concentrations. Labeled phosphorus cycles among the labeled compartments but undergoes the same rates (mineralization, settling, etc.) as unlabeled phosphorus. Only when calculating the algal growth and phosphorus uptake rates do the labeled and unlabeled systems need to communicate and correctly proportion the increase in algal biomass and the decrease in orthophosphate between the two systems.

The modification in the EUTRO model is the mass uptake rates of labeled and unlabeled orthophosphate in the water column:

$$\frac{dM_1}{dt} = a_{pc}G_p(C_1 + C_2)V\frac{P_1}{P_1 + P_2} \qquad (7\text{-}5)$$

$$\frac{dM_2}{dt} = a_{pc}G_p(C_1 + C_2)V\frac{P_2}{P_1 + P_2} \qquad (7\text{-}6)$$

where

M_1 = unlabeled orthophosphate mass uptake rate (mg/day)
M_2 = labeled orthophosphate mass uptake rate (mg/day)
a_{pc} = phosphorus to carbon ratio in phytoplankton biomass (mg P/mg C)
G_p = phytoplankton growth rate (day^{-1})
C_1 = unlabeled phytoplankton carbon concentration (mg/L)
C_2 = labeled phytoplankton carbon concentration (mg/L)
V = volume of the segment (L)
P_1 = unlabeled orthophosphate concentration (mg/L)
P_2 = labeled orthophosphate concentration (mg/L)

Note that the uptake rate is proportional to the concentrations of labeled and unlabeled orthophosphate in the water column. The unlabeled orthophosphate mass in Eq. 7-5 is then incorporated into the unlabeled phosphorus in the phytoplankton biomass. The labeled orthophosphate (Eq. 7-6) forms the labeled phosphorus in the phytoplankton biomass. In the EUTRO model, the phytoplankton biomass is calculated as carbon. A stoichiometric coefficient is used to convert carbon to chlorophyll *a*.

The original water quality model for the Upper Mississippi River and Lake Pepin has a total of 15 tributaries and point sources. The Metro Plant, the Mississippi River above Lock and Dam No. 1, and the Minnesota River are the three major phosphorus sources in the study area in 1988, and they are tracked separately using this numerical tagging model. The remaining phosphorus sources are lumped into one group, and they include the St. Croix River, Cannon River, three other minor tributaries, and seven small point sources. Additionally, the sediment phosphorus flux in Lake Pepin (an internal phosphorus source) has an important role in the model and is also tracked separately.

To implement this tagging technique, three additional subroutines to quantify the kinetics for labeled nonliving organic phosphorus, labeled orthophosphate, and labeled phytoplankton are added to the EUTRO model. The original EUTRO model has sediment flux "hardwired," or explicitly written, in the algorithms of the orthophosphate system (i.e., water quality variable 3). Since this internal source of phosphorus would be either labeled or not labeled, it is an easy task to rewire the code for the labeled flux condition. That is, for the model runs in which sediment phosphorus was tracked, the flux rate is turned off in system 3 (unlabeled orthophosphate) and turned on in system 9 (labeled orthophosphate).

Given the number of changes made to the original EUTRO source code, it is essential to verify the accuracy of the tagging model through a series of tests prior to conducting the numerical tagging analysis. First, the original model and the modified (i.e., tagging) model yield the same results (Lung, 1996b). Second, the splitting between the labeled and unlabeled orthophosphate components is proportioned correctly. The sum of the labeled and unlabeled phosphorus concentrations from the tagging model was the same as that from the original model. The same test is also successful for the chlorophyll *a* concentrations—a most crucial test (Lung, 1996b). The third test involves running the tagging model with various phosphorus sources labeled one at a time. The sums of these components matched well with the concentrations calculated in the original model application by Lung and Larson (1995).

Another key test is to examine the mass loading rates of a phosphorus component along the main channel. Figure 7-26 shows the Metro Plant mass rates for orthophosphate, total organic phosphorus, phosphorus in algae, and total phosphorus under the summer (June–September) 1988 condition. The mass rate is calculated by multiplying the phosphorus concentration by the advective outflow of each segment and totaling the rates for the top and bottom layers. Lateral flows between the main channel and the shallow area in Spring Lake are not included in this calculation. The important result in Figure 7-26 is that total phosphorus and orthophosphate loading rates are declining in the downstream direction, primarily due to settling in the water column. The amount of total phosphorus released by the Metro Plant is about 2,730 kg/day, with the majority (about 84%) in the orthophosphate form. The sharp decrease of orthophosphate and total phosphorus mass rates in the Spring Lake area is due to lateral flows and dispersion transporting mass out of the main channel. The segmentation and flow matrix for this area are complex, so the mass rates for the lateral segments could not be included. The mass rate in the main channel regains from the lateral area near Lock and Dam No. 2. The difference between the two curves in the middle panel of Figure 7-26 represents the mass rates of nonliving organic phosphorus.

Figure 7-26 follows the fate of phosphorus from the Metro Plant alone. The phosphorus mass rate in the phytoplankton increases following the Metro Plant input and attains a level of about 450 kg/day by Lock & Dam No. 2, with a sharpest rise in the Spring Lake area. This mass rate continues to grow, reaching a peak of 500 kg/day in the upper portion of Lake Pepin, and declines throughout Lake Pepin, maintaining a level of 90 kg/day at the outlet. The shallow depths in the Spring Lake area offer a favorable condition for algal growth, while the significant depths in Lake Pepin, particularly in the lower portion, drastically reduce the available light for algal

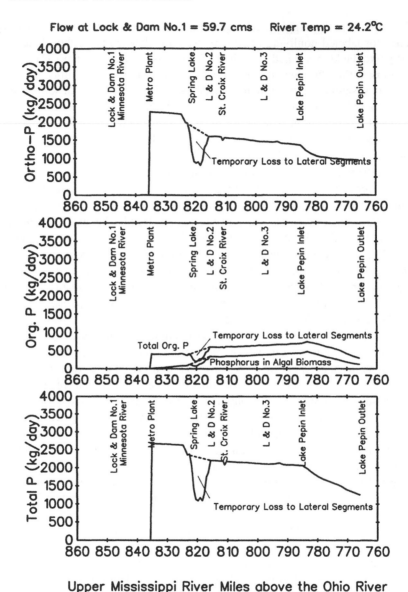

Upper Mississippi River Miles above the Ohio River

Figure 7-26 Fate and transport of orthophosphate, organic phosphorus, and total phosphorus from the Metro Plant in the Upper Mississippi River and Lake Pepin in summer 1988.

growth as the algae are mixed to greater depths. Such a difference in the phytoplankton mass rates between Spring Lake and Lake Pepin is particularly pronounced during low flow conditions with extremely shallow waters in the Spring Lake area.

There is a net loss of about 680 kg/day of Metro Plant orthophosphate between UM830 and UM810, of which 273 kg/day is converted to phytoplankton biomass.

The balance of 410 kg/day represents a net loss of phosphorus via the settling of algae and nonliving detritus (see the total phosphorus panel in Figure 7-26). While there is tremendous growth of algae in Spring Lake, more algal biomass settles into the sediment behind Lock & Dam No. 2 than makes it over the dam. As a result, the pooled area behind Lock & Dam No. 2 served as a phosphorus sink in the summer of 1988. Similarly, there is a net loss of approximately 410 kg/day orthophosphate between the inlet and outlet of Lake Pepin. Unlike Spring Lake, there is also a net loss in phosphorus from the algal biomass (about 360 kg/day) due to settling. Thus a total loss rate of 770 kg/day phosphorus is seen for Lake Pepin. In summary, the two major loss rates of phosphorus from the Metro Plant input are calculated in the pooled area behind Lock & Dam No. 2 (about 410 kg/day) and in Lake Pepin (about 770 kg/day).

Results from a similar analysis for upstream phosphorus sources (i.e., represented by the Mississippi River at Lock & Dam No. 1) are shown in Figure 7-27. The total phosphorus mass rate from Lock & Dam No. 1 is lower than that of the Metro Plant. A similar sharp decline and regain of orthophosphate and total phosphorus between UM825 and UM815 is also encountered for this phosphorus source—an artifact of the lateral segments. Phosphorus in algae from this upstream source continues to decrease in the downstream direction due to settling and recycling to organic phosphorus and orthophosphate. In fact, recycling minimizes the loss of phosphorus throughout the study area for this phosphorus source. The total reduction over the modeled reach is roughly 114 kg/day in summer 1988, representing about 20% of the total phosphorus mass rate at Lock & Dam No. 1. At the outlet of Lake Pepin, the majority of phosphorus is in orthophosphate form, resulting from recycling.

Results of the numerical tagging analysis for the summer 1988 condition are presented in the left-hand column of Figure 7-28, showing total phosphorus, orthophosphate, and chlorophyll *a* concentration profiles in the top layer of the main channel from the Mississippi River at Lock & Dam No. 1 to the outlet of Lake Pepin. Concentrations in the bottom-layer segments are similar to those in the top-layer segments as the water column is fairly well mixed in 1988. Figure 7-28 combines the results from five model runs, in which varying sources of phosphorus are tracked. First, results are plotted from the model run in which phosphorus from the Mississippi River at Lock & Dam No. 1 is labeled. Next, results from the run with the Minnesota River loads labeled are added. Subsequently, results from the model runs with phosphorus loads from the Metro Plant, other sources, and sediment flux are added individually. The model runs indicate that the total concentration profiles match the original model calibration results for these three water quality constituents, suggesting the successful operation of the tagging model.

Figure 7-28 shows the dominating effect of the Metro Plant discharge on the phosphorus concentrations in the water column during the summer months (June–September) of 1988. Under extremely low flow conditions, as seen in 1988, the relatively stable flow from the Metro Plant grows in proportion to diminishing flows from the tributaries. As a consequence, phosphorus loads from the Metro Plant have a greater influence on ambient river concentrations in low flow years than in higher flow years, as will be demonstrated in the 1990 results. Generally in 1988, the total phosphorus and orthophosphate concentrations are reduced in the downstream

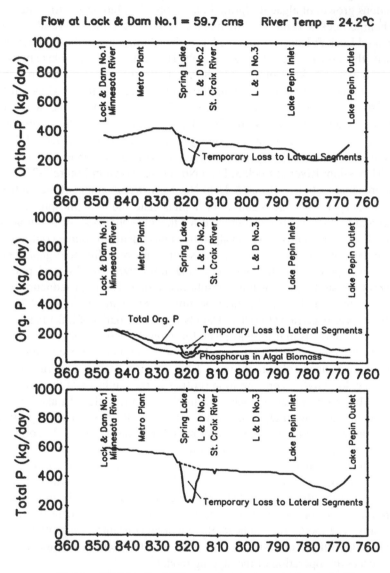

Figure 7-27   Fate and transport of orthophosphate, organic phosphorus, and total phosphorus from the upstream nonpoint source loads in the Upper Mississippi River and Lake Pepin in summer 1988.

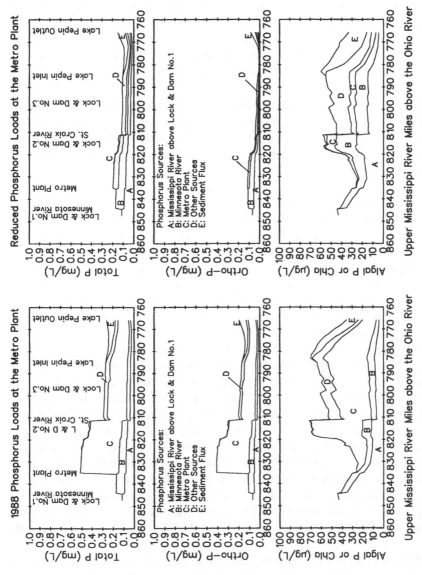

Figure 7-28 Model results showing the fate and transport of various phosphorus sources in the Upper Mississippi River and Lake Pepin under 1988 and reduced phosphorus loads at the Metro Plant.

249

direction, indicating loss of phosphorus along the river (due to algal uptake and settling) in addition to dilution by the St. Croix River.

Although the Metro Plant effluent contains little or no phytoplankton biomass,
phosphorus from the effluent is gradually taken up by the phytoplankton in the river,
leading to the band of viable chlorophyll *a* attributed to the Metro Plant in the lower
panel of Figure 7-28. As previously discussed, the numerical tagging model is
designed to proportion algal uptake of phosphorus according to the proportion of various sources of orthophosphate in the water column. Because the Metro Plant contributes the largest portion of orthophosphate in summer 1988, it is not surprising to
see a large amount of the water column phytoplankton biomass below the plant containing this Metro Plant nutrient source.

The model also uses a one-to-one ratio of phosphorus to chlorophyll *a* in the phytoplankton. Therefore, Figure 7-28 shows that roughly 50 µg/L of algal phosphorus
at the end of Pool 2 and 40 µg/L of algal phosphorus at the head of Lake Pepin are
derived from Metro Plant phosphorus in summer 1988. On the other hand, the role of
phosphorus from the headwaters (i.e., the Mississippi River at Lock & Dam No. 1) is
greatly diminished, showing 40 µg/L of algal phosphorus at the dam but only about
4 µg/L at the outlet to Lake Pepin.

One should not infer from Figure 7-28 that the Metro Plant phosphorus causes the
large increase in viable chlorophyll *a* in Pool 2. In fact, flow and light are the primary
controlling factors of algal growth in the Upper Mississippi River (Lung and Larson,
1995). High growth levels in Pool 2 are attributable to reduced flows and increased
light in the vicinity of Spring Lake. Also, given the nonlinear nature of algal dynamics,
it cannot be assumed that reductions in Metro Plant phosphorus will result in similar
reductions in phytoplankton biomass, as will be seen in the results of the numerical
tagging analysis using reduced Metro Plant loads.

The tagging model is then applied to quantify the phosphorus components under
a reduced Metro Plant load. Again, the 1988 summer steady-state model is used for
this analysis with low sediment phosphorus fluxes. This load reduction scenario calls
for a filtration process to lower the total phosphorus concentration in the Plant effluent to 0.4 mg/L, discharging 127 kg/day of orthophosphate and 253 kg/day of nonliving organic phosphorus, respectively. Loading rates for other water quality
constituents in the model are presented in a report by EnviroTech (1993). Of the five
possible phosphorus reduction scenarios considered for the Metro Plant, only the one
with the lowest phosphorus concentration in the effluent (0.4 mg/L) is analyzed in
this study. The selection of this scenario is intended to bracket the range of results for
the tagging analysis with high (1988 loads) and low (0.4 mg/L) Metro Plant loads.
Scenarios for the Metro Plant at 1.0 mg/L would fall somewhere in between the
results.

Model results of total phosphorus, orthophosphate, and chlorophyll *a* concentration profiles for the top layer of the water column under this load reduction scenario
are shown in the right-hand column of Figure 7-28. While the phosphorus loads from
the Metro Plant are significantly reduced under this scenario, the subsequent uptake
by the algal biomass in the water column is also greatly reduced compared with the
summer 1988 condition. In fact, the Metro Plant phosphorus would correspond to

only 10 µg/L of chlorophyll *a* behind Lock & Dam No. 2 and even less (about 5 µg/L) in Lake Pepin. On the other hand, phosphorus from all other sources increases its contribution to the phytoplankton biomass in the water column, almost making up the lost contribution from the Metro Plant. A noticeable source of algal phosphorus under this scenario is the sediment phosphate flux in Lake Pepin, and use of the higher flux rates would further increase this role. Despite the significant reduction of phosphorus loads from the Metro Plant, the phytoplankton biomass reduction in Lake Pepin amounts to only 10 µg/L, in agreement with the original modeling results (EnviroTech, 1993). Such an outcome results from the nonlinear relationship of algal dynamics.

Perhaps the above results can be put into perspective by quantifying the phosphorus loads to Lake Pepin. In compiling the point and nonpoint phosphorus loads in 1988, 1990, and 1991, Lung (1996b) reported that while the Metro Plant and other point source loads remain relatively stable over these three years, nonpoint loads increase sharply from 1988 (a low flow year) to 1991 (an extremely high flow year). Knowledge gained from this study suggests that sediment plays a major role in assessing the water quality of Lake Pepin. Phosphorus inputs to the lake are large, and even controlling the single largest point source (in a low flow year) is insufficient in abating eutrophication. While the effect of nutrient loading is exacerbated under low flow conditions, the true problem lies within year-to-year nutrient loadings into the lake.

The model results are particularly useful in quantifying the contribution of an individual phosphorus source or a group of sources to the phytoplankton biomass in the receiving water. Applying the numerical tagging technique on a watershed basis would yield helpful information for developing a sound water quality management strategy, particularly in terms of the trade-off between point and nonpoint loads. This study has demonstrated that the numerical tagging analysis can be instrumental in improving the overall TMDL development of a particular watershed.

Further, the EUTRO model can also be modified to track nitrogen components in the water column as demonstrated by Brown (1995) for the James River Estuary, Virginia. The mass uptake rates of the four forms of labeled and unlabeled inorganic nitrogen in the water column are as follows:

$$\frac{dM_1}{dt} = a_{nc}G_p(C_1 + C_2)V\alpha \frac{N_1}{N_1 + N_2} \tag{7-7}$$

$$\frac{dM_2}{dt} = a_{nc}G_p(C_1 + C_2)V\alpha \frac{N_2}{N_1 + N_2} \tag{7-8}$$

$$\frac{dM_3}{dt} = a_{nc}G_p(C_1 + C_2)V(1 - \alpha) \frac{N_3}{N_3 + N_4} \tag{7-9}$$

$$\frac{dM_4}{dt} = a_{nc}G_p(C_1 + C_2)V(1 - \alpha) \frac{N_4}{N_3 + N_4} \tag{7-10}$$

where

Gp = phytoplankton growth rate (day^{-1})
M_1 = unlabeled ammonia nitrogen mass uptake rate (mg/day)
M_2 = labeled ammonia nitrogen mass uptake rate (mg/day)
M_3 = unlabeled nitrite/nitrate nitrogen mass uptake rate (mg/day)
M_4 = labeled nitrite/nitrate nitrogen mass uptake rate (mg/day)
a_{nc} = nitrogen to carbon ratio in phytoplankton biomass (mg N/mg C)
C_1 = unlabeled phytoplankton carbon concentration (mg/L)
C_2 = labeled phytoplankton carbon concentration (mg/L)
V = volume of the water column
α = ammonia preference factor by algae
N_1 = unlabeled ammonia nitrogen concentration (mg/L)
N_2 = labeled ammonia nitrogen concentration (mg/L)
N_3 = unlabeled nitrite/nitrate nitrogen concentration (mg/L)
N_4 = labeled nitrite/nitrate nitrogen concentration (mg/L)

To incorporate the labeled nitrogen components, additional state variables must be included in the model. The modification to the program code is similar to that for the phosphorus tagging analysis.

7.6   MODELING SEDIMENT-WATER INTERACTIONS

The WASP/EUTRO modeling framework can also be used to simulate sediment-water interactions in receiving waters. This section presents an example for the Norwalk Harbor in Connecticut (Figure 7-29). The receiving water system is a relatively shallow, short tidal river off Long Island Sound. DO concentrations in the bottom waters are depressed to below 1 mg/L during the summer months. Under anaerobic conditions, significant sediment oxygen demand (SOD) rates and nutrient release fluxes have been observed.

The settling of particulate organic materials and the subsequent nutrient regeneration and release back to the overlying water may have profound impact on the water quality of Norwalk Harbor. On a whole system basis, the growth of phytoplankton in the shallow water is followed by settling of phytoplankton biomass in the water column. Subsequent decomposition of the organic material in the sediment exerts oxygen demands in the bottom layer of the water column. Additionally, nutrients such as orthophosphate and ammonia may be released from the sediment into the overlying water. On a volumetric basis, the fluxes from the sediment can be substantial nutrient sources or oxygen sinks to the overlying water column. Furthermore, the occurrence of anoxia, due in part to the SOD, may significantly increase these fluxes.

In the past, the sediment nutrient fluxes and oxygen demand were incorporated into eutrophication models as external inputs to the system. Similarly, zero-order kinetics are also used in formulating the EUTRO model. While direct inputs of sediment fluxes are useful in calibrating the water column kinetics, such an approach

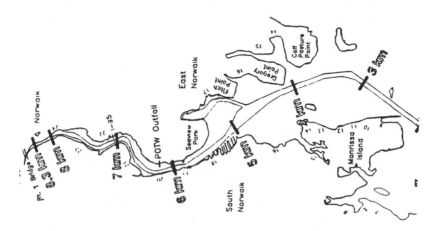

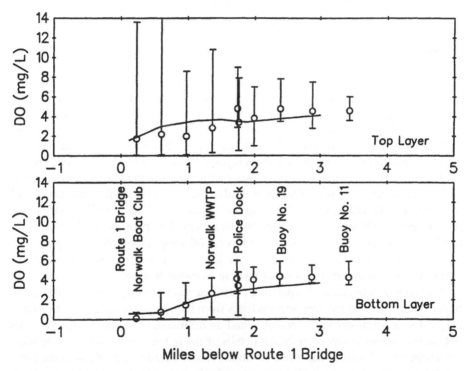

Figure 7-29 Calculated steady-state dissolved oxygen concentrations from WASP/EUTRO versus data in Norwalk Harbor in August 1988.

lacks the predictive capability necessary for models to simulate the response under nutrient controls. Recognizing this deficiency, an interactive sediment layer is necessary to address the question of long-term water quality impact.

One difficulty in sediment modeling is the availability of data. Although some SOD and nutrient flux data are available, no data on interstitial water is available. Further, sediment porosity and compaction information is lacking. Such data limitation does not warrant a comprehensive sediment model to be interfaced with the water column model. In addition, one desires to keep the computation to manageable proportions. Therefore, the attempt to model the sediment system is limited to the quantification of sediment-water fluxes. More refined computations such as multiple layers, which would include the effects of organism-induced mixing (bioturbation) and interstitial water diffusion, as well as porosity variations and compaction, are considered beyond the scope of this study.

The settling velocity of particulates across the sediment-water interface is the net velocity on a tidally averaged basis. This velocity can be thought of as the settling velocity, which represents the new flux to the sediment due to the difference between the downward settling flux and the upward resuspension flux. A more comprehensive treatment of the particle settling into the sediment would include not only a downward velocity but an upward resuspension velocity as well.

The thickness of the sediment layer for the Norwalk Harbor model is an important parameter. Such a thickness should adequately reflect the thickness of the active layer, the depth to which the sediment is influenced through exchange with the overlying water column. In addition, the thickness in the model formulation should characterize a reasonable time history or memory in the sediment layer. Note that the temporal and spatial scales are much smaller in the sediment system than the water column. As such, the memory of the sediment layer is much longer than the memory of the water column. Too thin a sediment layer will remember or be influenced by recent deposition of materials, occurring only within the last year or two of the period being analyzed. Conversely, if the layer used in the model is too thick, the model will average too long a history, not reflecting substantial reductions in sedimentary phosphorus resulting from reduced wastewater treatment plant phosphorus discharges. In the Norwalk Harbor study, the sediment layer depths, together with the assigned sedimentation velocities, provide for a 10-year detention time or memory, a reasonable approximation of the active sediment layer in Norwalk Harbor.

This approach is based on separating the initial reactions that convert sedimentary organic material into reactive intermediates from the remaining redox reactions that occur. Using a transformation variable and an orthogonality relationship, the mass balance equations independent of the details of the redox equations can be derived. As a result, the details of the redox equations can be avoided. Functions of the component concentrations are used to compute only the component concentrations, which can be treated identically to any other variable in the mass transport calculation. The convenient choice of components for the calculation is that which parallels the water quality variables in the water column, CBOD, and DO. Restricting the calculation to these components, however, eliminates the possibility of explicitly including the effects of other reduced species such as iron, manganese, and sulfide, which play a role in overall redox reactions and may also be involved in the generation of SOD. The decomposition reactions that drive the component mass balance equations, are the anaerobic decomposition of the algal carbon and the anaerobic breakdown of sedimentary organic carbon. Both reactions are sinks of the oxygen and rapidly drive its concentration *negative*, indicating that the sediment is *reduced* rather than oxidized. The negative concentrations computed can be thought of as the oxygen equivalents of the reduced end products produced by the chains of redox reactions occurring in the sediment. Since the calculated concentration of oxygen is positive in the overlying water, it is assumed that the reduced carbon species (negative oxygen equivalents) that are transported across the sediment-water interface combine with the available oxygen and are oxidized to CO_2 and H_2O, with a consequent reduction of oxygen in the overlying water column.

In a sediment system, detritus algae and zooplankton are decomposed to produce ammonia and organic nitrogen. Particulate organic nitrogen is hydrolyzed to ammonia by bacterial action within the sediment. Ammonia is also generated by the anaerobic decomposition of algae. Initially, however, the end product is not exclusively ammonia. A fraction of the algal nitrogen becomes particulate organic nitrogen, which must undergo hydrolysis before becoming ammonia. Ammonia produced by the hydrolysis of nonalgal organic nitrogen and the decomposition of detritus algal nitrogen may then be fluxed to the overlying water column via diffusive exchange. Organic nitrogen is also hydrolyzed to ammonia in a temperature dependent reaction. Ammonia is lost from the sediment through diffusion into the overlying water column. No nitrification occurs in the sediment due to the existing anaerobic conditions, yet nitrate is still present in the sediment because of diffusive exchange with the overlying water. Moreover, denitrification, the conversion of nitrate to nitrogen gas, may occur.

The EUTRO model has the capability of including a sediment layer and quantifying the SOD, ammonia, and orthophosphate rates internally. However, minor modifications of the source code are required to allow negative DO concentrations in the calculations. In this study, the EUTRO model is first configured to form a two-layer mass transport model using salinity as a tracer. The calibration mass transport pattern is then incorporated into the eutrophication model. Limited water quality data are available from the study area for the modeling analysis. The data collected during an intensive survey in August 1988 are used to calibrate the mass transport and eutrophication models.

The water column of Norwalk Harbor is first sliced into two layers, each of which is subdivided into eight segments in the longitudinal direction. Water quality constituents are considered to be completely mixed within each segment. The most recent hydraulic geometry data developed in this project are incorporated into this analysis. The surface layer volume is essentially the tidal prism volume with an average depth of 1.067 m. In addition, a sediment layer of eight segments, each of which is 10 cm thick, is incorporated.

The exchange coefficient between the pore water in the sediment and the bottom layer of the water column is 10^{-5} cm²/sec, and is an order of magnitude similar to that for molecular diffusion. Table 7-12 shows the configuration of water column diffusion as well as sediment diffusion coefficients for the 25-segment Norwalk Harbor

TABLE 7-12 Data Groups A and B for Norwalk Harbor EUTRO Model Input

Norwalk Harbor Two-Dimensional Eutrophication Model
August 1988 Calibration

NSEG	NSYS	ICRD	MFLG	IDMP	NSLN	INTY	ADFE		DD	HHMM		A:MODEL OPTIONS
24	8	0	1	1	1	0	0		1	0000		
2	8	12	14	16	17							
1												
0.012		20.										
1												
0.012		200.										
0	0	0	0	0	0	0	0					
2	0	+	*	+	*	+	*		+	*	+	*	B:EXCHANGES
8		1.0	1.00e-4						(surface water)			
1												
12090.		0.37	1	2								
2												
0.0		0.0		0.0	365.0							
1												
139580.		1.11	3	4								
2												
0.05		0.0		0.05	365.0							
1												
74080.		1.14	5	6								
2												
0.09		0.0		0.09	365.0							
1												
82860.		.998	7	8								
2												
0.10		0.0		0.10	365.0							
1												
100140.		1.393	9	10								
2												
1.0		0.0		1.0	365.0							
1												
153770.		1.18	11	12								
2												
1.0		0.0		1.0	365.0							
1												
374460.		1.21	13	14								
2												
1.0		0.0		1.0	365.0							
1												
626600.		1.21	15	16								
2												
1.0		0.		1.0	365.							
1		1.0	1.00e-4						(Pore water exchange)			
8												
12090.		0.10	2	17								
139580.		0.10	4	18								
74080.		0.10	6	19								
82860.		0.10	8	20								
100140.		0.10	10	21								
153770.		0.10	12	22								
374460.		0.10	14	23								
626600.		0.10	16	24								
2												
1.00e-5		0.0	1.00e-5		365.0							

model. A net sedimentation velocity of 1 cm/yr is assumed for the sediment layer to account for compaction of new sediments and sedimentation of old sediments into deeper layers. The EUTRO model configures this downward velocity as a flow out of the sediment segments, thereby requiring a boundary condition. To minimize the number of boundary conditions, a 25th segment is configured as a catchall segment for the eight sediment-layer segments. Thus boundary conditions for the sediment system are eliminated.

To allow negative DO concentrations in the sediment system in the model calculations, the EUTRO model input file is modified by setting the parameter, NSLN to 1. In addition, the WASP source code is modified to allow the display of negative DO concentrations on the screen during model executions.

The EUTRO model of the Norwalk Harbor is calibrated using the August 1988 data. Model results for DO in the water column are presented in Figure 7-29 for comparison with the data. Steady-state concentration profiles of these variables are plotted from Route 1 Bridge to Buoy No. 11 in the Norwalk Harbor. In general, the model results match the data well, particularly for the vertical stratification of DO.

The calculated steady-state concentration of DO in the sediment layer is about $-4,220$ mg/L, which yields a vertical flux of DO (i.e., SOD) of 3.65 gm $O_2/m^2/day$, a value within the range of 2 to 4.7 $O_2/m^2/day$ measured in the Norwalk Harbor in 1989. Such a result provides additional substantiation of the water quality model.

7.7 INCORPORATING NONPOINT SOURCE LOADS

The WASP/EUTRO model is also capable of accommodating nonpoint source loads. A sample application is the water quality model for the South Fork South Branch Potomac River in West Virginia (Figure 7-30). The river covers a total length of over 63 miles from Polo Alto to Moorefield, and is divided into 84 segments for the model. Major nutrient loads to the river are from this rural watershed where poultry farms are the primary contributors. Nonpoint nutrient loads are estimated for the one-month period of model simulation. In the input data file, a filename is assigned for nonpoint loads. This file containing all the time-variable loads is then called by the model during simulation. Table 7-13 shows the first 87 lines of the file. The first line indicates the filename and the number of segments that receive time-variable nonpoint loads (51 segments). The segments that receive nonpoint loads are then identified in the following 51 lines. Next, the water quality constituents that have nonpoint loads are indicated. Since the model is run time-variably to a steady-state condition for the summer months, the first date of the nonpoint source loads is Julian day 169. Nonpoint loads are entered on a daily basis for 33 consecutive days, the period of the time-variable simulation. For example, ammonium loads on the first day (Julian day 169) are entered on one line for all 51 segments in the order defined earlier in the input. (Due to length limitation, only the first five loads for segments 83, 82, 80, 79, and 78 are shown in Table 7-13.) Table 7-13 stops on day 171 while the actual data file continues to Julian day 202.

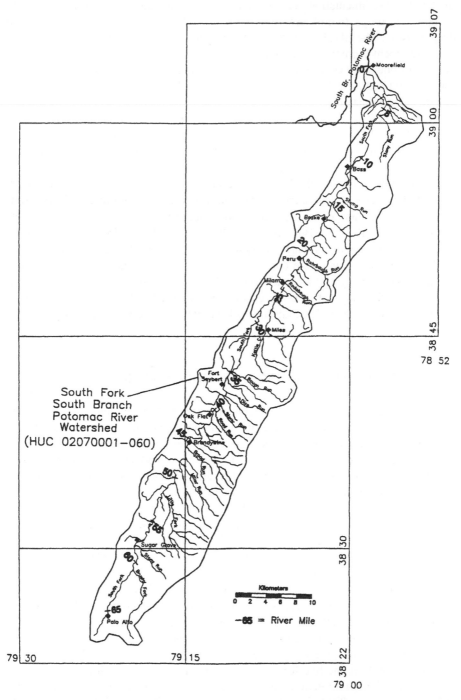

Figure 7-30 The South Fork South Branch of the Potomac River and its watershed in West Virginia.

**TABLE 7-13   Nonpoint Source File for EUTRO Model
(South Fork South Branch of the Potomac River)**

SFK_94.NPS   51   1   8

83	Seg# 83
82	Seg# 82
80	Seg# 80
79	Seg# 79
78	Seg# 78
77	Seg# 77
74	Seg# 74
70	Seg# 70
69	Seg# 69
68	Seg# 68
66	Seg# 66
65	Seg# 65
63	Seg# 63
62	Seg# 62
61	Seg# 61
59	Seg# 59
58	Seg# 58
57	Seg# 57
55	Seg# 55
52	Seg# 52
51	Seg# 51
49	Seg# 49
46	Seg# 46
42	Seg# 42
38	Seg# 38
37	Seg# 37
35	Seg# 35
34	Seg# 34
33	Seg# 33
31	Seg# 31
30	Seg# 30
28	Seg# 28
29	Seg# 29
27	Seg# 27
26	Seg# 26
24	Seg# 24
25	Seg# 25
21	Seg# 21
22	Seg# 22
23	Seg# 23
18	Seg# 18
17	Seg# 17
20	Seg# 20
16	Seg# 16

(continues)

TABLE 7-13 (*Continued*)

```
 15        Seg# 15
 11        Seg# 11
 10        Seg# 10
  7        Seg#  7
  6        Seg#  6
  1        Seg#  1
  3        Seg#  3
  1    2    3    4    5    6    7    8
NH3_N
NO_N
O_PO4_P
CHL_A
CBOD
OXYGEN
TON
TOP
169.00010   Time (days)      1 of   33
NH3_N          5.326E-03 3.064E-02 6.364E-03 1.245E-02 1.210E-02 6.654E-02
NO_N           2.397E-01 1.379E+00 2.864E-01 5.603E-01 5.447E-01 2.994E+00
O_PO4_P        2.663E-03 1.532E-02 3.182E-03 6.225E-03 6.052E-03 3.327E-02
CHL_A          2.663E-04 1.532E-03 3.182E-04 6.225E-04 6.052E-04 3.327E-03
CBOD           7.989E-01 4.596E+00 9.545E-01 1.868E+00 1.816E+00 9.981E+00
OXYGEN         2.130E+00 1.226E+01 2.545E+00 4.980E+00 4.842E+00 2.662E+01
TON            1.065E-01 6.128E-01 1.273E-01 2.490E-01 2.421E-01 1.331E+00
TOP            1.065E-02 6.128E-02 1.273E-02 2.490E-02 2.421E-02 1.331E-01
170.00000   Time (days)      2 of   33
NH3_N          5.326E-03 3.064E-02 6.364E-03 1.245E-02 1.210E-02 6.654E-02
NO_N           2.397E-01 1.379E+00 2.864E-01 5.603E-01 5.447E-01 2.994E+00
O_PO4_P        2.663E-03 1.532E-02 3.182E-03 6.225E-03 6.052E-03 3.327E-02
CHL_A          2.663E-04 1.532E-03 3.182E-04 6.225E-04 6.052E-04 3.327E-03
CBOD           7.989E-01 4.596E+00 9.545E-01 1.868E+00 1.816E+00 9.981E+00
OXYGEN         2.130E+00 1.226E+01 2.545E+00 4.980E+00 4.842E+00 2.662E+01
TON            1.065E-01 6.128E-01 1.273E-01 2.490E-01 2.421E-01 1.331E+00
TOP            1.065E-02 6.128E-02 1.273E-02 2.490E-02 2.421E-02 1.331E-01
171.00000   Time (days)      3 of   33
NH3_N          5.326E-03 3.064E-02 6.364E-03 1.245E-02 1.210E-02 6.654E-02
NO_N           2.397E-01 1.379E+00 2.864E-01 5.603E-01 5.447E-01 2.994E+00
O_PO4_P        2.663E-03 1.532E-02 3.182E-03 6.225E-03 6.052E-03 3.327E-02
CHL_A          2.663E-04 1.532E-03 3.182E-04 6.225E-04 6.052E-04 3.327E-03
CBOD           7.989E-01 4.596E+00 9.545E-01 1.868E+00 1.816E+00 9.981E+00
OXYGEN         2.130E+00 1.226E+01 2.545E+00 4.980E+00 4.842E+00 2.662E+01
TON            1.065E-01 6.128E-01 1.273E-01 2.490E-01 2.421E-01 1.331E+00
TOP            1.065E-02 6.128E-02 1.273E-02 2.490E-02 2.421E-02 1.331E-01
```

7.8   LINKING A HYDRODYNAMIC MODEL WITH THE WASP/EUTRO MODEL

The WASP model has a hydrodynamic module, DYNHYD, that is capable of simulating one-dimensional velocities and flow rates for streams, rivers, and estuaries. In addition, one-dimensional tributaries and branches can also be incorporated. In a recent mixing zone modeling study of the Salisbury wastewater treatment plant effluent, the receiving water, the Wicomico Estuary in Maryland (Figure 7-31), is divided into 97 segments from the dam upstream of Salisbury to the mouth of the river. Table 7-14 presents the input data file to run the DYNHYD model. The DYNHYD model output is next stored in a file (INPUT4.HYD in this case), which is then read by the WASP/EUTRO input data file. Table 7-15 shows data groups A and D of the WASP/EUTRO model input file for the Wicomico Estuary. Note the file name storing the DYNHYD model results is INPUT4.HYD, which is entered in the first line of data group D (see Table 7-15) in the WASP/EUTRO input file.

Since the DYNHYD model is limited to one-dimensional configurations, other hydrodynamic models are needed to generate the two- or three-dimensional flow fields in complicated waterbodies. In the case of the Loch Raven Reservoir, Maryland (Figure 7-32), the two-dimensional (longitudinal-vertical) CE-QUAL-W2 model (simply called W2 model) by Cole et al. (1998) is used to quantify the advective flows in both longitudinal and vertical directions. First, the reservoir is divided into 17 longitudinal segments, each of which is further divided into a number of 2-meter vertical layers, for a total of 112 segments in the water column (Figure 7-32). Note that a 4-meter surface layer is necessary to prevent the loss of the layer during the year's drought months. This extra thick surface layer is acceptable because vertical profiles of temperature and conductivity measured from the field show insignificant gradients in the surface 5 meters of the water column. This model configuration is important because the same spatial configuration (112 segments) is also used in the WASP/EUTRO model, thereby significantly reducing the effort in linking the W2 and WASP models together.

In this study, it was a difficult task to process the calculated velocities from the W2 model and to generate the flows for incorporation into the input data file to run the WASP model because the temporal two-dimensional flow fields are complicated with regard to the required format needed by the WASP model. Instead, the W2 model code is slightly modified to generate a hydrodynamic results file with WASP/EUTRO model format specifications. Thus the hydrodynamic file is easily read by the WASP/EUTRO model, just as if the hydrodynamic results were generated by the DYNHYD model.

An important checkpoint in linking the hydrodynamic and WASP/EUTRO models is to run the WASP model for a conservative substance (in this case, specific conductivity) using the flow field provided by the hydrodynamic model results. Results from the two models must match to insure a proper linkage. Figure 7-33 displays the model results for specific conductivity from the W2 and WASP models of Loch Raven Reservoir in 1991. Results from these two models match each other closely and also reproduce the same measured vertical profiles of specific conductivity in the

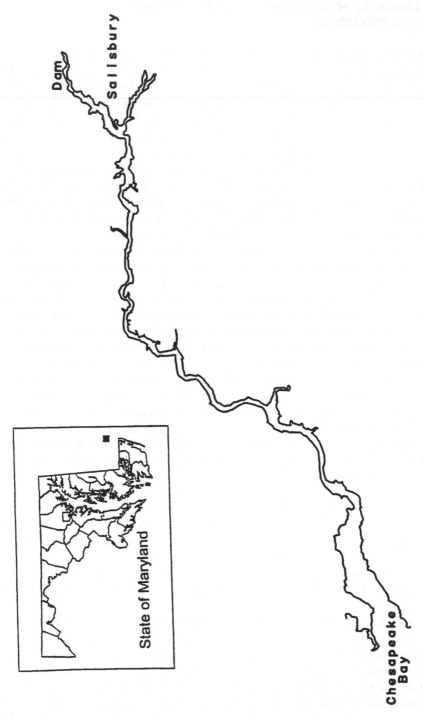

Figure 7-31   The Wicomico Estuary, Maryland, and town of Salisbury.

TABLE 7-14   Input Data File to Run the DYNHYD Model of Wicomico Estuary

```
1-D Wicomico Estuary Hydrodynamic Model
Linkage to WASP5
***** PROGRAM CONTROL DATA **************************************************
   98   97    0  25.    5    1  6 0    45 24 0
***** OUTPUT CONTROL DATA ***************************************************
0.0000E-01 6.000         98
    1    2    3    4    5    6    7    8    9   10   11   12   13   14   15   16
   17   18   19   20   21   22   23   24   25   26   27   28   29   30   31   32
   33   34   35   36   37   38   39   40   41   42   43   44   45   46   47   48
   49   50   51   52   53   54   55   56   57   58   59   60   61   62   63   64
   65   66   67   68   69   70   71   72   73   74   75   76   77   78   79   80
   81   82   83   84   85   86   87   88   89   90   91   92   93   94   95   96
   97   98
***** SUMMARY CONTROL DATA **************************************************
    1    1 12  0 12.0   60   10
***** JUNCTION DATA *********************************************************
    1       -0.3    39291.0      -4.0    1    0    0    0    0    0
    2       -0.3    40910.0      -4.0    1    2    0    0    0    0
    3       -0.3    42529.0      -4.0    2    3    0    0    0    0
    4       -0.3    56655.0      -4.0    3    4    0    0    0    0
    5       -0.3    69163.0      -4.0    4    5    0    0    0    0
    6       -0.3    69163.0      -4.0    5    6    0    0    0    0
    7       -0.3    69163.0      -4.0    6    7    0    0    0    0
    8       -0.3    81672.1      -4.0    7    8    0    0    0    0
    9       -0.3    94034.0      -4.0    8    9    0    0    0    0
   10       -0.3    93886.0      -4.0    9   10    0    0    0    0
   11       -0.3    93886.0      -4.0   10   11    0    0    0    0
   12       -0.3    93886.0      -4.0   11   12    0    0    0    0
   13       -0.3    93738.0      -4.0   12   13    0    0    0    0
   14       -0.3    70192.0      -4.0   13   14    0    0    0    0
   15       -0.3    46793.0      -4.0   14   15    0    0    0    0
   16       -0.3    46793.0      -4.0   15   16    0    0    0    0
   17       -0.3    46793.0      -4.0   16   17    0    0    0    0
   18       -0.3    46793.0      -4.0   17   18    0    0    0    0
   19       -0.3    46793.0      -4.0   18   19    0    0    0    0
   20       -0.3    46793.0      -4.0   19   20    0    0    0    0
   21       -0.3    40760.0      -4.0   20   21    0    0    0    0
   22       -0.3    40760.0      -4.0   21   22    0    0    0    0
   23       -0.3    46793.0      -4.0   22   23    0    0    0    0
   24       -0.3    46793.0      -4.0   23   24    0    0    0    0
   25       -0.3    46793.0      -4.0   24   25    0    0    0    0
   26       -0.3    52828.0      -4.0   25   26    0    0    0    0
   27       -0.3    56214.0      -4.0   26   27    0    0    0    0
   28       -0.3    53565.0      -4.0   27   28    0    0    0    0
   29       -0.3    53565.0      -4.0   28   29    0    0    0    0
   30       -0.3    53565.0      -4.0   29   30    0    0    0    0
```

(continues)

TABLE 7-14 (*Continued*)

31	-0.3	53565.0	-4.0	30	31	0	0	0	0
32	-0.3	50916.0	-4.0	31	32	0	0	0	0
33	-0.3	48047.0	-4.0	32	33	0	0	0	0
34	-0.3	47826.0	-4.0	33	34	0	0	0	0
35	-0.3	47826.0	-4.0	34	35	0	0	0	0
36	-0.3	47604.0	-4.0	35	36	0	0	0	0
37	-0.3	50989.0	-4.0	36	37	0	0	0	0
38	-0.3	54595.0	-4.0	37	38	0	0	0	0
39	-0.3	54595.0	-4.0	38	39	0	0	0	0
40	-0.3	54595.0	-4.0	39	40	0	0	0	0
41	-0.3	54595.0	-4.0	40	41	0	0	0	0
42	-0.3	54595.0	-4.0	41	42	0	0	0	0
43	-0.3	58200.0	-4.0	42	43	0	0	0	0
44	-0.3	60334.0	-4.0	43	44	0	0	0	0
45	-0.3	58863.0	-4.0	44	45	0	0	0	0
46	-0.3	58863.0	-4.0	45	46	0	0	0	0
47	-0.3	58863.0	-4.0	46	47	0	0	0	0
48	-0.3	58863.0	-4.0	47	48	0	0	0	0
49	-0.3	57391.0	-4.0	48	49	0	0	0	0
50	-0.3	57022.0	-4.0	49	50	0	0	0	0
51	-0.3	58126.0	-4.0	50	51	0	0	0	0
52	-0.3	58126.0	-4.0	51	52	0	0	0	0
53	-0.3	58126.0	-4.0	52	53	0	0	0	0
54	-0.3	58126.0	-4.0	53	54	0	0	0	0
55	-0.3	58126.0	-4.0	54	55	0	0	0	0
56	-0.3	58126.0	-4.0	55	56	0	0	0	0
57	-0.3	58126.0	-4.0	56	57	0	0	0	0
58	-0.3	59230.0	-4.0	57	58	0	0	0	0
59	-0.3	81230.1	-4.0	58	59	0	0	0	0
60	-0.3	95724.9	-4.0	59	60	0	0	0	0
61	-0.3	89324.8	-4.0	60	61	0	0	0	0
62	-0.3	89324.8	-4.0	61	62	0	0	0	0
63	-0.3	89324.8	-4.0	62	63	0	0	0	0
64	-0.3	89324.8	-4.0	63	64	0	0	0	0
65	-0.3	89324.8	-4.0	64	65	0	0	0	0
66	-0.3	89324.8	-4.0	65	66	0	0	0	0
67	-0.3	82923.0	-4.0	66	67	0	0	0	0
68	-0.3	72916.0	-4.0	67	68	0	0	0	0
69	-0.3	69311.0	-4.0	68	69	0	0	0	0
70	-0.3	65705.0	-4.0	69	70	0	0	0	0
71	-0.3	60186.0	-4.0	70	71	0	0	0	0
72	-0.3	58274.0	-4.0	71	72	0	0	0	0
73	-0.3	58274.0	-4.0	72	73	0	0	0	0
74	-0.3	58274.0	-4.0	73	74	0	0	0	0
75	-0.3	58274.0	-4.0	74	75	0	0	0	0
76	-0.3	87412.0	-4.0	75	76	0	0	0	0
77	-0.3	197780.0	-4.0	76	77	0	0	0	0

TABLE 7-14 (*Continued*)

78	-0.3	279011.0	-4.0	77	78	0	0	0	0	
79	-0.3	279011.0	-4.0	78	79	0	0	0	0	
80	-0.3	279011.0	-4.0	79	80	0	0	0	0	
81	-0.3	279011.0	-4.0	80	81	0	0	0	0	
82	-0.3	279011.0	-4.0	81	82	0	0	0	0	
83	-0.3	279011.0	-4.0	82	83	0	0	0	0	
84	-0.3	360244.0	-4.0	83	84	0	0	0	0	
85	-0.3	391592.0	-4.0	84	85	0	0	0	0	
86	-0.3	341709.0	-4.0	85	86	0	0	0	0	
87	-0.3	341709.0	-4.0	86	87	0	0	0	0	
88	-0.3	341709.0	-4.0	87	88	0	0	0	0	
89	-0.3	291818.0	-4.0	88	89	0	0	0	0	
90	-0.3	307266.0	-4.0	89	90	0	0	0	0	
91	-0.3	372605.0	-4.0	90	91	0	0	0	0	
92	-0.3	372605.0	-4.0	91	92	0	0	0	0	
93	-0.3	437943.0	-4.0	92	93	0	0	0	0	
94	-0.3	531977.1	-4.0	93	94	0	0	0	0	
95	-0.3	560673.0	-4.0	94	95	0	0	0	0	
96	-0.3	560673.0	-4.0	95	96	0	0	0	0	
97	-0.3	589369.0	-4.0	96	97	0	0	0	0	
98	-0.3	589369.0	-4.0	97	0	0	0	0	0	

TABLE 7-15   Input Groups A and D for EUTRO Model of Wicomico Estuary

```
Wicomico Estuary 1-D Mass Transport Model     March 1999  (Conductivity)
For Mixing Zone Modeling of Salisbury Wastewater Treatment Plant Effluent
 NSEG NSYS ICRD MFLG IDMP NSLN INTY ADFC    DD HHMM       A:MODEL OPTIONS
  97   01    0    1    1    0    0    0   11 0000
   1    2    3    4    5    6
   1
   0.001     90.0
   2
     1.0       10.       1.0      145.
   0    0    0    0    0    0    0    0
   1    0    +    *    +    *    +    *    +    *    +    *    B:EXCHANGES
  (77 groups of exchange coefficients)
   0    0    0    0    0    0    0    0
   2    0       1.0    +    *    +    *    +    *    +    *    C: VOLUMES
  (97 segment volumes)
   3    2 input4.hyd   +    *    +    *    +    *    +    *    D: FLOWS
   0
   0    0    0    0    0    0    0    0
```

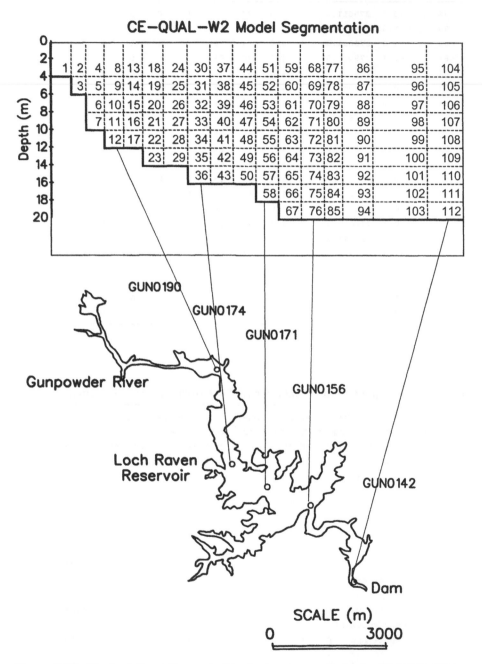

Figure 7-32   The Loch Raven Reservoir, Maryland, and segmentation for CE-QUAL-W2 hydrodynamic model and WASP/EUTRO model.

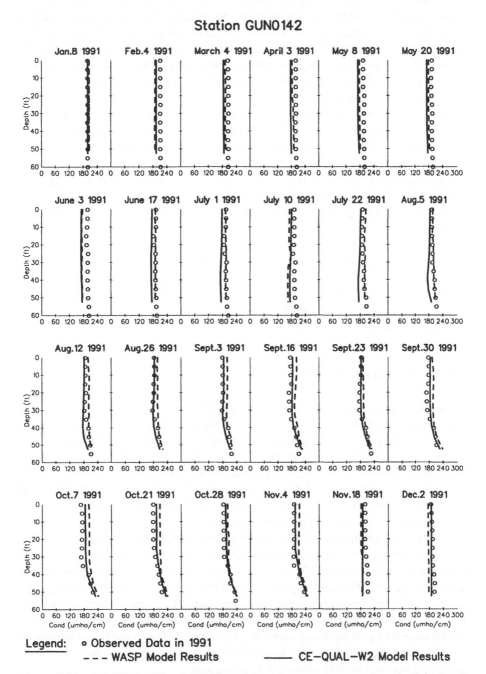

Figure 7-33 CE-QUAL-W2 and WASP model results versus measured conductivity levels at the dam in Loch Raven Reservoir, Maryland, 1991. Open circles are measured values. Solid and dashed lines are CE-QUAL-W2 and WASP model results, respectively.

reservoir. Perhaps another important result is that the slightly increased specific conductivity levels in the bottom waters during the summer months are reproduced by both models. The elevated conductivity levels are due to the release of iron and manganese ions under anaerobic conditions in the deep water of the reservoir. Results presented in Figure 7-33 further substantiate the validity of the mass transport calculations in the WASP model and indicate that the mass transport model is ready for water quality simulations.

Notice that to satisfy the numerical stability of the hydrodynamic model execution, the time steps used in the W2 (or DYNHYD model in one-dimensional applications) hydrodynamic calculations are often small. To illustrate, in the Loch Raven Reservoir study, a time step of 0.005 days is used. When directly linking a hydrodynamic model (DYNHYD, W2, or any other model) with WASP, this hydrodynamic time step is adopted in the WASP water quality model simulations. Any time steps entered in Group A of the WASP input data file will be ignored by the model simulation.

Since the W2 model is a real time hydrodynamic model, its output yields positive and negative flows. Special care must be exercised in working with the EUTRO code so that the magnitude of velocities is used in calculating the reaeration velocity for the surface layer of the model segmentation. In addition, transient dry and wet periods cause fluctuation of the surface elevation. Modification of the WASP code in this situation will prevent the loss of segments.

7.9   SUMMARY REMARKS

A number of case studies are presented in this chapter demonstrating successful applications of the WASP/EUTRO modeling framework. In emphasizing the model's diversity, configurations for a variety of natural water systems, including rivers, streams, impoundments, estuaries, and coastal waters, are made. Spatial configurations, including one- , two- , and three-dimensional segmentations, as well as steady-state and time-variable modeling analyses are illustrated. The WASP/EUTRO modeling framework is also shown to work well with conservative substances, BOD/DO, and nutrient/eutrophication applications.

One notable application of the modeling framework is mass transport modeling, indicating the model's diversity and flexibility. The Upper Mississippi River and Lake Pepin application has 174 segments with straightforward one-dimensional advective transport, while the Maryland Coastal Bay application has only 30 segments with complicated two-dimensional horizontal, advective, and dispersive transport patterns. The Patuxent Estuary application is perhaps the most sophisticated case of all the mass transport modeling applications presented. The spatially and temporally variable advective and dispersive mass transport coefficients are calibrated to reproduce a 38-segment salinity model for an entire year.

The WASP/EUTRO modeling framework is useful in tracking nonconservative substances on an individual basis by using the ammonia variable slot. Presented is an example for total phosphorus in the Upper Mississippi River and Lake Pepin. The model structure is designed to properly handle vertical settling from one layer to another in the water column. Combined, these cases demonstrate the flexibility of the modeling framework.

With careful modifications to the program code, the model can provide additional insights into the fate and transport of nutrients as demonstrated in the numerical tagging modeling. Both phosphorus and nitrogen components can be tagged by including additional system variables to represent the labeled and unlabeled components. Such a procedure is particularly useful in watershed analysis to track the fate and transport of a single nutrient source or a group of sources.

For user convenience, all the case studies presented are run on microcomputer systems. Like any modeling framework, however, WASP/EUTRO has its limitations. The hydrodynamic module, DYNHYD is limited to one-dimensional configurations, while mass transport coefficients for two- or three-dimensional circulation patterns require an independent quantification of the transport coefficients using a separate hydrodynamic model. Yet, the model extends far beyond the fixed, eight system variable applications. Its use in mass transport modeling is truly appreciated.

CHAPTER 8

USING THE HAR03 MODEL

The WASP/EUTRO model presented in the preceding chapter is capable of time-variable water quality simulations for one-, two-, and three-dimensional spatial configurations. The modeling framework can be used for steady-state water quality analyses by running the calculations to equilibrium. The HAR03 modeling framework described in this chapter, however, is a steady-state calculation code, solving the steady-state solutions explicitly. Although the HAR03 model is less prevalent, it originated long before the development of the WASP/EUTRO model (U.S. EPA, 1990).

HAR03 is a steady-state, multidimensional modeling framework that utilizes compartment (i.e., segment) modeling techniques. An orthogonal system of segmentation is used for each segment having up to six interfaces. The model includes the effect of net advection and dispersive tidal exchange. HAR03 models the biochemical oxygen demand/dissolved oxygen (BOD/DO) deficit system as a coupled reaction with first-order decay of carbonaceous BOD (CBOD). With minor modifications, the program may also be used to model variables analogous to the BOD/DO system such as ammonia-nitrate. Zero-order net photosynthetic and benthic oxygen demands can be user-supplied to the model and used in the DO balance.

8.1 MODEL CAPABILITIES

Numerous combinations of water quality constituents can be modeled using the HAR03 model (Chapra and Nossa, 1974):

1. *Conservative Substances:* Chlorides; total dissolved solids; specific conductivity; and salinity; or any substance that does not decay or can be approximated by zero decay such as persistent pesticides and herbicides.

2. *Single Reactive Substance:* CBOD; coliform bacteria; ammonia; total phosphorus; total suspended solids; or any substance that decays according to first-order kinetics.

3. *Coupled Reactive Substances:* CBOD/DO deficit; ammonia/nitrate; or a feed forward system of two substances reacting with first-order kinetics.

4. *Additive Coupled Substances:* CBOD; nitrogenous BOD (NBOD); and DO deficit.

5. *Estuarine Coupled Reactive Substances:* Chlorides, CBOD/DO deficit. A chloride run can precede the CBOD/DO deficit run so that the resulting chloride concentrations can be used to determine the saturation DO concentrations.

6. *Estuarine Additive Coupled System:* Chlorides; NBOD; and CBOD/DO deficit. The same as 5 but for the additive system.

Because HAR03 does not have the capability to perform water quality simulations beyond BOD/DO, its use in water quality modeling is limited. However, the framework's capability and simplicity in mass transport modeling render it an excellent tool in modeling the fate and transport of conservative substances in one -, two -, or three-dimensional configurations under steady-state conditions. As demonstrated later, the model input structure is simple and preparing the input data file requires a minimum level of effort, particularly when compared with that for the WASP/EUTRO model. Using the HAR03 model in mass transport modeling emphasizes the strength of this modeling framework. But since HAR03 is a mass transport model and not a hydrodynamic model, it requires the user to specify advective and dispersive flows. In some cases, hydrodynamic model results are processed to develop the advective and dispersive flows for HAR03. In other studies, field work such as a dye study is performed to independently quantify the dispersive coefficients in the receiving water for the model.

As in the preceding chapters, this chapter presents examples of applying this modeling framework to demonstrate its use in mass transport modeling. Two types of model application are discussed in this chapter: near-field mixing zone modeling and far-field ambient mass transport modeling under steady-state conditions.

8.2   INPUT DATA STRUCTURE AND MODEL CONFIGURATION

Table 8-1 displays the line-by-line description of the model input data structure of HAR03. Compared with the QUAL2E and WASP/EUTRO models, HAR03 has much simpler model input. Although the model has default units for the input parameters, such as length, area, flow, and dispersion coefficient, the user can override them with his/her preferred units and conversion (scale) factors. While the one-dimensional longitudinal dispersion coefficient in estuaries is usually expressed in mi^2/day, the vertical diffusion coefficient is found in cm^2/s in many scientific and engineering studies. Conversion factors are needed if the user prefers these original

TABLE 8-1 Input Data Structure for the HAR03 Model

Data Group	Column(s)	Variable	Description	Units	Format
1	(1–40)	Title	To be used to label the run		A40
	(41–43)	N	The number of segments in the model		I3
	(49)	IPRNT	An indicator: IPRNT = 1 for printing solution matrices		I1
			IPRNT = 0 to suppress the printing solution matrices		I1
	(54)	JCON	Set JCON = 1 for MASS TRANSPORT model runs		I1
	(55–59)	SCALE1	Area scale factor	ft^2/**ou**[a]	F5.0
	(60–68)	SCALE2	Dispersion coefficient scale factor	mi^2/day/**ou**	F9.0
	(69–73)	SCALE3	Flow scale factor	cfs/**ou**	F5.0
	(74–78)	SCALE4	Length scale factor	ft/**ou**	F5.0
2^b	(1–5)	AREA	Cross-sectional area of interface with section	ft^2	F15.0
	(6–10)	E	Dispersion coefficient of first interface with segment	mi^2/day	F5.0
	(11–17)	Q	Flow[c] across the first interface with segment	cfs	F7.0
	(18–21)	IARAY	Section forming the first interface with segment		I4
	(22–26)	AREA	Cross-sectional area of second interface	ft^2	F5.0
	(27–31)	E	Dispersion coefficient of second interface	mi^2/day	F5.0
	(32–38)	Q	Flow across the second interface	cfs	F7.0
	(39–42)	IARAY	Section forming the second interface		I4
	(43–47)	AREA	Cross-sectional area of third interface	ft^2	F5.0
	(48–52)	E	Dispersion coefficient of third interface	mi^2/day	F5.0
	(53–59)	Q	Flow across the third interface	cfs	F7.0
	(60–63)	IARAY	Section forming the third interface		I4

272

Card	Columns	Variable	Description	Units	Format
3^d	(3–5)	JI	1st segment that forms an interface with segment I		I3
	(6–15)	LA(I, JI)	Length of segment I with respect to segment JI	ft	F10.0
	(18–20)	JJ	2nd segment that forms an interface with segment I		I3
	(21–30)	LA(I, JJ)	Length of segment I with respect to segment JJ	ft	F10.0
	(33–35)	JK	3rd segment that forms an interface with segment I		I3
	(36–45)	LA(I, JK)	Length of segment I with respect to segment JK	ft	F10.0
	(48–50)	JL	4th segment that forms an interface with segment I		I3
	(51–60)	LA(I, JL)	Length of segment I with respect to segment JL	ft	F10.0
	(63–65)	JM	5th segment that forms an interface with segment I		I3
	(66–75)	LA(I, JM)	Length of segment I with respect to segment JM	ft	F10.0
4^e	(1–10)	H_1	Depth of 1st segment	ft	F10.0
	(11–20)	H_2	Depth of 2nd segment	ft	F10.0
	(21–30)	H_3	Depth of 3rd segment	ft	F10.0
	———	H_n			
5^f	(1–10)	T_1	Water temperature of 1st segment	°C	F10.0
	(11–20)	VOL_1	Volume of 1st segment	10^6 ft^3	F10.0
	(21–30)	T_1	Water temperature of 2nd segment	°C	F10.0
	(31–40)	VOL_1	Volume of 2nd segment	10^6 ft^3	F10.0
	———	T_n	Water temperature of nth segment	°C	F10.0
	———	VOL_n	Volume of nth segment	10^6 ft^3	F10.0
6	(1–2)	ICON	= 0 for MASS TRANSPORT runs		I2
7	(1–2)	INDIC	= –1 for MASS TRANSPORT runs		I2
8	(1–2)	NUMBC	Number of boundary conditions		I2

(continues)

TABLE 8-1 (*Continued*)

Data Group	Column(s)	Variable	Units	Description	Format
9[g]	(1–8)	BC(1)	mg/L	Concentration for 1st boundary	F8.0
	(9–12)	ICOL(1)		Segment adjoining 1st boundary	I4
	(13–20)	BC(2)	mg/L	Concentration for 2nd boundary	F8.0
	(21–24)	ICOL(2)		Segment adjoining 2nd boundary	I4
	(61–68)	BC(6)	mg/L	Concentration for 6th boundary	F8.0
	(69–72)	ICOL(6)		Segment adjoining 6th boundary	I4
10	(1–5)	FAC1	day^{-1}	Temperature correction factor for reaction rate	F5.0
	(6–10)	K(1)		Reaction rate (= 0.0 for MASS TRANSPORT runs)	F5.0
11[h]	(1–10)	LOAD	lb/day	Load to a segment (= 0.0 for MASS TRANSPORT runs)	F10.0
	(11–13)	ISEC		Segment to which the load is applied	I3
12	(4)	IEXIT		set IEXIT = 1 to terminate the model input	I2

[a]**ou** stands for original unit in which the parameter to be converted by the scale factor is expressed. The purpose of the scale factors is to allow the users to input parameters in units that are different from those specified on the following pages. For instance, according to this program length should be input as feet. However, it may be more convenient to enter it in miles and set SCALE4 to 5280. The program would then internally convert the length from miles to feet.

[b]The 4th, 5th, and 6th interface parameter entries are identical to above. If a segment has an interface that forms a boundary, an interface parameter must be input for it. To do this, input the appropriate AREA, E, and Q for the interface and input the segment's number as the IARAY. It is only necessary to input the parameters for a particular interface once. For example, after inputting the parameters of the interface of segment 1 with segment 2, it is not necessary to input the parameters of the interface of segment 2 with segment 1. Two lines for each segment are required; that is, a blank, second line is needed if the total number of interfaces is less than 4.

[c]Flow out of a segment is positive; flow into a segment is negative.

[d]Two lines are required for each segment. A blank, second line is needed if the total number of interfaces is less than 6.

[e]Eight depths per line; continue until all segment depths are complete.

[f]Four temperature and volume entries per line; continue until all segments are complete.

[g]Proceed with input until a total of NUMBC boundary conditions have been read.

[h]The load LOAD must be followed by a blank line to indicate the end of loads.

274

units to the default HAR03 units, emphasizing the importance of the model's flexibility.

Once the scale factors are entered, hydraulic geometry parameters, such as interfacial area, dispersion coefficient, and advective flow, are incorporated. Note that advective flows into and out of a given segment should carry a negative and positive sign, respectively. Table 8-1 shows that information for two interfacial cross sections occupies an input entry line. Since the HAR03 model allows a maximum of six interfaces, three lines of data must be entered. If a given segment has only two interfaces, only one line will have data, while the other two blank lines must also be included in the data entry. Data for any interface need only be entered once. That is, once the interface data is entered for a given segment, it is not needed to be entered again for the adjacent segment. Again, the user should pay special attention to the scale factors so that preferred units can be accommodated.

The mixing length between two adjacent segments is needed to quantify the dispersive flow in the following manner: dispersive flow = dispersion coefficient × interfacial area/mixing length. This value is approximated as the distance between the centers of two adjacent segments. Obviously, the horizontal mixing length is usually greater than the vertical mixing length by several orders of magnitude.

The maximum number of segments is specified in the HAR03 source code. The user can change this limit to any number and recompile the code for specific configurations.

8.3 NEAR-FIELD MIXING ZONE MODELING

Mixing zones are areas of a waterbody where an effluent discharge undergoes rapid initial dilution of the waste resulting from the hydraulic characteristics of the discharge and the receiving water. In a recent technical support document on water quality-based toxics control, the U.S. EPA (1991b) recognized the need for a mixing zone. Further, they suggested that acute and chronic water quality criteria could be exceeded within a mixing zone as long as adverse ecological impacts do not result from the discharge. Subject to various conditions, many states now allow mixing zones in water quality–based toxics control. States typically specify allowable spatial dimensions to limit the areal extent of the mixing zone. Mixing zones are then either allowed or rejected on a case-by-case basis.

When wastewater is discharged into the receiving water, its transport may be divided into two stages with distinct mixing characteristics. The initial discharge momentum in the first stage determines the mixing and dilution. The design of the discharge outfall should provide ample momentum to dilute the concentrations in the immediate contact area as quickly as possible. Many existing outfalls with small flows, which are being regulated for whole effluent toxicity (WET) limits, do not have much momentum for initial dilution. The second stage of mixing covers a more extensive area in which the effect of initial momentum is diminished and the waste is mixed primarily by ambient turbulence. Thus an allocated impact zone may also be allowed within a mixing zone.

Regulatory agencies are now focusing NPDES permits for existing domestic and industrial wastewater dischargers, often having small wastewater flow rates (i.e., as low as 1,000 gallons per day). Many of these discharges lack sufficient momentum to produce any initial dilution (i.e., dilution upon impact). For example, under the 1-day 10-year low flow condition for toxicity evaluation, many of these discharge pipes are above the receiving water surface, producing almost zero momentum-induced mixing. Thus initial dilution models such as CORMIX I, II, III are *not* applicable. Instead, natural turbulence-induced mixing would account for the dilution. A technical approach has been developed by Lung (1995) to address these small dischargers. Successful applications of this modeling analysis include rivers, lakes, and estuaries for the evaluation of WET, BOD, color, ammonia, and total chlorine residue levels in the mixing zone. One modeling framework used in this assessment is the HAR03 model. The following case studies demonstrate several applications of configuring HAR03 to mixing zones.

8.3.1   Simplified Mixing Zone Modeling of Rivers and Estuaries

Prior to applying the HAR03 model to mixing zone analysis, it is necessary to discuss a simplified modeling approach currently used in quantifying mixing zones. For tidal rivers and estuaries, a two-dimensional depth averaged, tidally averaged advection-dispersion equation, derived by Hamrick and Neilson (1989), is:

$$\frac{\partial C}{\partial t} + u\frac{\partial C}{\partial x} + v\frac{\partial C}{\partial y} = \frac{1}{h}\frac{\partial}{\partial x}\left(hD_{xx}\frac{\partial C}{\partial x} + hD_{xy}\frac{\partial C}{\partial y}\right) +$$

$$\frac{1}{h}\frac{\partial}{\partial y}\left(hD_{yx}\frac{\partial C}{\partial x} + hD_{yy}\frac{\partial C}{\partial y}\right) - K_d C \tag{8-1}$$

where
  C = depth and tidally averaged concentration
  u = depth and tidally averaged advective velocity in x direction
  v = depth and tidally averaged advective velocity in y direction
  h = tidally averaged depth
  D = dispersion coefficient tensor
  K_d = first-order decay coefficient

The first solution examined is for a continuous point source discharge at $x = y = 0$ on the shoreline of a channel of width, B. In this case, x is positive in the direction of net flow (seaward) and y is positive across the channel toward the opposite shore at $y = B$. If the velocity field is unidirectional and the x coordinate is aligned in that direction, the transverse velocity v, and the off-diagonal dispersion coefficients, D_{xy} and D_{yx}, may be set to zero. Furthermore, under steady-state conditions, Eq. 8-1 becomes:

$$hu\frac{\partial C}{\partial x} = \frac{\partial}{\partial x}\left(hD_x\frac{\partial C}{\partial x}\right) + \frac{\partial}{\partial y}\left(hD_y\frac{\partial C}{\partial y}\right) - hK_d C \tag{8-2}$$

where $D_x = D_{xx}$ and $D_y = D_{yy}$. Hamrick and Neilson (1989) provided detailed methods for determining the dispersion coefficients.

Equation 8-2 has closed form analytical solutions only if the coefficients h, u, D_x, D_y, and K_d are constant (Hamrick and Neilson, 1989). Although this is seldom true for actual situations, it is possible to choose representative values that will give reasonable results in the vicinity of the outfall. The solution is:

$$ C = \frac{M}{\pi h (D_x D_y)^{1/2}} \exp\left[\frac{u}{(4 K_d D_x)^{1/2}} \left(\frac{K_d}{D_x} \right)^{1/2} x \right] + $$

$$ \sum_{i=-\infty}^{\infty} K_0 \left[\left(1 + \frac{u^2}{4 K_d D_x} \right)^{1/2} \left(\frac{K_d x^2}{D_x} + \frac{K_d}{D_y} (y + 2iB)^2 \right)^{1/2} \right] \tag{8-3} $$

where

M = mass discharged/unit time
K_0 = modified Bessel function of the second kind of order zero
B = depth-averaged width

If the estuary channel is sufficiently wide, satisfying:

$$ B^2 >>> (D_y \,/\, K_d) \qquad \text{or} \qquad B^2 > 300(D_y \,/\, K_d) $$

the series in Eq. 8-3 may be truncated at the $i = 0$ term. Further, for conservative substances, $K_d = 0$, Eq. 8-3 becomes:

$$ C = \frac{M}{\pi h (D_x D_y)^{1/2}} \exp\left(\frac{ux}{2 D_x} \right) K_0 \left[\frac{u}{2 D_x^{1/2}} \left(\frac{x^2}{D_x} + \frac{y^2}{D_y} \right)^{1/2} \right] \tag{8-4} $$

The application of Eq. 8-4 requires approximating the actual site conditions with idealized geometry, topography, and current fields. While this results in simple analytical solutions to the transport equations, the uncertainty must be incorporated into the dispersion coefficients. On the other hand, more complex models allowing spatial variations of advection and dispersion may be constructed by solving Eq.8-1 in a box-model configuration for mass transport (Beltaos, 1978a,b). In practice, tracer (dye) dispersion studies are used to determine characteristics of the ambient mixing for the mass transport model. At times, dispersion coefficient values reported in the literature (Holley and Jirka, 1989) may be used in lieu of a tracer study. Subsequent model sensitivity analyses substantiate and fine-tune the dispersion coefficients.

8.3.2 Applying the Simplified Modeling Analysis

The original outfall of the Falling Creek wastewater treatment plant (WWTP) discharged effluent into Falling Creek, a tributary of the James River (Figure 8-1). A

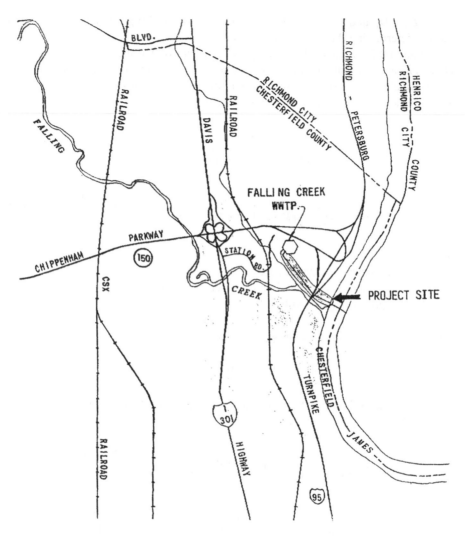

Figure 8-1 Falling Creek and James Estuary near Richmond, Virginia (tidal fresh portion).

new outfall was designed to discharge the effluent directly into the James Estuary. A new outfall design, consisting of a 48-in. pipe on the western bank of the James River, would change the discharge site from Falling Creek to the James River. The proposed discharge site is still under tidal influence, although the salinity level is close to that of freshwater. The acute impact of the discharge on the aquatic community in the vicinity of the proposed outfall became a primary concern to the regulatory agency, Virginia State Water Control Board (SWCB).

SWCB proposed an acute WET limit of 1 TU_a at the end of the outfall for the Falling Creek WWTP, that is, one allowing no dilution. However, the discharger conducted a modeling study to examine whether the proposed acute toxicity limit of 1 TU_a would be too stringent for the discharger, or could lethality to passing organisms still be prevented with an even stronger effluent, thereby allowing dilution in the mixing zone?

The Federal guideline requires that the acute toxicity impact be evaluated under the 1-day 10-year (1Q10) low flow condition at the study site. The designed outfall is above the water surface under the 1Q10 low flow condition, resulting in zero initial dilution. Thus the discharge-induced mixing was neglected, and mixing between the effluent and the river water was analyzed under the condition solely provided by ambient water motion. (Momentum-induced mixing, if any, would provide additional relief, thereby making the result conservative.)

Interestingly, the initial dilution caused by momentum-induced mixing near the outfall is insignificant for the Falling Creek plant discharge. Considering a momentum jet in a crossflow, one can evaluate the dilution ratio in the jet momentum-dominated regime (or momentum-dominated near field, MDNF) using the following equation:

$$s = 0.17 \frac{M_o^{0.5} y}{Q_o} \tag{8-5}$$

where

Q_o = discharge flow
M_o = momentum flux = $Q_o u_o$
u_o = effluent velocity

For the Falling Creek plant, the following data are used:

Wastewater flow, $Q_o = 10$ mgd = 0.438 m^3/s
Cross-sectional area of pipe, $A = 1.167$ m^2
Effluent velocity, $u_o = Q_o/A = 0.375$ m/s
Momentum flux, $M_o = 2 Q_o u_o = 0.329$ m^4/s^2

The initial dilution at the edge of allocated impact zone ($y = 27$ ft = 8.23 m) is 0.916, which is below 1.0 (no dilution!). Thus the only dilution would be from the ambient turbulence in the river.

In the amendments to the Water Quality Standards proposed by the Virginia Water Control Board, Section VR680-21-01.2.C, mixing zones are allowed under specific conditions. The mixing zone shall not constitute more than one-half of the width of the receiving watercourse nor shall it constitute more than one-third of the area of any cross section of the receiving watercourse. In addition, it shall not extend downstream at any time a distance more than five times the width of the receiving watercourse at the point of discharge.

The dimensions of the allocated impact zone within which the criterion maximum concentration (CMC) is met depend on the size of the regulatory mixing zone as specified and are calculated based on the Federal guidelines and State specifications as follows:

1. The length of the regulatory mixing zone = 2,700 ft (five times the river width)
2. The width of the regulatory mixing zone = 270 ft (one-half of the river width)
3. The CMC should be met within 10% of the lateral distance from the edge of the outfall structure to the edge of the mixing zone = 27 ft (10% of 270 ft)
4. The CMC should be met within a distance of 50 times the discharge length scale in any spatial direction = 177 ft (50 times 3.55 ft, which is the discharge length scale for a 4-ft diameter pipe)
5. The CMC should be met within a distance of five times the local water depth = 125 ft (5 times 25 ft)

Based on the above limitations, the size of the allocated impact zone is 27 ft in any spatial direction. Thus the CMC of 0.3 TU_a must be met at the edge of this zone.

Equation 8-4 is applied with the following site-specific data:

- Total wastewater flow = 10 mgd
- Effluent acute toxicity = 1.0 TU_a
- River velocity = 0.045 ft/s (with a 1Q10 low flow of 605 cfs)
- Longitudinal dispersion coefficient = 10 ft^2/s
- Lateral dispersion coefficient = 2.0 ft^2/s
- River depth = 24.9 ft

In lieu of a dye dispersion study, literature data on the mixing coefficients in the James Estuary were used. Also, the longitudinal and lateral dispersion coefficients were derived from the hydrodynamic modeling study of the James River Estuary by Hamrick and Neilson (1989).

The model results shown in Figure 8-2a display five isopleth toxicity contours. Contour No. 1 has a toxicity of 0.3 TU_a and contour No. 5 represents the toxicity level of complete mixing (i.e., 0.025 TU_a). The other three isopleths are 0.033 TU_a, 0.05 TU_a, and 0.10 TU_a, respectively. The model results show that the 0.3 TU_a isopleth is within the allocated impact zone (shown in the dot-dot/dashed line), thus meeting the water quality standard for whole effluent toxicity.

The ambient mixing model was then used to evaluate a higher effluent toxicity limit, such as 2 TU_a. The results of the model calculations are shown in Figure 8-2b, using the same value of D_y (2.0 ft^2/s). It is seen that the 0.3 TU_a isopleth is still within the allocated impact zone (shown in the dot-dot/dashed line), suggesting that a 50% dilution of the effluent may be allowed. Model prediction results were submitted to and subsequently approved by the SWCB and Environmental Protection Agency (EPA) Region III. Thus a mixing zone is allowed in lieu of the 1 TU_a acute toxic limit for the Falling Creek Plant.

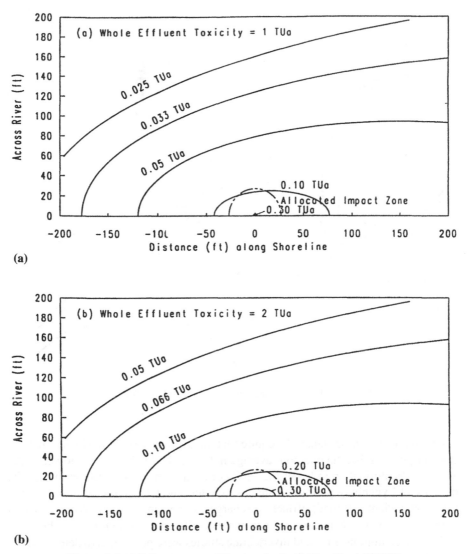

Figure 8-2   Mixing zone modeling results for Falling Creek WWTP.

8.4   MODELING MIXING ZONE IN THE JAMES RIVER USING HAR03

8.4.1   Total Residual Chlorine in Lynchburg Plant's Effluent

The NPDES permit for the City of Lynchburg requires that dechlorination facilities be constructed and operated to reduce the total residual chlorine (TRC) in the effluent to below 0.1 mg/L to meet the water quality standards in the James River (Figure 8-3). (Note that the James River in the Lynchburg area is a free-flowing river, without

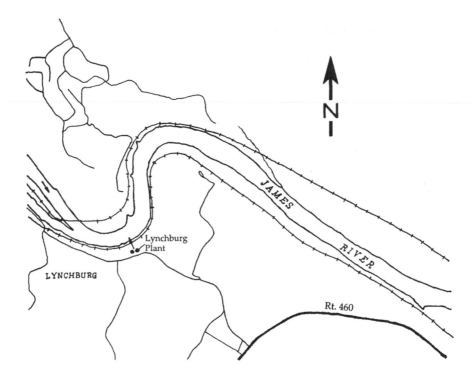

Figure 8-3 Lynchburg Plant and James River, Virginia.

any tidal effect.) These standards require that the average TRC in freshwater not exceed 11 ppb and that the one-hour average not exceed 19 ppb. In order to verify the need for dechlorination to lower the TRC concentrations to required permit levels; the City of Lynchburg performed studies to determine the ecological impact of the Lynchburg Plant (10 mgd) effluent discharge on the James River. Work included both biological and chemical studies to evaluate effluent toxicity for both laboratory and *in situ* conditions. Fish and invertebrate studies were performed to determine the acute toxicity of the effluent for resident aquatic species. In addition, TRC measurements were made in the outfall plume to evaluate the mixing zone under the flow conditions experienced during the testing.

Since it is impractical to test the outfall plume over a wide range of flow and effluent quality conditions, a TRC model was developed to estimate both the zone of initial dilution and the mixing zone. Model calculations provided a basis for evaluating the impact of a wide range of effluent flow and chlorine concentrations on the receiving water under various stream flow conditions. These results were then used to establish rational guidelines for effluent TRC levels, based on the predicted mixing zone and the toxicity levels established.

8.4.2 Field Data to Support Modeling

The field program consisted of three components: hydraulic geometry measurements, a dye dispersion study, and receiving water measurements of TRC levels. The hydraulic geometry measurements included velocity, depth, and width of the river at a number of locations (transects). The dye dispersion study collected data needed to calibrate the mixing in the James River. The receiving water measurements of TRC were required to perform final verification of the mixing zone model.

To assist in the design and planning of the dye dispersion study, Eq. 8-4 was used to approximate the mixing zone extent and to locate sampling transects:

$$C(x,\ y) = \frac{M}{du(4\pi D_y x\ /\ u)^{1/2}}\ \exp\left(\frac{-y^2 u}{4D_y x}\right) \qquad (8\text{-}6)$$

where:

M = point source mass discharge rate (kg/day)
u = average river velocity (m/s)
D_y = lateral dispersion coefficient (m²/s)
x = distance downstream (m)
y = distance across the river (m)
d = average river depth (m)

Results of the analysis using Eq. 8-6 suggested that eight transects should be established along the river. One transect upstream of the Lynchburg Plant outfall served as a control station and the seven downstream transects were used to define the plume characteristics. The transects were spaced to extend approximately 4,000 ft downstream of the outfall. The relative elevation of each transect was established to determine the river slope and water level. The river depth and velocity were measured at selected locations across the transects to determine the flow characteristics.

During the dye dispersion study, Rhodamine WT dye was continuously injected into the Lynchburg Plant's effluent at a constant rate. The Rhodamine WT red dye was chosen because it is a relatively stable material and not expected to decay during the sampling period. The low turbidity and suspended solids levels in the James River observed during the survey did not raise any concern of dye adsorption. The dye concentration in the effluent was 400 µg/L. Figure 8-4 displays the dye concentrations along each transect and shows a rapid decline of dye concentration downstream from the Lynchburg Plant. Further, total residual chlorine was measured on a different day and its concentrations are also presented in Figure 8-4. The TRC concentration in the plant effluent is 0.42 mg/L (420 µg/L). Note the rapid decline of TRC levels along each transect when compared with the dye data. Such a rapid decline of TRC concentration is due to the decay of chlorine in the receiving water.

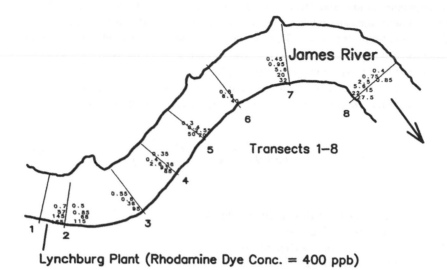

Measured Rhodamine WT Dye Conc. (ppb) on 1–14–89

Lynchburg Plant (Rhodamine Dye Conc. = 400 ppb)

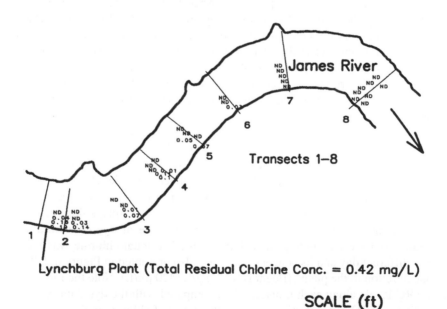

Measured Total Residual Chlorine Conc. (mg/L) on 1–27–89

Lynchburg Plant (Total Residual Chlorine Conc. = 0.42 mg/L)

SCALE (ft)

0   1000

Figure 8-4   Rhodamine WT red dye and residual chlorine data in James River.

8.4.3   Modeling Rhodamine WT Red Dye and Total Chlorine Residue in the Mixing Zone

The simple plume model in Eq. 8-1 has limitations that prevent an accurate calculation of a mixing zone. First, the lateral dispersion coefficient, D_y, characterizing the spread rate of water quality constituents, is assumed constant in space, which may not be totally valid along a 4,000-ft distance. Second, the ambient river flow, which has a direct impact on the mixing zone size, is only indirectly incorporated into the calculations through the use of an average velocity. These points make it clear that a more rigorous model is needed to address the mixing of the Lynchburg Plant effluent with the James River. Thus HAR03 was configured for the study area using the field data to calibrate the mixing zone model. Model calculated dye concentrations were then compared with the data to fine-tune the lateral dispersion coefficients.

The James River study area is divided into 48 segments (8 segments along the river and 6 segments across the river). Figure 8-5 shows the segmentation map. The HAR03 input data file is listed in Table 8-2. The advective flows in Table 8-2 are derived from velocity and hydraulic geometry measurements along the river. To better illustrate the flow pattern, the advective flows (in cfs) are shown in Figure 8-6. Lateral advective flows are derived from flow balance for each segment. Figure 8-6 also displays the lateral dispersion coefficients (in ft²/s) entered in Table 8-2. Note that the dispersion coefficients range from 0.52 ft²/s to 1.50 ft²/s, comparable to the values used in the James Estuary for the Falling Creek plant (see Section 8.3.2) A scale factor of 0.0031 is entered to convert all lateral dispersion coefficient values in ft²/s to mile²/day. As mentioned earlier, advective flows entering a given segment are negative (see Table 8-2). Since the areas and lengths are entered in ft² and ft, respectively,

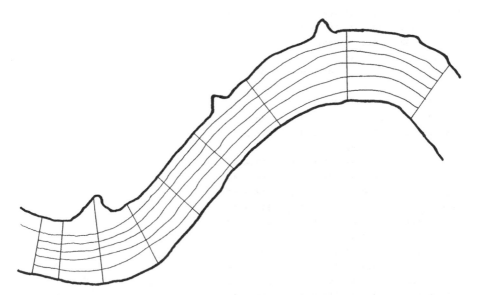

Figure 8-5   HAR03 model segmentation for mixing zone modeling of Lynchburg Plant.

TABLE 8-2   Input Data for Modeling Rhodamine WT Red Dye

James River Calibration 1/14/89						48	1	1. 0.0031		1.	1.
32.4	0.0	-25.9	1	35.5 0.73	11.6	2	37.3	0.0	29.8	7	
32.4	0.0	-20.0	2 116.8 0.73		1.0	3	39.5	0.0	30.6	8	
197.2	0.0	-249.6	3 209.7 0.73		0.	4	197.2	0.0	250.6	9	
222.2	0.0	-414.8	4 217.2 0.73		0.	5	222.2	0.0	414.8	10	
318.3	0.0	-528.5	5 168.0 0.73		0.	6	318.3	0.0	528.5	11	
185.3	0.0	-270.7	6 185.3 0.00		270.7 12						
412.1	0.71	-5.0	8 46.4 0.00		34.8 13						
983.3	0.71	-12.0	9 50.0 0.0		37.6 14						
1715.0	0.71	-5.0	10 150.0 0.0		243.6 15						
1766.9	0.71	0.0	11 214.5 0.0		409.8 16						
1152.4	0.71	0.0	12 355.0 0.0		528.5 17						
222.3	0.0	270.7 18									
458.4	0.67	0.0	14 46.4 0.00		34.8 19						
1015.9	0.67	0.0	15 50.0 0.0		37.6 20						
1707.3	0.67	0.0	16 201.5 0.0		243.6 21						
1679.7	0.67	0.0	17 214.5 0.0		409.8 22						
1052.6	0.67	0.0	18 355.0 0.0		528.5 23						
222.3	0.0	270.7 24									
458.4	0.58	0.0	20 46.4 0.0		34.8 25						
1136.9	0.58	0.0	21 50.0 0.0		37.6 26						
1695.4	0.58	0.0	22 187.5 0.0		243.6 27						
1458.6	0.58	0.0	23 150.0 0.0		409.8 28						
1054.9	0.58	0.0	24 315.1 0.0		528.5 29						
282.1	0.0	270.7 30									
523.8	0.52	-22.0	26 75.0 0.0		56.8 31						
1009.4	0.52	-44.0	27 75.0 0.0		59.6 32						
1407.8	0.52	-44.0	28 162.5 0.0		243.6 33						
1224.0	0.52	-20.0	29 162.5 0.0		385.8 34						
962.6	0.52	0.0	30 285.0 0.0		508.5 35						
170.5	0.0	270.7 36									
876.0	0.62	-34.0	32 100.0 0.0		90.8 37						
1405.0	0.62	-130.0	33 115.0 0.0		155.6 38						
1762.5	0.62	-130.0	34 221.0 0.0		243.6 39						
1533.9	0.62	-90.0	35 159.0 0.0		345.8 40						
1281.0	0.62	0.0	36 258.2 0.0		418.5 41						
218.2	0.0	270.7 42									
1210.0	1.43	0.0	38 100.0 0.0		90.8 43						
1863.0	1.43	0.0	39 125.0 0.0		155.6 44						
2450.8	1.43	0.0	40 187.2 0.0		243.6 45						
2176.5	1.43	0.0	41 230.2 0.0		345.8 46						
1846.2	1.43	0.0	42 305.5 0.0		418.5 47						
395.9	0.0	270.7 48									
1622.0	1.50	0.0	44 100.0 0.0		90.8 43						
2583.0	1.50	0.0	45 155.6 0.0		155.6 44						
3364.0	1.50	0.0	46 250.0 0.0		243.6 45						
3281.4	1.50	0.0	47 212.5 0.0		345.8 46						

TABLE 8-2 (*Continued*)

```
2603.9 1.50    0.0 48 400.0 0.0    418.5 47
450.0   0.0   270.7 48
   1   50.0     2   50.0      7   268.8
   1   50.0     2   50.0      3   50.0      8   256.3
   2   50.0     3   50.0      4   50.0      9   246.9
   3   50.0     4   50.0      5   62.5     10   238.1
   4   62.5     5   50.0      6   75.0     11   216.3
   5   75.0     6   50.0     12  187.5
   1  268.8     8   50.0     13  487.5
   7   50.0     2  256.3      9   50.0     14   462.5
   8   50.0     3  246.9     10   50.0     15   443.8
   9   50.0     4  238.1     11   65.6     16   426.3
  10   65.6     5  216.3     12   82.5     17   382.5
  11   82.5     6  187.5     18  325.0
   7  487.5    14   50.0     19  483.8
  13   50.0     8  462.5     15   50.0     20   466.3
  14   50.0     9  443.8     16   50.0     21   450.6
  15   50.0    10  426.3     17   71.9     22   435.6
  16   71.9    11  382.5     18   97.5     23   405.0
  17   97.5    12  325.0     24  361.2
  13  483.8    20   50.0     25  452.5
  19   50.0    14  466.3     21   50.0     26   447.5
  20   50.0    15  450.6     22   50.0     27   441.3
  21   50.0    16  435.6     23   82.5     28   435.0
  22   82.5    17  405.0     24  117.5     29   426.3
  23  117.5    18  361.2     30  411.3
  19  452.5    26   50.0     31  451.3
  25   50.0    20  447.5     27   50.0     32   455.0
  26   50.0    21  441.3     28   50.0     33   460.0
  27   50.0    22  435.0     29   82.5     34   463.8
  28   82.5    23  426.3     30  116.3     35   470.0
  29  116.3    24  411.3     36  485.0
  25  451.3    32   50.0     37  507.5
  31   50.0    26  455.0     33   50.0     38   523.8
  32   50.0    27  460.0     34   50.0     39   545.0
  33   50.0    28  463.8     35   72.5     40   563.8
  34   72.5    29  470.0     36   96.3     41   591.3
  35   96.3    30  485.0     42  652.5
  31  507.5    38   50.0     43  591.3
  37   50.0    32  523.8     39   50.0     44   630.0
  38   50.0    33  545.0     40   50.0     45   668.8
  39   50.0    34  563.8     41   72.5     46   710.0
  40   72.5    35  591.3     42  115.0     47   771.3
  41  115.0    36  652.5     48  903.8
  37  591.3    44   50.0     43  645.0
  43   50.0    38  630.0     45   50.0     44   697.5
  44   50.0    39  668.8     46   50.0     45   742.5
```

(*continues*)

TABLE 8-2 *(Continued)*

45	50.0	40	710.0	47	75.0	46	795.0
46	75.0	41	771.3	48	151.3	47	875.0
47	151.3	42	903.8	48	1047.5		
0.75	0.79	3.90	4.44	4.24	2.47	0.75	1.00
4.03	4.29	3.55	2.46	0.75	1.00	4.03	4.29
3.55	2.46	0.75	1.00	3.41	3.37	2.91	2.10
0.75	1.00	2.87	2.66	2.58	1.93	0.75	1.00
3.69	3.02	2.82	1.99	0.75	1.00	4.09	3.89
2.97	2.27	0.75	1.00	3.66	4.00	3.43	1.84
15.	0.002021	15.	0.004130	15.	0.009860	15.	0.011110
15.	0.015915	15.	0.009265	15.	0.061350	15.	0.153409
15.	0.176920	15.	0.186141	15.	0.257533	15.	0.132466
15.	0.015915	15.	0.009265	15.	0.061350	15.	0.153409
15.	0.176920	15.	0.186141	15.	0.257533	15.	0.132466
15.	0.015915	15.	0.009265	15.	0.061350	15.	0.153409
15.	0.002021	15.	0.004130	15.	0.009860	15.	0.011110
15.	0.015915	15.	0.009265	15.	0.061350	15.	0.153409
15.	0.176920	15.	0.186141	15.	0.257533	15.	0.132466
15.	0.015915	15.	0.009265	15.	0.061350	15.	0.153409
15.	0.176920	15.	0.186141	15.	0.257533	15.	0.132466
15.	0.015915	15.	0.009265	15.	0.061350	15.	0.153409

```
 0
-1
 6
0.40      1   0.40    2    0.40   3   0.40    4   0.40    5   0.40    6
1.      0.0
3336 0.0     1

   1
/*
```

the scale factors for these two parameters are 1.0 (see the first line of the input data file). Also note that each segment takes two lines to enter the hydraulic geometry data. A blank line is needed for the second line of each segment.

The depths and temperatures are not needed in the mass transport modeling of the Rhodamine WT dye. Their being entered simply affects the completeness of the model input even though the values are not used by the model. Segments 1 to 6 are the upstream segments and therefore require the specification of a boundary concentration. In this case, 0.40 µg/L is used to reflect a nonzero but very minimal background level of dye. The dye concentration of 400 µg/L in the Lynchburg Plant effluent is multiplied with the wastewater flow rate (10 mgd) to generate a point source dye loading rate of 33.36 lb/day, entering segment No.1 (see the 4th line from the end of the data file in Table 8-2). Note that the point source load entered is actually 33,360, so that the calculated concentration values are in µg/L. Also, the plant

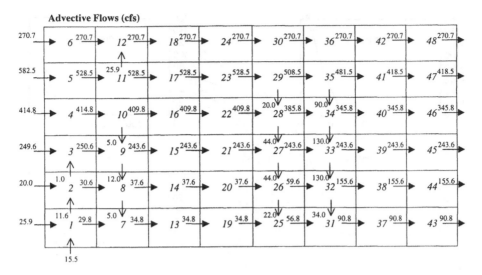

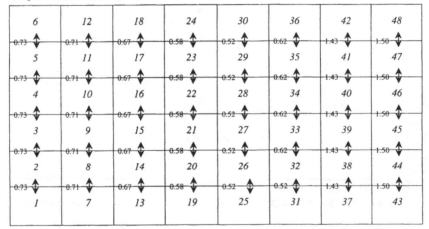

Note: Segment Numbers in *italic*

Figure 8-6 Advective flows and lateral dispersion coefficients in the James River at Lynchburg, Virginia.

flow of 10 mgd (= 15.5 cfs) is not needed in the input data file because the Rhodamine WT dye loading rate is entered. The model output message indicates a flow unbalance of 15.5 cfs, which is exactly the plant flow.

Figure 8-7 shows model calculated dye concentrations versus data at all eight transects. In general, the dye concentration gradually reduces from Transect 1 to Transect 8 along the river. At each transect, high concentrations remain close to the shore, with a sharp lateral concentration gradient across the river. The model is able to mimic the lateral concentration profiles at all transects, suggesting that the mass transport coefficients (advective flows and lateral dispersion coefficients) are correct

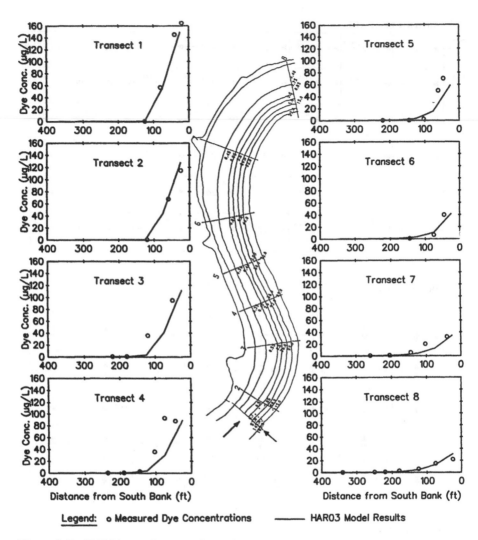

Figure 8-7 HAR03 model results of lateral Rhodamine WT dye concentration profiles versus data in the James River.

in reproducing the two-dimensional concentration distribution in the James River. Figure 8-8 presents the longitudinal concentration profiles in the first three tubes (125 ft, 75 ft, and 25 ft from the shore) in the model compared with the data. Again, the results demonstrate that the mass transport model results match the data quite well. The calibrated mass transport pattern is then used to simulate the residual chlorine concentrations in the James River. Figures 8-9 and 8-10 show the lateral and longitudinal concentration profiles versus data. Note that the plant effluent residual chlorine concentration is 0.42 mg/L and a decay rate of 4.733 day^{-1} is used for chlorine in the receiving water.

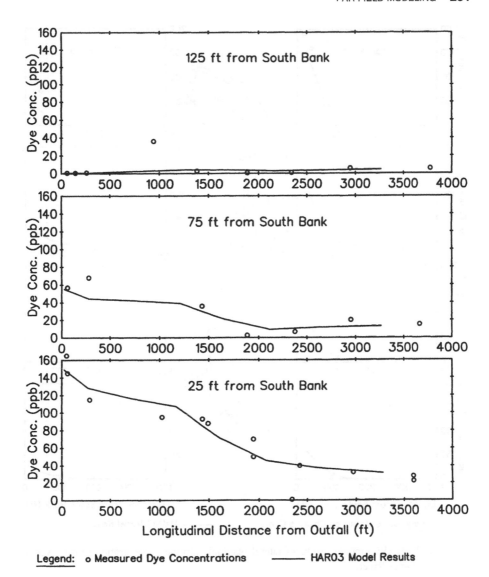

Figure 8-8 Longitudinal concentration profiles of Rhodamine WT dye in the James River: HAR03 model results versus data.

8.5 FAR-FIELD MODELING

8.5.1 Quantifying Mass Transport in Stratified Embayments and Estuaries

Following the *Exxon Valdez* oil spill in March 1989, fertilizers were sprayed on the contaminated beach in the summer of 1989 to accelerate bacterial growth in a bioremediation effort to clean up the beaches. As a first step, a field demonstration project

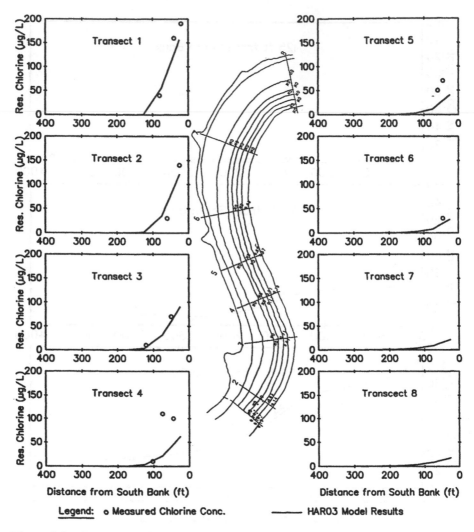

Figure 8-9 HAR03 model results of lateral residual chlorine concentration profiles versus data in the James River.

was conducted to determine whether nutrient (in the fertilizer) addition to contaminated beaches would stimulate hydrocarbon breakdown by indigenous bacteria (McCutcheon, 1989). One concern associated with nutrient application was whether the added nutrients would cause excessive algal growth during the growing season in the study area. The federal and state agencies on the Shoreline Committee raised the concern during the bioremediation experiments planning and later when Exxon proposed a wide-scale application of nutrients. Another concern was whether there would be an adverse long-term quality impact after the fertilizer application was completed and discontinued.

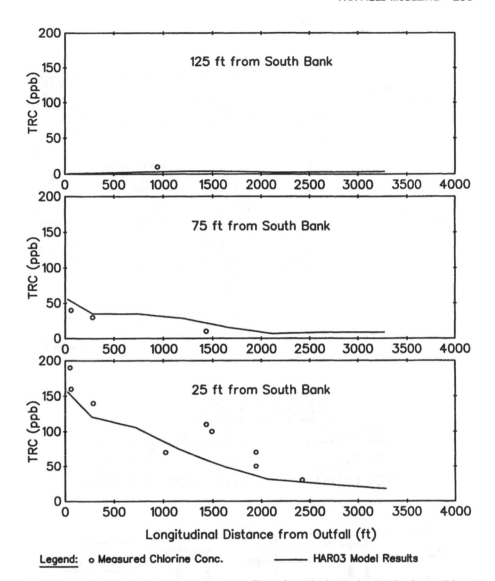

Figure 8-10 Longitudinal concentration profiles of residual chlorine in the James River: HAR03 model results versus data.

One of the studied sites is Snug Harbor, Alaska, located on the southeastern side of Knight Island (Figure 8-11). Major sources of freshwater runoff in this area are from precipitation and snowmelt, typical of islands in Prince William Sound. Although some shorelines in Snug Harbor were heavily contaminated with oil, it appeared that little oil was being released to the water, thus minimizing the prospect of reoiling on the beaches chosen for treatment.

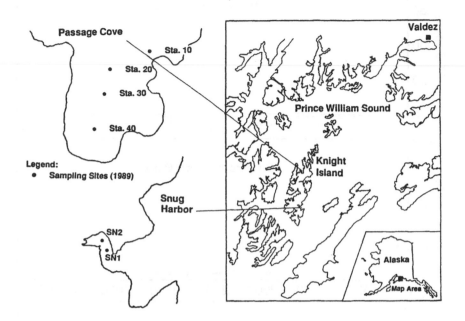

SNUG HARBOR

SEG. NO.	1	2	3	4
DEPTH (m)	10.0	11.0	10.0	36.0
VOL. ($10^6 m^3$)	3.21	7.20	16.2	116.8

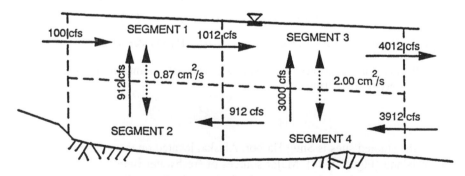

Figure 8-11   Snug Harbor, Alaska, and two-layer mass transport under tidally averaged conditions.

Mixing and circulation play a major role in determining the eutrophication potential of these small embayments. Flushing time and residence time of a conservative substance are good indicators of how long a contaminant will remain in the embayment. Two methods were utilized to determine the flushing times: the fraction of freshwater and the tidal prism methods (Dyer, 1973). Table 8-3 shows the results using these two methods to calculate the Snug Harbor flushing time. Based on the small freshwater inflows, the flushing time is long (slow flushing rates). On the other hand, strong tides provide efficient mixing, resulting in a very short flushing time. Note that salinity distributions in Snug Harbor suggest incomplete mixing in the vertical direction. Thus the results in Table 8-3 serve as a bound analysis for mixing in this embayment system. It is expected that the actual mixing rate falls between these bounds and must be quantified using a more rigorous analysis.

The methodology developed by Lung and O'Connor (1984) is used to calculate the two-layer mass transport in Snug Harbor. The water column in Snug Harbor is divided into four segments in a two-layer configuration (Figure 8-11). Also shown in Figure 8-11 are the transport coefficients quantified by the Lung and O'Connor method. They are incorporated into the mass transport model using HAR03.

The HAR03 modeling framework is also configured to form a two-layer mass transport model for the Patuxent Estuary, in Maryland. Along the 100-km length of the estuary, each layer in the water column is divided into 19 segments (see Figure 7-14 for the segmentation map). The freshwater flow at the head of the estuary (approximately 100 km from the river mouth) is 255 cfs. The mass transport coefficients, including the advective flows and vertical dispersion coefficients, are derived using the Lung and O'Connor (1984) method. Results of the two-layer model simulation are presented in Figure 8-12, showing model calculated salinity levels versus data for both layers. This moderate freshwater flow pushes the salinity profiles in the downstream direction, resulting in a salinity intrusion ended at approximately 55 km from the mouth.

TABLE 8-3 Flushing Time Calculation for Snug Harbor, Alaska

Parameter	Snug Harbor
Tidally averaged volume (m^3/s)	143.421×10^6
Tidal range (m)	4.57
Surface area (m^2)	5.961×10^6
Tidal prism (m^3)	27.263×10^6
Low tide volume (m^3)	129.790×10^6
Freshwater flow (m^3/tidal cycle)	127,932.8
Open sea salinity, S_i (ppt)	32.3
Salinity in embayment, S (ppt)	30.0
Flushing time (tidal cycles[a]):	
Fraction of freshwater method	8974
Tidal prism method	5.76

[a]1 tidal cycle = 12.54 hours.

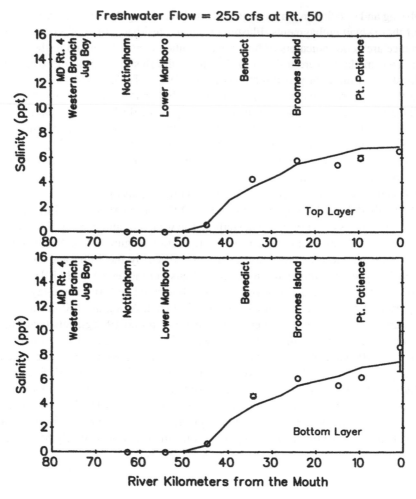

Figure 8-12 Quantifying two-layer mass transport in the Patuxent Estuary by simulating salinity concentrations.

8.5.2 Modeling Consumer Product Chemicals in Estuaries

Estuaries throughout the United States receive chemical inputs from a variety of sources. In particular, about 25 to 30% of all U.S. publicly owned treatment works (POTW) discharges enter estuaries and coastal waters. The fate and transport of these chemicals in a receiving estuary is an important aspect in the overall exposure analysis in the environment. Processes affecting the fate and transport of consumer product chemicals include mass transport, biodegradation, photolysis, volatilization, sorption, and subsequent settling/sedimentation.

To predict the fate and transport of the chemicals and to quantify the chemical exposure levels in estuaries, a modeling framework called ESTUARY has been developed in which several of these processes are incorporated with physical characteristics such as estuarine circulation and mass transport. The modeling development was conducted by Lung et al. (1990) with support from the Procter & Gamble Co.

In general, usage rates of consumer product chemicals were relatively stable and supported by over 10 years of influent wastewater monitoring data by the Procter & Gamble Co. (Rapaport and Eckhoff, 1990). These data suggested that tidally averaged mass transport patterns using a steady-state fate-transport solution are appropriate for examining the exposure levels of these chemicals. The steady-state mass transport was analyzed under three flow conditions: 7-day 10-year low flow, summer low flow, and annual mean flow. To account for varying mixing characteristics and salinity distributions observed in a given estuary, the modeling framework includes one-dimensional and two-dimensional (either longitudinal/vertical or horizontal) mass transport algorithms. To date, the modeling framework has been applied to 17 estuaries.

In addition to mass transport, loss of chemicals from the water column (e.g., by biodegradation) is accounted for by a first-order kinetics algorithm. Each POTW is classified as one of four groups: primary, trickling filter, lagoon, and activated sludge in terms of treatment level. For validation purposes, BOD concentrations are also predicted based on industrial as well as municipal organic carbon loads. In several estuaries, predicted in-stream BOD levels have been compared to measured BOD data, resulting in successful model validation. The validation models have been used to accurately predict linear alkylbenzene sulfonate (LAS) concentrations in estuaries (Lung et al., 1990).

One of the 17 estuaries modeled was Boston Harbor, where the HAR03 model was applied in a two-dimensional (horizontal) configuration. A hydrodynamic model of the Boston Harbor by Hydroscience (1973) was used to develop the circulation pattern. The hydrodynamic model results were processed to derive the advective flows under a steady-state condition. Figure 8-13 shows model results of fecal coliform using the mass transport model under tidally averaged conditions.

8.5.3 Modeling Color of Lake Martin Using HAR03

Russell Corporation and Avondale Mills discharge textile dyeing and finishing wastewater to the Sugar Creek Plant located in Alexander City, Alabama. The major flows to the Plant are from Russell Corporation (4 mgd) and Avondale Mills (1 mgd). The wastewater is biologically treated with the extended aeration process and the plant effluent is discharged into Sugar Creek, which then flows into Lake Martin. Although significant color reduction is achieved by the treatment process, relatively high levels of color persist in the discharge to Sugar Creek in the range of 800 to 1000 American Dye Manufacturer's Institute (ADMI) color units. The highly colored

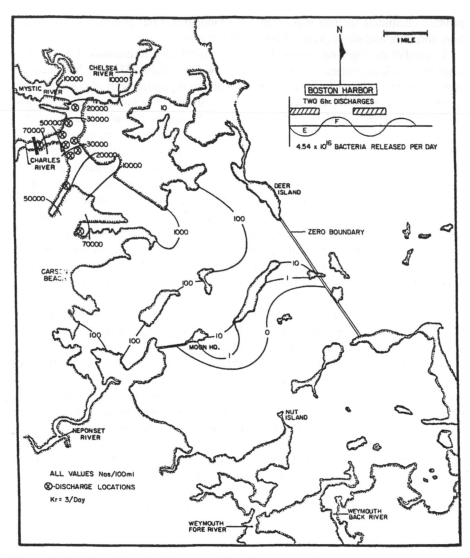

Figure 8-13 Two-dimensional segmentation of Boston Harbor and mass transport model results.

discharge produces visible color in both Sugar Creek and some parts of Lake Martin. Due to public concerns regarding the aesthetic quality of Lake Martin with respect to visible color, the Alabama Department of Environment Management (ADEM) indicated that color limits would be established on the effluent from the Sugar Creek Plant. A three-dimensional mass transport model was developed to quantify the advective and dispersive transports affecting the fate and transport of colored effluent from the Sugar Creek Plant through the backwaters of Lake Martin.

Lake Martin is located in Tallapoosa County, Alabama, with branches extending into Coosa and Elmore counties. The northern end lies between Alexander City and Dadeville, and the southernmost branch ends about 5 miles from Eclectic (Figure 8-14). The lake was formed in 1926 when a dam was built across the Tallapoosa River, flooding several creek channels and a few roadbeds. The water is used primarily for recreation (boating and fishing), and is classified as swimming water by ADEM. The lake is quite large, with depths up to 155 ft at the dam, covering 44,000 acres, and draining approximately 2 million acres. The primary tributary to the lake is the Tallapoosa River. The pool elevation varies seasonally, with lake drawn down in winter to provide flood control storage volume. Due to the relatively small flow in Sugar Creek, very little dilution of the effluent occurs prior to the confluence with Lake Martin. The color in Sugar Creek as it enters Lake Martin is thus comparable to the color in the plant.

Color limits are becoming increasingly prevalent on effluent discharge permits for both pretreatment and direct dischargers within the textile industry. Although federal guidelines have not established specific color limits, permit writers at the state level are addressing aesthetic concerns by establishing discharge color levels based on water quality standards. ADEM establishes water quality standards for color based on water quality criteria and water use. Current objective color limits established by ADEM are 300 ADMI for agricultural and industrial waters and 80 ADMI for fish and wildlife waters. The primary objective of this modeling study was to estimate color levels within Lake Martin as a result of a 300 ADMI discharge from the Sugar Creek Plant and to define the mixing zone within Lake Martin.

Field data to support the modeling analysis were collected from eight water quality surveys from October 1990 to May 1991. In each survey, 22 transects were visited for sampling and data collection. Ten of these transects were divided laterally in order to capture cross-sectional variations in the parameters being measured. Hydraulic geometry of the study area was determined from the depth measurements taken in the field and from maps of both the shoreline of the lake and of bottom bathymetry. Although color is the parameter modeled in this study, specific conductivity was measured in all samples and color only in selected samples. This was done because there is a very strong correlation between color and conductivity and because taking color measurements is a very time-consuming process. Color was thus indirectly determined in all samples by using the correlation between measured conductivity and color values for each sample set. The specific gravity of the Sugar Creek Plant effluent is higher than that of the receiving water due to the relatively high salt content in the effluent. This causes the wastewater to sink to the lowest elevation in the lake channel, resulting in density stratification in the water column as reflected in the color and conductivity data.

Figure 8-14 shows the measured conductivity levels in the upper branch of Martin Lake, strongly indicating significantly higher concentrations in the bottom waters of the reservoir. The vertical gradients of conductivity levels are strong throughout the upper branch. Much attenuation of the vertical gradients is achieved through the upper branch and the gradients gradually diminish in the lower branch, more or less reaching the background conductivity level (Figure 8-15). However, in the Raintree area (station 13), the bottom water still shows high conductivity (and color) levels.

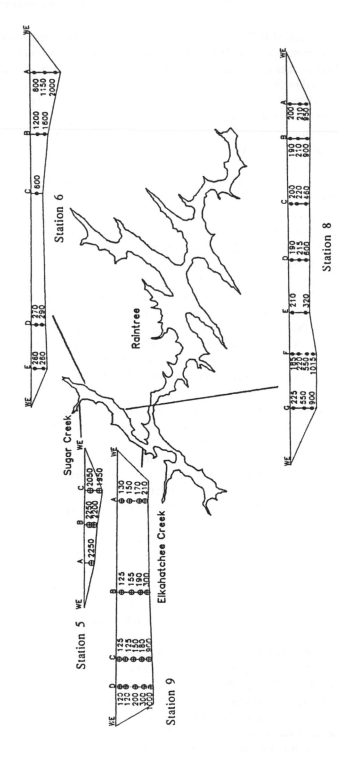

Figure 8-14 Measured specific conductivity (μmho/cm) in upper branch of Martin Lake.

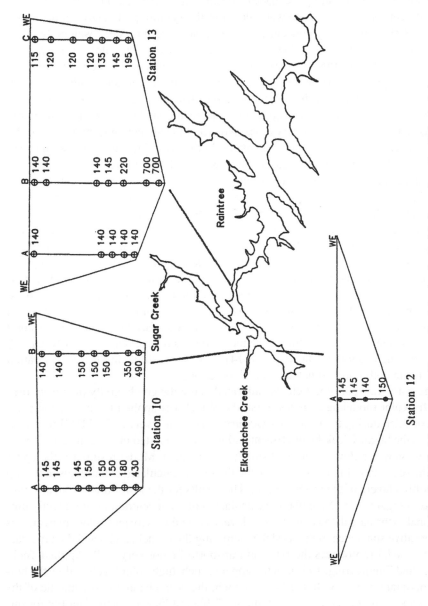

Figure 8-15 Measured specific conductivity (μmho/cm) in lower branch of Martin Lake.

To model the conductivity and color in Martin Lake, the water column is divided into 60 horizontal segments. With three layers in the water column, the total number of segments in the system is 180. In lieu of a hydrodynamic model, initial estimates of the advective flow routing are made by observing the correlation between the conductivity concentration and the water depth in the system, particularly in the upper branch. High conductivity concentrations are associated with deep waters. In addition, a photograph taken in 1963 shows the dense flow (with color) entering from Sugar Creek, crossing the sand berm to reach the deep channel on the right-hand side of the upper branch. The dense flow stays in the deep trough and gradually moves from right to left, due to the bottom topographic changes. Eventually, the dense, high conductivity water reaches the deep channel on the left-hand side. The advective flow was routed based on these observations. Note that the advective flow pattern in the bottom layer follows the bottom bathymetry. In the lower branch, a narrow and deep channel exists and receives the advective flow in the bottom layer.

Model calculated and measured specific conductivity concentrations for the upper branch are shown in Figure 8-16. The model calculation indicates that the high concentration in the surface layer rapidly decreases in the area adjacent to the Sugar Creek entrance due to sinking into the lower layers. Further, very little lateral concentration gradient (from one bank to the other) is shown, which is consistent with the field observations. In fact, the longitudinal concentration gradients are only significant in the immediate vicinity of the inflow and become much less further downstream from the input. The conductivity concentration in the surface layer almost reaches background levels at Station 9 (see the location in Figure 8-14). The model results for the middle layer are similar to those for the surface layer, with only a slight increase in overall concentrations. The model results for the bottom layer (Figure 8-16) match the observed data (Figure 8-14) very well, mimicking the concentration gradients in both longitudinal and lateral directions.

The mass transport coefficients calibrated using the conductivity data were then used for color modeling analysis. First, the model was applied to analyze the color data collected during the October 1990 survey. A color level of 824 ADMI was measured in the Sugar Creek Plant effluent and was incorporated into the model. The average wastewater flow during that survey is 6.17 mgd. Model results for the upper branch are given in Figure 8-16, showing the model calculated color levels in the top and bottom layers of the water column. The results for the lower branch are also presented in Figure 8-16. Note that the calculated color concentrations are directly proportional to the calculated conductivity levels as conductivity and color are treated as conservative substance in the model. Comparing the model results and data indicate that the model reproduces the fate and transport of color very well (Applied Technology and Engineering, 1991). In the upper branch, high color levels follow the bottom topography closely. In the lower branch, the deep channel in the middle of the branch shows higher color concentrations. Table 8-4 lists the input data file for this modeling analysis.

(Text continues on page 313.)

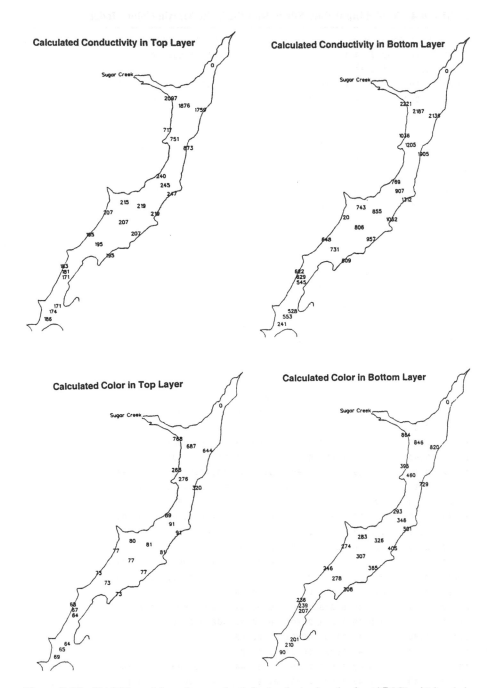

Figure 8-16 HAR03 model results: conductivity (μmho/cm) and color (ADMI unit) levels in upper branch of Martin Lake.

TABLE 8-4   Model Input Data File to Run the Lake Martin Color Model

```
Martin Lake Color Model (3-D)              180   0   4      1000.0 0.0334     1.0      1.0
0.181   0.00    -3.2  1 0.832 0.10     2.1    2 0.215 0.001   0.0    5
92.51   0.0      1.1 61
0.832   0.10     1.6  3 0.372 0.001   0.0     6 83.03   0.0    0.5   62
0.455   0.009    0.0  4 0.383 0.10    1.1     7118.57   0.0    0.5   63
195.8   0.0      0.0 64
1.306   0.05     0.00  6 0.215 0.10    0.00    8 45.98   0.0   0.00   65
0.924   0.10     0.0  7 0.372 0.10    0.00    9 79.77   0.0   0.00   66
0.268   0.10     0.75 10 82.00 0.00    0.35   67
1.646   0.12     0.00  9 0.324 2.50    0.00  11 93.51   0.0   0.00   68
1.646   0.12     0.00 10 0.415 2.50    0.00  12 119.6   0.0   0.00   69
77.508  0.00     0.25 70 0.355 2.50    0.50  13
1.017   0.12    -0.10 12 0.543 2.5     0.07  14 125.6   0.0    0.03  71
1.017   0.12    -0.12 13 0.503 2.5     0.00  15 67.68   0.0    0.02  72
0.428   2.50     0.21 16 57.58 0.0     0.17  73
1.405   0.12    -0.05 15 0.472 2.5     0.10  17 83.52   0.0   0.02   74
1.405   0.12    -0.07 16 0.596 2.5     0.00  18 98.61   0.0   0.02   75
0.552   2.50     0.07 19 158.1 0.0     0.07  76
1.612   0.12    -0.04 18 0.472 2.5     0.11  20 68.82   0.0   0.03   77
1.612   0.12    -0.05 19 0.525 2.5     0.00  21 86.89   0.0   0.01   78
0.288   2.50     0.0 22 80.60 0.0     0.02   79
2.028   0.12     0.0 21 0.715 2.5     0.08  24 108.5   0.0   0.03   80
2.028   0.12     0.0 22 0.525 2.5     0.00  23 79.75   0.0   0.00   81
0.288   2.50     0.0 23 0.288 2.5      0.0  26 43.79   0.0    0.0   82
1.971   0.10     0.00 24 0.947 2.5      0.0  26 93.03.00001   0.0   83
1.971   0.10     0.00 25 0.947 2.5     0.08  27 93.03.00001   0.0   84
93.03 .00001     0.00 85
3.092   0.10     0.00 27 1.187 2.50    0.00  28 85.75.00001   0.0   86
1.781   2.50     0.08 29 1.055 0.10    0.00  54 185.2.00001   0.0   87
3.092   0.10     0.00 29 1.187 2.50    0.00  31 85.75.00001   0.0   88
3.092   0.10     0.00 30 0.291.00001   0.08  32 85.75.00001   0.0   89
1.187   2.50     0.00 33 85.75.00001   0.00  90 1.187 2.50    0.0   55
8.138   0.10     0.00 32 1.307 2.50    0.00  34 124.4.00001   0.0   91
8.138   0.10     0.00 33 0.291 2.50    0.08  35 27.65.00001   0.0   92
1.307   2.50     0.0 36 124.4.00001   0.00  93
20.29   0.10     0.0 35 3.095 2.50    0.00  37 494.3.00001   0.0   94
20.29   0.10     0.0 36 0.688 2.50    0.08  38 109.8.00001   0.0   95
3.095   2.50     0.0 39 494.3.00001    0.0  96
23.77   0.10     0.0 38 8.205 2.50    0.00  41 1390..00001   0.0   97
23.77   0.10     0.0 39 1.823 2.50    0.08  42 308.9.00001   0.0   98
8.205   2.50     0.0 43 1390..00001   0.00  99
16.53   0.10     0.0 41 2502..00001    0.0 100
15.69   0.10     0.0 42 8.532 2.50     0.0  44 1370..00001   0.0 101
15.69   0.10     0.0 43 1.896 2.50    0.08  45 304.5.00001   0.0 102
8.532   2.50     0.0 46 1370..00001    0.0 103
35.19   0.10     0.0 45 8.532 2.50     0.0  47 1152..00001   0.0 104
35.19   0.10     0.0 46 1.896 2.50    0.08  48 256.0.00001   0.0 105
```

TABLE 8-4 (*Continued*)

```
8.532   2.50    0.0 49 1152..00001   0.0 106
35.19   0.10    0.0 48 5.710 2.50    0.0  50 1152..00001   0.0 107
35.19   0.10    0.0 49 1.269 2.50    0.08 51 256.0.00001   0.0 108
5.710   2.50    0.0 52 1152..00001   0.0 109
5.710   2.50    0.0 50 625.0.00001   0.0 110
1.269   2.50    0.08 51 138.9.00001  0.0 111
5.710   2.50    0.0 52 625.0.00001   0.0 112
39.61   0.10    0.0 54 4.747 2.50    0.0  56 348.4.00001   0.0 113
12.35   0.10    0.0 59
39.61   0.10    0.0 55 1.055 2.50    0.0  57 77.42.00001   0.0 114
4.747   2.50    0.0 58 348.4.00001   0.0 115
7.596   0.00    0.0 56 422.6.00001   0.0 116
1.688   0.00    0.0 57 93.92.00001   0.0 117
7.596   0.00    0.0 58 422.6.00001   0.0 118
10.80   2.50    0.0 60 210.0.00001   0.0 119
8.096   0.50    0.0 60 509.6.00001   0.0 120
0.181   0.00   -3.20 61 0.832 0.10    3.3 62 0.215  0.00   0.0  65
92.52   0.00    1.0 121
0.832   0.10    3.5 63 0.373 0.001    0.0 66 83.02 .0000   0.3 122
0.455   0.00    0.0 64 0.383 0.10    2.8 67 118.5 .0000   1.2 123
195.83  .0000   0.0124
1.306   0.10    0.0 66 0.215 0.10     0.0 68 45.98 .0000   0.0 125
0.924   0.10    0.0 67 0.372 0.10    0.00 69 79.77 .0000 0.00 126
0.268   0.10   2.15 70 82.01 .0000   1.0 127
1.646   0.12   0.00 69 0.391 2.5     0.00 71 93.31 .0000 0.00 128
1.646   0.12   0.00 70 0.415 2.5     0.00 72 119.6 .0000 0.00 129
77.51   .0000   0.8 130 0.355 2.5    1.60 73
1.017   0.12  -0.42 72 0.543 2.5     0.30 74 125.6 .0000 0.15 131
1.017   0.12  -0.60 73 0.503 2.5     0.00 75 67.68 .0000 0.20 132
0.428   2.50    0.6 76 57.58 .0000  0.57133
1.405   0.12  -0.18 75 0.472 2.5     0.34 77 83.52 .0000 0.16 134
1.405   0.12  -0.25 76 0.596 2.5     0.00 78 98.61 .0000 0.09 135
0.552   2.50    0.2 79 158.1 .0000  0.22136
1.612   0.12  -0.13 78 0.472 2.5     0.33 80 68.82 .0000 0.17 137
1.612   0.12  -0.15 79 0.525 2.5     0.00 81 86.89 .0000 0.03 138
0.288   2.50    0.0 82 80.60 .0000  0.07139
2.028   0.12    0.0 81 0.715 2.5     0.24 84 108.5 .0000 0.12 140
2.028   0.12    0.0 82 0.525 2.5     0.00 83 79.75 .0000 0.00 141
0.288   2.50    0.0 83 0.288 2.5      0.0 86 43.79 .0000  0.0 142
1.971   0.10    0.0  84 0.947 2.50    0.0  86 93.03.00001 0.00 143
1.971   0.10    0.0  85 0.947 2.50   0.24 87 93.03.00001 0.00 144
93.03   .00001  0.0 145
3.092   0.10    0.0  87 1.187 2.50    0.0  88 85.75.00001 0.00 146
1.187   2.50   0.24 89 1.055 0.10     0.0 114 185.2.00001 0.00 147
3.092   0.10    0.0  89 1.038 2.50    0.0  91 85.75.00001  0.0 148
3.092   0.10    0.0  90 0.291 2.50   0.24  92 85.75.00001 0.0  149
1.187   2.50    0.0  93 85.75.00001   0.0 150 1.187 2.50    0.0 115
```

(*continues*)

TABLE 8-4 (*Continued*)

```
8.138  0.10   0.0  92 1.308 2.50   0.0  94 124.4.00001  0.0 151
8.138  0.10   0.0  93 0.291 2.50  0.24  95 27.65.00001  0.0 152
1.308  2.50   0.0  96 124.4.00001   0.0 153
20.29  0.10   0.0  95 3.095 2.50   0.0  97 494.3.00001  0.0 154
20.29  0.10   0.0  96 0.688 2.50  0.24  98 109.8.00001  0.0 155
3.095  2.50   0.0  99 494.3.00001   0.0 156
23.77  0.10   0.0  98 8.205 2.50   0.0 101 1200..00001  0.0 157
23.77  0.10   0.0  99 1.823 2.50  0.24 102 266.8.00001  0.0 158
8.205  2.50   0.0 103 1200..00001   0.0 159
16.53  0.10   0.0 101 2502..00001   0.0 160
15.69  0.10   0.0 102 8.532 2.50   0.0 104 1370..00001  0.0 161
15.69  0.10   0.0 103 1.896 2.50  0.24 105 304.5.00001  0.0 162
8.532  2.50   0.0 106 1370..00001   0.0 163
35.19  0.10   0.0 105 8.530 2.50   0.0 107 1152..00001  0.0 164
35.19  0.10   0.0 106 1.896 2.50  0.24 108 256.0.00001  0.0 165
8.530  2.50   0.0 109 1152..00001   0.0 166
35.19  0.10   0.0 108 5.710 2.50   0.0 110 1152..00001  0.0 167
35.19  0.10   0.0 109 1.269 2.50  0.24 111 256.0.00001  0.0 168
5.710  0.10   0.0 112 1152..00001   0.0 169
5.710  2.50   0.0 110 625.3.00001   0.0 170
1.269  2.50  0.24 111 138.9.00001   0.0 171
5.710  2.50   0.0 112 625.3.00001   0.0 172
39.61  0.10   0.0 114 4.747 2.50   0.0 116 348.4.00001  0.0 173
12.35  0.10   0.0 119
39.61  0.10   0.0 115 1.055 2.50   0.0 117 77.42.00001  0.0 174
4.747  2.50   0.0 118 348.4.00001   0.0 175
7.596  0.00   0.0 116 422.6.00001   0.0 176
1.688  0.00   0.0 117 93.92.00001   0.0 177
7.596  0.00   0.0 118 422.6.00001   0.0 178
8.096  2.50   0.0 120 210.0.00001   0.0 179
8.096  0.50   0.0 120 509.6.00001   0.0 180
0.181  0.00 -3.20 121 0.645 0.10  4.20 122 0.168   0.00 0.00 125
0.645  0.10   4.5 123 0.269 0.001   0.0 126
0.392  0.00   0.0 124 0.267 0.10   5.7 127
0.615  0.01  0.00 126 0.168 0.10   0.0 128
0.690  0.01  0.00 127 0.261 0.10   0.0 129
0.166  0.10   6.7 130
0.938  0.02  0.00 129 0.271 2.5    0.0 131
1.035  0.02   0.0 130 0.261 2.5    0.0 132
0.222  2.50   7.5 133
0.555  0.02  -4.2 132 0.348 2.5   4.35 134
0.480  0.02  -4.0 133 0.274 2.5    0.0 135
0.270  2.50  4.07 136
0.540  0.02 -2.09 135 0.220 2.5    6.6 137
0.698  0.02 -2.00 136 0.269 2.5    0.0 138
0.239  2.5   2.29 139
```

TABLE 8-4   (*Continued*)

```
0.705   0.02 -2.39 138 0.220 9.5    9.16 140
0.728   0.02 -2.36 139 0.216 2.5     0.0 141
0.134   2.50    0.0 142
0.833   0.20    0.0 141 0.263 4.5    9.28 144
0.735   0.02    0.0 142 0.216 2.5     0.0 143
0.134   2.50    0.0 143 0.134 2.5     0.0 146
0.660   0.10    0.0 144 0.272 2.5     0.0 146
0.660   0.00    0.0 145 0.317 2.5    9.28 147
0.709   9.10    0.0 147 0.272 2.5     0.0 148
0.272   2.50   9.28 149 0.068 2.5     0.0 174
0.709   9.03    0.0 149 0.317 0.2     0.0 151
0.709   0.03    0.0 150 0.047 45.    9.28 152
0.317   2.50    0.0 153 0.306 2.5     0.0 175
1.320   0.00    0.0 152 0.212 2.5     0.0 154
1.320   0.00    0.0 153 0.047 4.5    9.28 155
0.212   2.50    0.0 156
2.700   0.00    0.0 155 0.412 2.5     0.0 157
2.700   0.00    0.0 156 0.092 4.5    12.0 158
0.412   2.50    0.0 159
2.798   0.00    0.0 158 0.966 2.5     0.0 161
2.798   0.00    0.0 159 0.215 4.5    12.0 162
0.966   2.50    0.0 163
1.192   0.10    0.0 161
1.688   0.00    0.0 162 0.792 2.5     0.0 164
1.688   0.00    0.0 163 0.176 4.5    12.5 165
0.792   2.50    0.0 166
3.270   0.00    0.0 165 0.762 2.5     0.0 167
3.270   0.00    0.0 166 0.176 4.5    12.5 168
0.762   2.50    0.0 169
3.270   0.00    0.0 168 0.762 2.5     0.0 170
3.270   0.00    0.0 169 0.176 4.5    13.5 171
0.762   2.50    0.0 172
0.458   2.50    0.0 170 3.075 0.0     0.0 171
0.102   2.50   13.5 171 3.075 0.0     0.0 172
0.458   2.50    0.0 172
2.558   2.10    0.0 174 0.306 0.5     0.0 176 0.735  4.5  0.0  179
2.558   0.01    0.0 175 0.068 0.5     0.0 177
0.306   0.50    0.0 178
0.466   0.00    0.0 176
0.104   0.00    0.0 177
0.466   0.00    0.0 178
0.466   2.50    0.0 180
0.466   0.50    0.0 180
   1 430.0       2 430.0      5  215.0     61  1.9
   3 193.0       6 430.0     62    1.9
   4 430.0       7 430.0     63    1.9
```

(continues)

TABLE 8-4 (*Continued*)

64	1.7					
6	103.0	8	445.0	65	2.1	
7	179.0	9	445.0	66	2.1	
10	445.0	67	2.1			
9	164.0	11	690.0	68	2.4	
10	174.0	12	690.0	69	2.4	
70	2.4	13	690.0			
12	340.0	14	370.0	71	2.7	
13	183.0	15	370.0	72	2.7	
16	370.0	73	2.7			
15	180.0	17	465.0	74	3.0	
16	212.0	18	465.0	75	3.0	
19	465.0	76	3.0			
18	142.0	20	485.0	77	3.3	
19	179.0	21	485.0	78	3.3	
22	485.0	79	3.3			
21	195.0	24	555.0	80	3.7	
22	144.0	23	555.0	81	3.7	
23	555.0	26	79.0	82	3.7	
24	211.3	26	293.5	83	4.5	
25	211.3	27	440.0	84	4.5	
85	4.5					
27	181.3	28	472.5	86	6.5	
29	472.5	54	362.8	87	6.5	
29	181.3	31	472.5	88	6.5	
30	181.3	32	472.5	89	6.5	
33	472.5	55	472.5	90	6.5	
32	141.3	34	880.0	91	9.2	
33	31.4	35	880.0	92	9.2	
36	880.0	93	9.2			
35	274.5	37	1800.0	94	11.3	
36	61.0	38	1800.0	95	11.3	
39	1800.0	96	11.3			
38	643.9	41	1865.0	97	12.7	
39	143.1	42	1865.0	98	12.7	
43	1865.0	99	12.7			
41	3150.0	100	20.8			
42	1218.6	44	1125.0	101	13.9	
43	270.1	45	1125.0	102	13.9	
46	1125.0	103	13.9			
45	528.3	47	2180.0	104	16.1	
46	117.4	48	2180.0	105	16.1	
49	2180.0	106	16.1			
48	528.3	50	2180.0	107	16.1	
49	117.4	51	2180.0	108	16.1	
52	2180.0	109	16.1			
50	2050.0	110	18.7	51	305.1	

TABLE 8-4 (*Continued*)

51	2050.0	111	18.7	52	67.8		
52	2050.0	112	18.7				
54	204.3	56	1705.0	113	23.2	59	204.3
55	45.4	57	1705.0	114	23.2		
58	1705.0	115	23.2				
57	310.9	56	1360.0	116	24.4		
58	69.1	57	1360.0	117	24.4		
58	1360.0	118	24.4				
60	490.0	119	25.2				
60	1640.0	120	25.2				
61	430.0	62	430.0	65	215.0	121	1.9
63	193.0	66	430.0	122	1.9		
64	430.0	67	430.0	123	1.9		
124	1.7						
66	112.0	68	410.0	125	2.1		
67	179.0	69	445.0	126	2.1		
70	445.0	127	2.1				
69	181.0	71	690.0	128	2.4		
70	174.0	72	690.0	129	2.4		
130	2.4	73	690.0				
72	340.0	74	370.0	131	2.7		
73	183.0	75	370.0	132	2.7		
76	370.0	133	2.7				
75	180.0	77	465.0	134	3.0		
76	212.0	78	465.0	135	3.0		
79	465.0	136	3.0				
78	142.0	80	485.0	137	3.3		
79	179.0	81	485.0	138	3.3		
82	485.0	139	3.3				
81	195.0	84	555.0	140	3.7		
82	144.0	83	555.0	141	3.7		
83	555.0	86	79.0	142	3.7		
84	211.3	86	293.5	143	4.5		
85	211.3	87	440.0	144	4.5		
145	4.5						
87	181.3	88	472.5	146	6.5		
89	472.5	114	362.8	147	6.5		
89	181.3	91	472.5	148	6.5		
90	181.3	92	472.5	149	6.5		
93	472.5	115	472.5	150	6.5		
92	141.3	94	880.0	151	9.2		
93	31.4	95	880.0	152	9.2		
96	880.0	153	9.2				
95	275.5	97	1800.0	154	11.3		
96	61.0	98	1800.0	155	11.3		
99	275.5	156	11.3				
98	644.0	101	1865.0	157	12.7		

(*continues*)

TABLE 8-4 (*Continued*)

99	143.1	102	1865.0	158	12.7	
103	1800.0	159	12.7			
101	3150.0	160	20.8			
102	1218.6	104	1129.0	161	13.9	
103	270.8	105	1129.0	162	13.9	
106	1129.0	163	13.9			
105	528.3	107	2180.0	164	16.1	
106	117.4	108	2180.0	165	16.1	
109	2180.0	166	16.1			
108	528.3	110	2180.0	167	16.1	
109	117.4	111	2180.0	168	16.1	
112	2180.0	169	16.1			
111	305.1	110	2050.0	170	18.7	
112	67.8	111	2050.0	171	18.7	
112	2050.0	172	18.7			
114	204.3	116	1705.0	173	23.2	119 204.3
115	45.4	117	1705.0	174	23.2	
118	1705.0	175	23.2			
117	311.0	116	1360.0	176	24.4	
118	69.1	117	1360.0	177	24.4	
118	1360.0	178	24.4			
120	490.0	179	25.2			
120	1640.0	180	25.2			
121	430.0	122	430.0	125	215.0	
123	193.0	126	430.0			
124	430.0	127	430.0			
126	103.0	128	445.0			
127	179.0	129	445.0			
130	445.0					
129	164.0	131	690.0			
130	174.0	132	690.0			
133	690.0					
132	340.0	134	370.0			
133	183.0	135	370.0			
136	370.0					
135	180.0	137	465.0			
136	212.0	138	465.0			
139	465.0					
138	142.0	140	485.0			
139	179.0	141	485.0			
142	485.0					
141	195.0	144	555.0			
142	144.0	143	555.0			
143	555.0	146	79.0			
144	211.3	146	293.5			
145	211.3	147	440.0			
147	181.3	148	472.5			

TABLE 8-4 (*Continued*)

```
149   472.5   174   362.7
149   181.3   151   472.5
150   181.3   152   472.5
153   945.0   175   945.0
152   141.3   154   880.0
153    31.4   155   880.0
156   880.0
155   274.5   157 1800.0
156    61.0   158 1800.0
159 1800.0
158   644.0   161 1865.0
159   143.1   162 1865.0
163 1865.0
161 3150.0
162 1218.6   164 1125.0
163   270.8   165 1125.0
166  1125.0
165   528.3   167 2180.0
166   117.4   168 2180.0
169  2180.0
168   528.3   170 2180.0
169   117.4   171 2180.0
172  2180.0
171   305.1   170 2050.0
172    67.8   171 2050.0
172  2050.0
174   204.3   176 1705.0   179   204.3
175    45.4   177 1705.0
178  1705.0
176  1360.0   177   311.0
177  1360.0   178    69.1
178  1360.0
180   490.0
180  1640.0
```

1.90	1.90	1.90	1.70	2.10	2.10	2.10	2.40
2.40	2.40	2.70	2.70	2.70	3.00	3.00	3.00
3.33	3.33	3.33	3.70	3.70	3.70	4.50	4.50
4.50	6.5	6.5	6.5	6.5	6.5	9.2	9.2
9.2	11.3	11.3	11.3	12.7	12.7	12.7	20.8
13.9	13.9	13.9	16.1	16.1	16.1	16.1	16.1
16.1	18.7	18.7	18.7	23.2	23.2	23.2	24.4
24.4	24.4	25.2	26.1	1.90	1.90	1.90	1.70
2.10	2.10	2.10	2.40	2.40	2.40	2.70	2.70
2.70	3.00	3.00	3.00	3.33	3.33	3.33	3.70
3.70	3.70	4.50	4.50	4.50	6.5	6.5	6.5
6.5	6.5	9.2	9.2	9.2	11.3	11.3	11.3
12.7	12.7	12.7	20.8	13.9	13.9	13.9	16.1

(*continues*)

TABLE 8-4 (*Continued*)

16.1	16.1	16.1	16.1	16.1	18.7	18.7	18.7
23.2	23.2	23.2	24.4	24.4	24.4	25.2	26.1
1.50	1.50	1.50	1.50	1.50	1.50	1.50	1.50
1.50	1.50	1.50	1.50	1.50	1.50	1.50	1.50
1.50	1.50	1.50	1.50	1.50	1.50	1.50	1.50
1.50	1.50	1.50	1.50	1.50	1.50	1.50	1.50
1.50	1.50	1.50	1.50	1.50	1.50	1.50	1.50
1.50	1.50	1.50	1.50	1.50	1.50	1.50	1.50
1.50	1.50	1.50	1.50	1.50	1.50	1.50	1.50
1.50	1.50	1.50	1.50				
23.	0.179005	23.	0.160582	23.	0.229337	23.	0.341118
23.	0.095520	23.	0.165638	23.	0.170268	23.	0.223956
23.	0.287160	23.	0.245918	23.	0.345455	23.	0.185998
23.	0.158246	23.	0.252512	23.	0.298048	23.	0.477803
23.	0.228818	23.	0.288827	23.	0.267913	23.	0.396561
23.	0.291447	23.	0.160029	23.	0.416729	23.	0.416729
23.	0.416729	23.	0.538671	23.	1.212009	23.	0.538671
23.	0.538671	23.	0.538671	23.	1.150686	23.	0.255708
23.	1.150686	23.	5.571252	23.	1.238056	23.	5.571252
23.	19.53005	23.	4.340011	23.	19.53005	23.	52.07186
23.	19.12117	23.	4.249150	23.	19.12117	23.	18.59816
23.	4.132925	23.	18.59816	23.	18.59816	23.	4.132925
23.	18.59816	23.	11.70457	23.	2.601015	23.	11.70457
23.	8.093901	23.	1.798645	23.	8.093901	23.	10.33121
23.	2.295802	23.	10.33121	23.	5.291538	23.	13.27811
23.	0.179005	23.	0.160582	23.	0.229337	23.	0.341118
23.	0.095520	23.	0.165638	23.	0.170268	23.	0.223956
23.	0.287160	23.	0.245918	23.	0.345455	23.	0.185998
23.	0.158246	23.	0.252512	23.	0.298048	23.	0.477803
23.	0.228818	23.	0.288827	23.	0.267913	23.	0.396561
23.	0.291447	23.	0.160029	23.	0.416729	23.	0.416729
23.	0.416729	23.	0.538671	23.	1.212009	23.	0.538671
23.	0.538671	23.	0.538671	23.	1.150686	23.	0.255708
23.	1.150686	23.	5.571252	23.	1.238056	23.	5.571252
23.	19.53005	23.	4.340011	23.	19.53005	23.	52.07186
23.	19.12117	23.	4.249150	23.	19.12117	23.	18.59816
23.	4.132925	23.	18.59816	23.	18.59816	23.	4.132925
23.	18.59816	23.	11.70457	23.	2.601015	23.	11.70457
23.	8.093901	23.	1.798645	23.	8.093901	23.	10.33121
23.	2.295802	23.	10.33121	23.	5.291538	23.	13.27811
23.	0.209487	23.	0.124542	23.	0.177866	23.	0.293751
23.	0.068982	23.	0.119666	23.	0.123011	23.	0.169466
23.	0.180275	23.	0.116262	23.	0.188516	23.	0.101529
23.	0.086381	23.	0.125291	23.	0.147924	23.	0.237138
23.	0.103235	23.	0.130340	23.	0.120902	23.	0.162750
23.	0.119637	23.	0.065691	23.	0.139561	23.	0.139561
23.	0.139561	23.	0.123475	23.	0.277819	23.	0.123475

TABLE 8-4 (*Continued*)

23.	0.123475	23.	0.123475	23.	0.186638	23.	0.041475
23.	0.186638	23.	0.741515	23.	0.164781	23.	0.741515
23.	2.085289	23.	0.463398	23.	2.085289	23.	3.753378
23.	2.056040	23.	0.456898	23.	2.056040	23.	1.728025
23.	0.384055	23.	1.728025	23.	1.728025	23.	0.384055
23.	1.728025	23.	0.937866	23.	0.208415	23.	0.937866
23.	0.522592	23.	0.116132	23.	0.522592	23.	0.633966
23.	0.140881	23.	0.633966	23.	0.315023	23.	0.764399

```
 0
-1
 4
824.0    1  824.0   61  894.0  121   50.0  171
1.04 0.000
```

8.6 SUMMARY

A number of practical applications of the HAR03 model to both near-field and far-field mass transport modeling were presented in this chapter. They range from two-dimensional horizontal, two-dimensional vertical, to three-dimensional configurations for streams, estuaries, and lakes. The applications show that the HAR03 modeling framework is a very versatile tool for calibrating mass transport. The computation is straightforward and efficient. The limitation of this modeling framework is that it requires the assignment of advective flows and dispersion coefficients. In practice, the advective flows may be derived from a hydrodynamic model or for simple configurations, developed via model calibration to reproduce the distribution of a conservative substance. In estuaries and coastal waters, salinity is used as the conservative tracer. In freshwater streams and lakes, chloride or specific conductivity is usually used as the tracer.

Note that running the WASP model (as presented in Chapter 7) for conservative substances to steady state can accomplish the same task of quantifying mass transport as the HAR03 model. However, the HAR03 model yields steady-state solutions directly, thereby requiring considerably shorter run time and its input data file structure is much simpler than that of the WASP model.

REFERENCES AND FURTHER READINGS

Allen, J. S., P. A. Newberger, and J. Federiuk. 1995. Upwelling Circulation on the Oregon Continental Shelf. *J. Phys. Oceanogr.*, 35: 1843–1889.

Ambrose, R. B., T. A. Wool, and J. L. Martin. 1993a. The Water Quality Analysis Simulation Program, WASP5, Part A: Model Documentation. U.S. EPA Center for Exposure Assessment Modeling, Athens, GA.

Ambrose, R. B., T. A. Wool, and J. L. Martin. 1993b. The Water Quality Analysis Simulation Program, WASP5, Part B: The WASP5 Input Dataset. U.S. EPA Center for Exposure Assessment Modeling, Athens, GA.

APHA (American Public Health Association). 1992. Standard Methods for the Examination of Water and Wastewater, 18th ed., Washington, DC.

Applied Technology and Engineering. 1989. Mixing Zone Analysis of the James River at the Lynchburg Regional Wastewater Treatment Plant. Report prepared for City of Lynchburg, VA, 35 pp.

Applied Technology and Engineering. 1991. Results of Water Quality Modeling to Evaluate the Fate of Color in Martin Lake. Report prepared for Russell Corporation Avondale Mills, City of Alexander, AL.

Banks, R. B. and E. F. Herrera. 1977. Effect of Wind and Rain on Surface Reaeration. *J. Environ. Eng.*, 103(EE3): 489–504.

Bedford, K. W. 1985. Selection of Turbulence and Mixing Parameterizations for Estuary Water Quality Models. U.S. Army Engineers Waterways Experiment Station, Miscellaneous Paper EL-85-2, Vicksburg, MS.

Beltaos, S. 1978a. Mixing Processes in Natural Streams. In *Transport Processes and River Modeling Workshop*. Canada Centre for Inland Waters, Burlington, ON.

Beltaos, S. 1978b. Transverse Mixing in Natural Streams. Transportation and Surface Water Eng. Div., Alberta Research Council, Report No. SWE-78/01, Edmonton, Alberta.

Blumberg, A. F. 1977. Numerical Model of Estuarine Circulation. *J. Hydraul. Div.*, ASCE, 103: 295.

Blumberg, A. F. and D. M. Goodrich. 1990. Modeling of Wind-Induced Destratification in Chesapeake Bay. *Estuaries*, 13: 1236–1249.

Blumberg, A. F. and H. J. Herring. 1987. Circulation Modeling Using Orthogonal Curvilinear Coordinates. In *Three-Dimensional Models of Marine and Estuarine Dynamics*, J. C. J. Nihoul and B. M. Jamart, Eds., Elsevier, New York, pp. 55–88.

Blumberg, A. F. and G. L. Mellor. 1980. A Coastal Ocean Numerical Model. In *Mathematical Modelling of Estuarine Physics, Proceedings of an International Symposium*, Hamburg, August 24–26, 1978, J. Sundermann and K. P. Holz, Eds., Springer-Verlag, Berlin.

Blumberg, A. F. and G. L. Mellor. 1983. Diagnostic and Prognostic Numerical Circulation Studies of the South Atlantic Bight. *J. Geophys. Res.*, 88: 4579–4592.

Blumberg, A. F. and G. L. Mellor. 1985. A Simulation of the Circulation in the Gulf of Mexico. *Isr. J. Earth Sci.*, 34: 122–144.

Bowie, G. L., W. B. Mills, D. B. Porcella, C. L. Campbell, J. K. Pagenkopf, G. L. Rupp, K. M. Johnson, P. W. H. Chan, and S. A. Gherini. 1985. Rates, Constants and Kinetics Formulations in Surface Water Quality Modeling, 2nd ed., EPA/600/3-85/040. U.S. Environmental Protection Agency, Environmental Research Laboratory, Athens, GA.

Boynton, W. and W. M. Kemp. 1985. Nutrient Regeneration and Oxygen Consumption by Sediments along an Estuarine Salinity Gradient. *Mar. Ecol. Prog. Ser.*, 23, 45.

Boynton, W., W. M. Kemp, and C. Osborne. 1980. Nutrient Fluxes across the Sediment-Water Interface in the Turbid Zone of a Coastal Plain Estuary. In *Estuarine Perspectives*, V. Kennedy, Ed., Academic Press, New York, p. 93.

Broecker, H. C., J. Petermann, and W. Siems. 1978. The Influence of Wind on CO_2 Exchange in a Wind-Wave Tunnel. *J. Marine. Res.*, 36(4): 595–610.

Brown, L. C. and T. O. Barnwell, Jr. 1987. The Enhanced Stream Water Quality Models QUAL2E and QUAL2E-UNCAS: Documentation and User Manual, Athens, GA, EPA/600/3-87/007, 189 pp.

Brown, T. 1995. Numerical Tagging of Nitrogen Components in the James River Estuary. M.S. Thesis, Department of Civil Engineering, University of Virginia, Charlottesville, VA.

Butts, T. A. and R. L. Evans. 1983. Effects of Channel Dams on Dissolved Oxygen Concentrations in Northeastern Illinois Streams. Circulars 132, State of Illinois, Dept. of Reg. And Educ., Illinois Water Survey, Urbana, IL.

Cerco, C. F. 1985a. Sediment-Water Column Exchanges of Nutrients and Oxygen in the Tidal James and Appomattox Rivers. Virginia Institute of Marine Science, Gloucester Point, VA.

Cerco, C. F. 1985b. Effect of Temperature and Dissolved Oxygen on Sediment-Water Nutrient Flux. Virginia Institute of Marine Science, Gloucester Point, VA.

Cerco, C. F. 1988. Sediment Nutrient Fluxes in a Tidal Freshwater Embayment. *Water Resources Bulletin*, 24(2): 255–267.

Cerco, C. F. 2001. Phytoplankton Kinetics in the Chesapeake Bay Eutrophication Model. Manuscript submitted to *Water Qual. Ecosyst. Model.* for publication.

Cerco, C. F. and T. Cole, T. 1993. Three-Dimensional Eutrophication Model of Chesapeake Bay. *J. Environ. Eng.*, 119(6):1006–1025.

Chapra, S. C. 1997. *Surface Water-Quality Modeling*. McGraw-Hill, New York, NY, pp. 382–383.

Chapra, S. C. and R. P. Canale. 1991. Long-Term Phenomenological Model of Phosphorus and Oxygen in Stratified Lakes. *Water Res.*, 25(6):707–715.

Chapra, S. C. and D. M. Di Toro. 1991. The Delta Method for Estimating Community Production, Respiration, and Reaeration in Streams. *J. Environ. Eng.*, 117(5):640–655.

Chapra, S. C. and G. A. Nossa. 1974. Documentation for HAR03—A Computer Program for the Modeling of Water Quality Parameters in Steady-State Multi-Dimensional Natural Aquatic Systems. U.S. EPA, Region II, New York.

Chen, C., R. C. Beardsley, and R. Limeburner. 1995. A Numerical Study of Stratified Tidal Rectification over Finite-Amplitude Banks. Part II: Georges Bank. *J. Phys. Oceanogr.*, 25: 2111–2128.

Churchill, M. A., H. L. Elmore, and R. A. Buckingham. 1962. The Prediction of Stream Reaeration Rates. *J. Sanitary Eng.*, 88(SA4):1–46.

Cole, T. and E. Buchak. 1995. CE-QUAL-W2: A Two-Dimensional, Laterally Averaged, Hydrodynamic and Water Quality Model, Version 2.0. U.S. Army Engineers Waterways Experiment Station, Tech. Report EL-95-May 1995, Vicksburg, MS.

Cover, A. P. 1976. Selecting the Proper Reaeration Coefficient for Use in Water Quality Models. In *Environmental Modeling and Simulation*, W. R. Ott, Ed., EPA Office of Research and Development and Office of Planning and Management, Washington, DC, pp. 340–343.

Defant, A. 1961. *Physical Oceanography*, Volume 1, Pergamon Press, New York.

Di Toro, D. M. 1975. Algae and Dissolved Oxygen. Summer Institute in Water Pollution Notes, Manhattan College, Bronx, NY.

Di Toro, D. M. 1978. Optics of Turbid Estuarine Waters: Approximations and Applications. *Water Res.*, 12:1059–1068.

Di Toro, D. M. 1986. A Diagenetic Oxygen Equivalents Model of Sediment Oxygen Demand. In *Sediment Oxygen Demand*, K. Hatcher, Ed., Institute of Natural Resources, University of Georgia, Athens, p. 171.

Di Toro, D. M., J. J. Fitzpatrick, and R. V. Thomann. 1983. Water Quality Analysis Simulation Program (WASP) and Model Verification Program (MVP) Documentation. Report submitted by Hydroscience, Inc. to EPA Environmental Research Laboratory, Duluth, MN.

Di Toro, D. M., P. R. Paquin, K. Subburamu, and D. A. Gruber. 1990. Sediment Oxygen Demand: Methane and Ammonia Oxidation, *J. Environ. Eng.*, 116(5):945–986.

Di Toro, D. M., R. V. Thomann, and O'Connor. 1971. A Dynamic Model of Phytoplankton Population in the Sacramento-San Joaquin Delta. In *Advances in Chemistry* Series 106: Nonequilibrium Systems in Natural Water Chemistry, R. F. Gould, Ed., American Chemical Society, Washington, DC, p.131.

Domotor, D. K., M. S. Haire, N. M. Panday, and R. M. Summers. 1989. Patuxent Estuary Water Quality Assessment: Special Emphasis 1983–1987. Maryland Department of the Environment, Technical Report No. 104, Baltimore, MD.

Donigian, A. S., Jr., J. C. Imhoff, B. R. Bicknell, and J. L. Kittle, Jr. 1984. Application Guide for Hydrological Simulation Program Fortran (HSPF). U.S. EPA, EPA-600/3-84-065, Athens, GA.

Dyer, K. R. 1973. *Estuaries: A Physical Introduction*. John Wiley and Sons, Inc., New York.

Dyer, K. R. 1986. *Coastal and Estuarine Sediment Dynamics*, John Wiley and Sons, Inc., New York.

Eckenfelder, W. W., Jr., and D. J. O'Connor. 1961. *Biological Waste Treatment*. Pergamon, New York.

EnviroTech Associates, Inc. 1993. Mississippi River Phosphorus Study, Section 6, Part B: Developing Mass Transport for WASP Models of the Upper Mississippi River. Final report prepared for Metropolitan Waste Control Commission, St. Paul, MN.

Eppley, R. W. 1972. Temperature and Phytoplankton Growth in the Sea. *Fishery Bulletin*, 70(4): 1063–1085.

Ezer, T. and G. L. Mellor. 1992. A Numerical Study of the Variability and the Separation of the Gulf Stream, Induced by Surface Atmosphere Forcing and Lateral Boundary Flows. *J. Phys. Oceanogr.*, 22: 660–682.

Falch, J. A., P. J. Beilke, and H. G. Stefan. 1979. Travel Time and Longitudinal Dispersion in the Upper Mississippi River between Anoka and Lock and Dam No. 2 near Hastings, Minnesota and in the Lower Minnesota River between Jordan and Mendota, Minnesota. University of Minnesota St. Anthony Falls Hydraulic Laboratory, Minneapolis, MN, External Memorandum No. 164, 90 pp.

Fischer, H. B. 1968. Dispersion Predictions in Natural Streams. *J. Sanitary Eng.*, 94(SA5): 927–944.

Fischer, H. B., E. J. List, R. C. Y. Koh, J. Imberger, and N. H. Brooks. 1979. *Mixing in Inland and Coastal Waters*. Academic, New York.

Fisher, J. S., J. D. Ditmars, and A. T. Ippen. 1972. Mathematical Simulation of Tidal Time-Averages of Salinity and Velocity Profiles in Estuaries. Ralph M. Parsons Laboratory, Massachusetts Institute of Technology, Report No. 151, Cambridge, MA.

FWPCA (Federal Water Pollution Control Administration) 1965. Report on Hydrographic Studies of the Mississippi, Minnesota, and St. Croix Rivers, St. Paul, MN.

Gelda, R. K., M. T. Auer, S. W. Effler, S. C. Chapra, and M. L. Storey. 1996. Determination of Reaeration Coefficients: A Whole Lake Approach. *J. Environ. Eng.*, 122(4): 269–275.

Grant, R. S. 1976. Reaeration Coefficient Measurements of Ten Small Streams in Wisconsin Using Radioactive Tracers. U.S. Geological Survey Water Resources Investigations, pp. 76–96.

Haffely, G. 1997. Long-Term BOD Data on the Upper Mississippi River, Minnesota River, St. Croix River, and the Metro Plant. Personal communications.

Haffely, G. and L. Johnson. 1994. Biodegradation of TCMP (N-Serve) Nitrification Inhibitor in the Ultimate BOD Test. Proc. WEFTEC 1994, 64th Ann. Conf., Chicago, IL.

Hall, J. C. and R. J. Foxen. 1983. Nitrification in BOD Test Increases POTW Noncompliance. *J. Water Pollution Contr. Fed.*, 55(12):1461–1469.

Hamrick, J. M. 1992a. A Three-Dimensional Environmental Fluid Dynamics Computer Code: Theoretical and Computational Aspects. The College of William and Mary, Virginia Institute of Marine Science, Special Report 317, Gloucester Point, VA, 63 pp.

Hamrick, J. M. 1992b. Estuarine Environmental Impact Assessment Using a Three-Dimensional Circulation and Transport Model. In *Estuarine and Coastal Modeling*, Proceedings of the 2nd International Conference, M. L. Spaulding, Ed., ASCE, New York, pp. 292–303.

Hamrick, J. M. 1994. Evaluation of Island Creation Alternatives in the Hampton Flats of the James River. Report submitted by Virginia Institute of Marine Science to U.S. Army Corps of Engineers, Norfolk District.

Hamrick, J. M. 1995. Evaluation of the Environmental Impacts of Channel Deepening and Dredge Spoil Disposal Site Expansion in the Lower James River, Virginia. The College of William and Mary, Virginia Institute of Marine Science. Gloucester Point, VA.

Hamrick, J. M. and B. J. Neilson. 1989. Determination of Marine Buffer Zones Using Simple Mixing and Transport Models. Report submitted by Virginia Institute of Marine Science for Virginia Department of Health, Gloucester Point, VA, 68 pp.

Harleman, D. R. F., J. E. Dailey, M. L. Thatcher, T. O. Najarian, D. N. Brocard, and R. A. Ferrara. 1977. User's Manual for the M.I.T. Transient Water Quality Network Model—Including Nitrogen Cycle Dynamics for Rivers and Estuaries. R. M. Parsons Laboratory for Water Resources and Hydrodynamics, Cambridge, MA.

Hartman, B. and D. E. Hammond. 1985. Gas Exchange in San Francisco Bay. *Hydrobiologia*, 129:59.

Hatcher, K. J. 1986. Introduction of Sediment Oxygen Demand Modeling. In Sediment Oxygen Demand, Processes, Modeling, and Measurement, K. J. Hatcher, Ed., Institute of Natural Resources, University of Georgia, Athens, GA, pp. 113–138.

Holley, E. R. and G. H. Jirka. 1989. Mixing in Rivers. Army Engineers Waterways Experiment Station, Technical Report E-86-11, Vicksburg, MS.

Holmes, R. W. 1970. The Secchi Disk in Turbid Coastal Waters. *Limnol. Oceanogr.*, 15: 668–694.

Houck, O. A. 1999. The Clean Water Act TMDL Program: Law, Policy, and Implementation. Environmental Law Institute, Washington, DC.

Hubbard, E. F., F. A. Kilpatrick, L. A. Martens, and J. F. Wilson. 1982. Measurement of Time of Travel and Dispersion in Streams by Dye Tracing. Book 3, Application of Hydraulics, Chapter A9, U.S. Geological Survey, Washington, DC, 44 pp.

HydroQual, Inc. 1981. Water Quality Analysis of the Patuxent River. Report prepared for Maryland Office of Environmental Programs, Baltimore, MD.

HydroQual, Inc. 1982. Grand Calumet River Wasteload Allocation Study. Report submitted to Indiana State Board of Health, Baltimore, MD.

HydroQual, Inc. 1986. Water Quality Analysis of the James and Appomattox Rivers. Report prepared for Richmond Regional Planning District Commission, Richmond, VA.

HydroQual, Inc. 1987. Evaluation of Sediment Oxygen Demand in the Upper Potomac Estuary. Report submitted to the Metropolitan Washington Council of Governments, Washington, DC.

HydroQual, Inc. 1992. An Evaluation of the Effect of CSO Controls on Anacostia River Sediment Oxygen Demand. Report submitted to Metropolitan Washington Council of Governments, Washington, DC.

Hydroscience, Inc. 1973. Development of Hydrodynamic and Time Variable Water Quality Models of Boston Harbor. Report submitted to Commonwealth of Massachusetts Water Resources Commission, Boston, MA.

Hydroscience, Inc. 1978. NYC 208 Task Report on Seasonal Steady-State Modeling. Report submitted to Hazen & Sawyer, Engineers, New York.

Hydroscience, Inc. 1979. Upper Mississippi River 208 Grant Water Quality Modeling Study. Report prepared for Metropolitan Waste Control Commission, St. Paul, MN.

James, W. F. and J. W. Barko. 1992. Internal Phosphorus Loading in Lake Pepin (Minnesota-Wisconsin): Data Summary for Phase II—In Situ Rates of Nutrient Release from the Sediments and Nutrient Concentrations in the Water Column. Eau Galle Limnological Laboratory, Spring Valley, WS, and U.S. Army Engineers Waterways Experiment Station, Vicksburg, MS.

James, W. F., J. W. Barko, and H. L. Eakin. 1992. Internal Phosphorus Loading in Lake Pepin (Minnesota-Wisconsin): Data Summary for Phase I—Sediment Composition and Rates of P Release from the Sediment. Eau Galle Limnological Laboratory, Spring Valley, WS, and U.S. Army Engineers Waterways Experiment Station, Vicksburg, MS.

Jassby, A. and T. Powell. 1975. Vertical Patterns of Eddy Diffusion during Stratification in Castle Lake, California. *Limnol. & Oceanogr.*, 20: 530–543.

Johnson, B. H., K. W. Kim, R. E. Heath, B. B. Hsieh, and H. L. Butler. 1993. Validation of Three-Dimensional Hydrodynamic Model of Chesapeake Bay. *J. Hydraul. Eng.*, 119(1): 2–20.

Krenkel, P. A. and G. T. Orlob. 1963. Turbulent Diffusion and the Reaeration Coefficient. *Am. Soc. Civil Engrs.*, 128: 293–334.

Krenkel, P. and R. J. Ruane. 1979. Basic Approach to Water Quality Modeling. In *Modeling of Rivers*, H. W. Shen, Ed., Wiley Interscience, John Wiley & Sons, Inc., New York, Chapter 18.

Langbien, W. B. and W. H. Durum. 1967. The Aeration Capacity of Streams. U.S. Geological Survey Circular 542, Washington, DC.

Larson, C. E. 1993. Derivation of Kinetic Coefficients for the Time Variable WASP/EUTRO5 Model of the Upper Mississippi River. Metropolitan Waste Control Commission, St. Paul, MN.

Leo, M. W., R. V. Thomann, and T. W. Gallagher. 1984. Before and After Case Studies: Comparisons of Water Quality Following Municipal Treatment Plant Improvements. Report prepared by HydroQual, Inc. for U.S. EPA Office of Water Program Operations, Washington, DC.

Leonard, B. P. 1979. A Stable and Accurate Convection Modelling Procedure Based on Quadratic Upstream Interpolation, *Comput. Meth. Appl. Mechan. Eng.*, 19: 59–98.

Lerman, A. and M. Stiller. 1969. Vertical Eddy Diffusion in Lake Tiberias. *Verh. Int. Ver. Limnol.*, 17: 323–333.

Lung, W. S. 1986. Advective Acceleration and Mass Transport in Estuaries. *J. Hydraul. Eng.*, 112(9): 874–878.

Lung, W. S. 1987a. Water Quality Modeling Using Personal Computers. *J. Water Pollution Contr. Fed.*, 59(10): 909–913.

Lung, W. S. 1987b. Lake Acidification Model: A Practical Tool. *J. Environ. Eng.*, 113(4): 900–915.

Lung, W. S. 1990. Development of a Water Quality Model for the Patuxent Estuary. In *Coastal and Estuarine Studies, Vol. 36: Estuarine Water Quality Management*, W. Michaelis, Ed., Springer-Verlag, Berlin, pp. 49–54.

Lung, W. S. 1992. A Water Quality Model for the Patuxent Estuary. Report submitted by University of Virginia to Maryland Department of the Environment, Baltimore, MD.

Lung, W. S. 1993. *Water Quality Modeling, Vol. III. Application to Estuaries.* CRC Press, Boca Raton, FL.

Lung, W. S. 1994. Water Quality Modeling of the St. Martin River, Assawoman and Isle of Wight Bays. Report submitted by University of Virginia to Maryland Department of the Environment, Baltimore, MD, 111 pp.

Lung, W. S. 1995. Mixing Zone Modeling for Toxic Waste-Load Allocations. *J. Environ. Eng.*, 121(11): 839–842.

Lung, W. S. 1996a. Post Audit of the Upper Mississippi River BOD/DO Model. *J. Environ. Eng.*, 122(5): 350–358.

Lung, W. S. 1996b. Fate and Transport Modeling Using a Numerical Tracer. *Water Resources Res.*, 32(1): 171–178.

Lung, W. S. 1998. Trends in BOD/DO Modeling for Wasteload Allocations. *J. Environ. Eng.*, 124(10): 1004–1007.

Lung, W. S. 2000. Modeling Sediment Oxygen Demand and Methane Production Across the Sediment-Water Interface in the Anacostia River Using WASP/EUTRO5. Report prepared for the Interstate Commission of the Potomac River Basin, Rockville, MD.

Lung, W. S. and R. P. Canale. 1976. Phosphorus Models for Eutrophic Lakes. *Water Res.*, 10: 1101–1114.

Lung, W. S. and C. C. Hwang. 1989. Integrating Hydrodynamic and Water Quality Models for Patuxent Estuary. *Estuarine and Coastal Modeling*, ASCE, New York, NY, pp. 420–429.

Lung, W. S. and C. E. Larson. 1995. Water Quality Modeling of Upper Mississippi River and Lake Pepin. *J. Environ. Eng.*, 121(10): 691–699.

Lung, W. S. and D. J. O'Connor. 1978. Assessment of the Effect of Proposed Submerged Sill on the Water Quality of Western Delta–Suisan Bay, Report submitted by Hydroscience, Inc. to U.S. Army Corps of Engineers, Sacramento District, Sacramento, CA.

Lung, W. S. and D. J. O'Connor. 1984. Two-Dimensional Mass Transport in Estuaries. *J. Hydraul. Eng.*, 110(10): 1340–1357.

Lung, W. S. and R. G. Sobeck. 1999. Renewed Use of BOD/DO Models in Water Quality Management. *J. Water Resources Plan. Manage.*, 125(4), 222–227.

Lung, W. S. and N. Testerman. 1989. Modeling Fate and Transport of Nutrients in the James Estuary. *J. Environ. Eng.*, 115(5): 978–991.

Lung, W. S., J. L. Martin, and S. C. McCutcheon. 1982. Eutrophication Analysis of Embayments in Prince William Sound, Alaska. *J. Environ. Eng.*, 119(5): 811–824.

Lung, W. S., R. A. Rapaport, and A. C. Franco. 1990. Predicting Concentrations of Consumer Product Chemicals in Estuaries, *Environ. Toxicol. Chem.*, 9(9): 1127–1136.

Lung, W. S., J. Tsatsaros, B. Landin, J. Harcum, M. Morton, W. Lane, and P. Corcker. 2000. Enhancement of EPA's WASP/EUTRO Model for Diurnal Dissolved Oxygen and pH to Simulate Periphyton Growth in Effluent-Dominated Streams. Presentation at the SETAC 21st Annual Meeting, November 12–16, 2000, Nashville, TN.

Manhattan College, 1976. Summer Institute Notes on Natural Water Quality Modeling Workshop, Bronx, NY.

Martin, J. L. and S. C. McCutcheon. 1999. *Hydrodynamics and Transport for Water Quality Modeling*. Lewis Publishers, Boca Raton, FL.

McBride, G. B. 1987. QUICKEST Algorithms for Estuarine Solute Transport Model. Hamilton Science Centre Internal Report No. IR/87/5, Ministry of Works and Development, Hamilton, New Zealand, 20 pp.

McCutcheon, S. C. 1985. Water Quality and Streamflow Data for the West Fork Trinity River in Fort Worth, TX. U.S. Geological Survey, Water Resources Investigation Report 84-4330, Washington, DC.

McCutcheon, S. C. 1989. *Water Quality Modeling, Volume I: Transport and Surface Exchange in Rivers*. CRC Press, Boca Raton, FL.

McCutcheon, S. C. 2000. Modeling Concepts for Mixing Analysis. Presentation before the mixing zone panel as part of the Modeling and Management of Emerging Environmental Issues Workshop 2000 sponsored by DuPont, July 25, 2000, Malvern, PA.

McCutcheon, S. C., D. Zhu, and S. Bird. 1990. Model Calibration, Validation, and Use. In Technical Guidance Manual for Performing Waste Load Allocations, Book III Estuaries. U.S. EPA, Athens, GA.

McKeown, J. J., L. C. Brown, and C. H. Martone. 1981. Ultimate BOD Estimation in Receiving Water Quality Modeling. *Wat. Sci. Tech.*, 13: 363–370.

Melching, C. S. and H. E. Flores. 1999. Reaeration Equations Derived from U.S. Geological Survey Database. *J. Environ. Eng.*, 125(5): 407–414.

MPCA (Minnesota Pollution Control Agency) 1981. Mississippi River Waste Load Allocation Study. Report prepared by Division of Water Quality, St. Paul, MN.

Murphy, P. J. and D. B. Hicks. 1986. *In-Situ* Method for Measuring Sediment Oxygen Demand. In *Sediment Oxygen Demand*, K. J. Hatcher, Ed., Institute of Natural Resources, University of Georgia, Athens, GA.

NCASI. 1982. A Review of Ultimate BOD Estimation and Its Kinetic Formulation for Pulp and Paper Mill Effluents. National Council of the Paper Industry for Air and Stream Improvement, Inc., Technical Bulletin No. 382, Medford, MA, 35 pp.

Neely, W. B. 1982. The Definition and Use of Mixing Zones. *Environ. Sci. & Technol.*, 16: 518A.

Nemura, A. D. 1992. *Modeling Sediment Oxygen Demand and Nutrient Fluxes in the Tidal Anacostia River*, 2 volumes. Metropolitan Washington Council of Governments, Washington, DC.

O'Connor, D. J. 1960. Oxygen Balance of an Estuary. *J. Sanitary Eng.*, 86(SA3): 35–55.

O'Connor, D. J. 1962. Organic Pollution of New York Harbor—Theoretical Considerations. *J. Water Pollution Fed.*, 34: 905.

O'Connor, D. J. 1965. Estuarine Distribution of Nonconservative Substances. *J. Sanitary Eng.*, 91: 23.

O'Connor, D. J. and W. E. Dobbins. 1958. Mechanism of Reaeration in Natural Streams. *ASCE Trans.*, 123: 641–684.

O'Connor, D. J. and W. S. Lung. 1981. Suspended Solids Analysis of Estuarine Systems. *J. Environ. Eng.*, 107: 101.

Oey, L. Y., G. L. Mellor, and R. I. Hires. 1985a. Tidal Modeling of the Hudson-Raritan Estuary. *Estuarine Coastal Shelf Sci.*, 20: 511–527.

Oey, L. Y., G. L. Mellor, and R. I. Hires. 1985b. A Three-Dimensional Simulation of the Hudson-Raritan Estuary. Part I: Description of the Model and Model Simulations. *J. Phys. Oceanogr.*, 15: 1676–1692.

Oey, L. Y., G. L. Mellor, and R. I. Hires. 1985c. A Three-Dimensional Simulation of the Hudson-Raritan Estuary. Part II: Comparison with Observations. *J. Phys. Oceanogr.*, 15: 1693–1709.

Officer, C. B. 1976. *Physical Oceanography of Estuaries (and Associated Coastal Waters)*. Wiley-Interscience, New York, 465 pp.

Officer, C. B. 1977. Longitudinal Circulation and Mixing Relations in Estuaries. In *Estuaries, Geophysics, and the Environment*. National Academy of Sciences, Washington, DC.

Officer, C. B. 1980. Box Models Revisited. In *Estuarine and Wetland Processes*, P. Hamilton and K. B. Macdonald, Eds. Plenum Press, New York, p. 65.

Orlob, G. T. 1972. Mathematical Modeling of Estuarial Systems. In *Modeling of Water Resources Systems*, A. K. Biswas, Ed. Harvest House, p. 87.

Owens, M., R. W. Edwards, and J. W. Gibbs. 1964. Some Reaeration Studies in Streams. *Int. J. Air Water Poll.* 8: 469–486.

Pritchard, D. W. 1952. Salinity Distribution and Circulation in the Chesapeake Bay Estuaries System. *J. Marine Res.*, 11: 106–123.

Pritchard, D. W. 1956. The Dynamic Structure of a Coastal Plain Estuary. *J. Marine Res.*, 15: 33–42.

Pritchard, D. W. 1958. The Equations of Mass Continuity and Salt Continuity in Estuaries. *J. Marine Res.*, 17: 412–423.

Pritchard, D. W. 1969. Dispersion and Flushing of Pollutants in Estuaries. *J. Hydraul. Div.*, 95, 115.

Rapaport, R. A. 1988. Prediction of Consumer Product Chemical Concentrations as a Function of Publicly Owned Treatment Works Treatment Type and Riverine Dilution. *Environ. Toxicol. Chem.*, 7: 107.

Rapaport, R. A. and W. S. Eckhoff. 1990. Monitoring Linear Alkylbenzene Sulfonate in the Environment: 1973–1986. *Environ. Toxicol. Chem.*, 9(10): 1245–1257.

Rathbun, R. E. 1977. Reaeration Coefficients of Streams—State-of-the-Art. *J. Hydraul. Div.*, ASCE, 103(HY4): 409–424.

Rathbun, R. E. 1998. Transport, Behavior, and Fate of Volatile Organic Compounds in Streams. U.S. Geological Survey Professional Paper 1589, Washington, DC, 151 pp.

Rattray, M. and D. V. Hansen. 1962. A Similarity Solution for Circulation in an Estuary. *J. Marine Res.*, 20: 121–133.

Rutherford, J. C. 1994. *River Mixing*. John Wiley & Sons, Inc., New York.

Sampou, P. 1990. Sediment-Water Exchanges and Disgenesis of Anacostia River Sediments. Report prepared by Center for Environmental and Estuarine Studies, Horn Point Environmental Laboratory, University of Maryland, Cambridge, MD.

Schroeder, S. H. 1981. Pollution Control: Changing Approaches. *Environ. Sci. & Technol.*, 15: 1287.

Seitzinger, S. P. 1985. The Effect of Oxygen Concentration and pH on Sediment-Water Nutrient Fluxes in the Potomac River. Report 85-2, The Academy of Natural Sciences, Philadelphia, PA.

Smith, D. J. 1978. WQRRS, Generalized Computer Program for River-Reservoir Systems. User's Manual 401-100, 100A, U.S. Army Engineers Hydrological Engineering Center, Davis, CA, 210 pp.

Smayda, T. J. 1970. The Suspension and Sinking of Phytoplankton in the Sea. *Annu. Rev. Oceanogr. Marine Biol.*, 8: 353–414.

Stefan, H. and K. Anderson. 1977. Analysis of Flow Through Sturgeon Lake and Backwater Channels of Mississippi River Pool No. 3 near Red Wing, Minnesota. Report submitted by University of Minnesota St. Anthony Falls Hydraulic Laboratory to Northern States Power Company, Project No. 165, Minneapolis, MN, 51 pp.

Stefan, H. and A. Demetracopoulos. 1979. A Model for Water Circulation and Solute Transport in Pool No. 2 of the Mississippi River. Report submitted by University of Minnesota St. Anthony Hydraulic Laboratory to Metropolitan Waste Control Commission, St. Paul, MN, 102 pp.

Stefan, H. and A. Wood. 1976. Field Investigations of Water Temperature Stratification and Wind Effects on Dissolved Oxygen in Pool No. 2 of the Mississippi River. Report prepared by University of Minnesota St. Anthony Hydraulic Laboratory, Minneapolis, MN, 116 pp.

St. John, J. P., T. W. Gallagher, and P. R. Paquin. 1984. The Sensitivity of the Dissolved Oxygen Balance to Predictive Reaeration Equations. In *Gas Transfer at Water Surface*, Brutsaert, W. and G. Jirka, eds., D. Reidel Publishing Co., Dordrecht, the Netherlands.

Streeter, N. W. 1936. The Rate of Atmospheric Reaeration of Sewage Polluted Streams. *Trans. ASCE*, 89: 1351.

Streeter, N. W. and E. B. Phelps. 1925. *Public Health Bull*. 146, U.S. Public Health Service.

Sundaram, T. R., C. C. Easterbrook, K. R. Piech, and G. Rudinger. 1969. An Investigation of the Physical Effects of Thermal Discharges into Cayuga Lake. Cornell Aeronautical Laboratory Report VT-2626-0-2, Ithaca, NY, 360 pp.

SWCB (State Water Control Board). 1976. Roanoke River Basin Water Quality Management Plan. Commonwealth of Virginia, Richmond, VA, 330 p.

Thomann, R. V. 1973. Principles of Waste Load Allocations. In Quality Models of Natural Water Systems, Manhattan College Summer Institute in Water Pollution Control, Bronx, NY.

Thomann, R. V. 1980. Measures of Verification. In Workshop on Verification of Water Quality Models, EPA-600/9-80-016, Washington, DC, pp. 37–61.

Thomann, R. V. 1987. System Analysis in Water Quality Management—A 25 Year Retrospect. In *Systems Analysis in Water Quality Management*, M. B. Beck, Ed. Pergamon Press, Tarrytown, NY, pp. 1–14.

Thomann, R. V. 1998. The Future "Golden Age" of Predictive Models for Surface Water Quality and Ecosystem Management. Simon W. Freese Lecture at the ASCE North American Water and Environment Congress '96, Anaheim, CA, June 24, 1996. *J. Environ. Eng.*, 124(2): 94–103.

Thomann, R. V. and J. J. Fitzpatrick. 1982. Calibration and Verification of the Potomac Estuary Model. HydroQual, Inc. final report prepared for the Washington, DC, Department of Environmental Services.

Thomann, R. V. and J. A. Mueller. 1987. *Principles of Surface Water Quality Modeling and Control*, Harper & Row, New York.

Tsivoglou, E. C. and L. A. Neal. 1976. Tracer Measurements of Reaeration: III. Predicting the Reaeration Capacity of Inland Streams. *J. Water Pollution Contr. Fed.*, 48(12): 2669–2689.

Tully, J. P. 1949. Oceanography and Prediction of Pulp Mill Pollution in Alberni Inlet. *Bull. Fisheries Res. Board Canada*, 83: 1–169.

U.S. EPA. 1971. Simplified Mathematical Modeling of Water Quality. Report prepared by Hydroscience, Inc. for U.S. Environmental Protection Agency, Washington, DC.

U.S. EPA. 1984. Handbook of Advanced Treatment Review Issues. Office of Water Program Operations, Washington, DC.

U.S. EPA 1985. Rates, Constants, and Kinetics Formulations in Surface Water Quality Modeling, 2nd ed., EPA 600/3-85/040, Washington, DC.

U.S. EPA 1990. Technical Guidance Manual for Performing Waste Load Allocations, Book III: Estuaries, Part 1: Estuaries and Waste Load Allocation Models, Washington, DC.

U.S. EPA 1991a. Guidance for Water Quality-Based Decisions: The TMDL Process. Office of Water (WH-553), EPA 440/4-91-001, Washington, DC.

U.S. EPA. 1991b. Technical Support Document for Water Quality-Based Toxics Control. EPA/505/2-90-001, Washington, DC.

U.S. EPA. 1995. Technical Guidance Manual for Developing Total Maximum Daily Loads, Book 2: Streams and Rivers, Part 1: Biochemical Oxygen Demand/Dissolved Oxygen and Nutrients/Eutrophication. EPA 823-B-95-007, Washington, DC.

U.S. EPA. 2000. Develop Modeling Approaches for Total Maximum Daily Loads (TMDLs) in Support of the Clean Water Act. Ecosystem Research Division, National Exposure Research Laboratory, Athens, GA.

Vigil, H. 1992. Literature Review of Sediment Oxygen Demand and Nutrient Flux Measurement Techniques. In Proceedings, U.S. Army Corps of Engineers Workshop on Sediment Oxygen Demand; Providence, RI, 21–22, August 1990, pp. 4–32.

Wanninkhof, R., J. R. Ledwell, and J. Crusius. 1991. Gas Transfer Velocities on Lakes Measured with Sulfur Hexafluoride. In Symposium Volume of the Second International Conference on Gas Transfer at Water Surface, S. C. Wilhelms and J. S. Gulliver, Eds. Minneapolis, MN.

Wells, S. A. and T. Cole. 2000. Hydrodynamic Modeling with Application to CE-QUAL-W2. Workshop Notes, Portland State University, Portland, OR, August 21–25, 2000.

Whittemore, R. 1986. Problems with *in situ* and Laboratory SOD Measurements and Their Implementation in Water Quality Modeling Studies. In Sediment Oxygen Demand, Processes, Modeling, and Measurement, K. J. Hatcher, Ed. Institute of Natural Resources, University of Georgia, Athens, GA, pp. 331–342.

Wilson, J. F., E. D. Cobb, and F. A. Kilpatrick. 1986. Fluorometric Procedures for Dye Tracing. In *Application of Hydraulics, Book 3*. U.S. Geological Survey, Chapter A12, Washington, DC, p. 34.

Wright, R. M. and A. J. McDonnel. 1979. In Stream Deoxygenation Rate Prediction. *J. Env. Eng.*, 105(EE2): 323–335.

Wright, R. M., P. M. Nolan, D. Pincumbe, and E. Hartman. 1998. Blackstone River Initiative: Water Quality Analysis of the Blackstone River Under Wet and Dry Weather Conditions. Department of Civil and Environmental Engineering, University of Rhode Island, Kingston, RI.

Yotsukura, N. and E. D. Cobb. 1972. Transverse Diffusion of Solutes in Natural Streams. U.S. Geological Survey Professional Paper 582-C, Washington, DC.

ABOUT THE AUTHOR

Wu-Seng Lung received his MS degree in Hydrology/Hydraulics in 1970 from the University of Minnesota, and his PhD degree in Environmental Engineering, with a specialty in water quality modeling, from the University of Michigan in 1975. After working for 8 years at several consulting firms, where he applied water quality modeling to various studies for regulatory agencies, industries, and law firms, he joined the Civil Engineering Department at the University of Virginia in 1983, where he is a Professor of Environmental Engineering.

At Virginia, he has been working on estuarine modeling of eutrophication and toxic substances. The modeling framework that he developed for the Procter & Gamble Co. to predict contaminant concentrations in 17 major estuaries in the United States has been adopted by the Environmental Protection Agency's (EPA's) Office of Pollution Prevention and Toxics and Office of Water for use in their fate and transport assessment analysis. In 1991, he was appointed by EPA to the model review panel for the EPA Chesapeake Bay Program, providing guidance to water quality modeling work of the Chesapeake Bay. His work on estuarine modeling was later synthesized into a book entitled *Water Quality Modeling: Application to Estuaries*, published by CRC Press in 1993. He was a major contributor to EPA's 1995 guidance manual on TMDL for stream and river BOD/DO and nutrients/eutrophication.

Through the years, Dr. Lung has served as a water quality-modeling consultant to a number of organizations including U.S. EPA, Army Engineers Waterways Experiment Station, and various states regulatory agencies, consulting firms, and industries, adding to his extensive background. Dr. Lung is a past Associate Editor for the *Journal of Environmental Engineering* (1994 to 1998), responsible for the area of water quality modeling. He is currently the Editor-in-Chief for *Water Quality and Ecosystem Modeling* by Kluwer Academic Publishers. Other professional affiliations include the EPA Science Advisory Board, where Dr. Lung has been a member since 1998.

INDEX

Acute toxicity, 278
Aeration, *see also* Reaeration, atmospheric, 89
  hypolimnetic, 96
  in-stream, 92
Alberni Inlet, British Colombia, 47
Anacostia River, Washington, DC, modeling sediment diagenesis, 119–124
Anaerobic DO model:
  estuary, 139
  lake, 97
  river, 83, 155
Appomattox River, Virginia, tributary to James River Estuary, 33
Area, cross-sectional:
  average over reach, 153
  river/stream, 27, 151

Bacteria, fecal coliform, 157
Banister River, Virginia, mixing modeling, conductivity, 35
Benthal oxygen demand, *see* Sediment oxygen demand
Biochemical oxygen demand (BOD), *see* CBOD and NBOD

Blackstone River, Massachusetts and Rhode Island:
  $CBOD_5$ concentration profile, 182
  mass transport model, 181
  QUAL2E model, 172, 185–187
Boston Harbor, Massachusetts, total coliform bacteria, 298
Bottle CBOD rate, 76, 80

Carbon, organic:
  chlorophyll *a* ratio, 223
  in phytoplankton model, 243
Carbonaceous BOD, CBOD:
  5-day, $CBOD_5$, 76–79
  bottle rate, k_1, 76, 80
  $CBOD_u$ to BOD_5 ratio, 77
  $CBOD_u$ to $CBOD_5$ ratio, 72, 76–79
  in-stream deoxygenation rate, k_d, 81, 83, 84, 86
  in-stream removal rate, k_r, 81
  long-term CBOD tests, 75, 85, 164
  ultimate, $CBOD_u$, 76–82
  unfiltered, 83
CE-QUAL-W2 model:
  application to Loch Raven Reservoir, 138, 266

CE-QUAL-W2 model (*cont'd*)
  application to Patuxent Estuary,
    65–66, 142
Chickahominy River, Virginia, tributary
    to James Estuary, 33
Chloride, model, Blackstone River,
    Massachusetts and Rhode Is-
    land, 29, 181
Chlorides:
  mass transport in river, 181
  oxygen saturation, 99
Chlorophyll *a*:
  algal carbon-chlorophyll *a* ratio, *see*
    Carbon
  algal phosphorus to carbon ratio, 249
  DO photosynthetic production, 104
Coliform bacteria, *see* Bacteria
  models:
    coastal water, 298
    river, 157
Combined sewer overflow, CSO, 119
Conductivity:
  distribution of, Lake Okeechobee,
    Florida, 45
  models:
    Loch Raven Reservoir, Maryland,
      267
    Martin Lake, 303
    mixing zone of the Banister River,
      35
    Upper Mississippi River, 31, 205
Conservative substances:
  chloride, *see* Chloride
  conductivity, *see* Conductivity
  salinity, *see* Salinity
  total dissolved solids, *see* Total dis-
    solved solids
Courant number, 69

Dam reaeration, 93
Deficit, DO, *see* Oxygen deficit
Delaware Estuary:
  CBOD$_u$ to CBOD$_5$ ratio in, 79
  DO model of, 109
  Rhodamine dye dispersion in, 36–37

Denitrification, rate coefficient, 227
Deoxygenation rate, *see* Carbonaceous
    BOD
Depth:
  CBOD deoxygenation rate relation-
    ship, 86
  reaeration coefficient relationship,
    91
  river flow relationship, 24, 175,
    196–197
Die-away of bacteria, *see* Bacteria
Diffusivity, vertical, 38, 40, 48
Dilution at outfall:
  estuary, 279
  river, 283
Dispersion:
  estuary, estimation of, 33
  lateral, river, 34
  numerical, 69
  river, estimation of, 30
  vertical, lake, 39–41
Dissolved oxygen, *see* Oxygen
Dye study:
  estuary, 36–37
  river, 284

EFDC model, 67
Effluent dominated streams, benthic
    (attached) algae, 108
Elevation, effect on oxygen solubility,
    99
Embayment, coastal, 294
Epilimnion, lake, 97
Error, root mean square, 209
Estuary, models:
  1D finite segment, 33, 261
  2D lateral finite segment, 298
  2D vertical finite segment, 56–61,
    217–220
  BOD/DO, 109, 124, 139–140
  conservative substances, 33, 56–61,
    217–220
  distributed sources, SOD, 123
  dye dispersion, 37, 46
  multiple sources, 109, 139–140,

nonconservative substances, 129, 139–140
Eutrophication, models:
    coastal bay phytoplankton, 141
    complete kinetics, 243
    estuary phytoplankton, 139–140
    lake phosphorus, 97, 118
    stream periphyton, 108
    stream phytoplankton, 185–187, 232–234, 236
Extinction coefficient:
    from light measurements, 114
    from Secchi depth, 113
    from suspended solids, 113, 115–116
    phytoplankton growth and, 112

Finite segment model:
    2D lateral, estuary, 213
    2D vertical, estuary, 218
    BOD/DO, 127
    integration, time variable, 203–208, 232–234
    numerical dispersion, 69
    one-dimensional, estuary, 5, 33
    phytoplankton, 139–140, 185–187, 232–234
    relative error, 238
    segment size, estimation of, 190
    stability, 261
    temperature, 138
    two-layer lake, 97–118
Flow:
    1Q10, 276
    7Q10, 28, 62, 151, 166
    depth relationship, 24
    velocity relationship, 24
Flume, Waterways Experiment Station, Vicksburg, Mississippi, two-dimensional velocity measurements, 52

Genessee River, New York, temperature distribution, 54
Geographical Information System, GIS, in TMDL work, 16, 133–136

Grand Calumet River, Illinois, lithium dispersion, 32
Growth rate, phytoplankton:
    in DO production, 104
    maximum, 111
    temperature, light, nutrient effect on, 111–116
Gunpowder River, Maryland, GIS for TMDL, 133
Gunston Cove, Virginia, sediment nutrient flux rates, 121

HEC-2 model, Quinebaug River, Massachusetts, 27
Hudson Estuary, two-layer salinity model, 58
Hunting Creek, Virginia, sediment nutrient flux rates, 121
Hydraulic geometry:
    Blackstone River, Massachusetts and Rhode Island, 175
    Quinebaug River, Massachusetts, 27
    Rock Creek, Pennsylvania, 151
    South Fork South Branch Potomac River, 24
    Upper Mississippi River, 197
Hydraulic model, Western Delta, California, 44–46
Hypolimnion, oxygen depletion rate, 97

Kalamazoo River, Michigan, 7Q10 low flow and time of travel, 28

Lake, models:
    2D stratified, 97, 118
    conductivity, 267
    temperature, 138
    total phosphorus, 118
Lake Okeechobee, conductivity distribution, 45
Lake Pepin, Minnesota:
    eutrophication model, 232–234
    numerical tagging, 249
    sediment oxygen demand, 101
Light, *see* Solar radiation

Light and dark bottle, photosynthesis respiration measurement, 104
Loch Raven Reservoir, Maryland:
  conductivity modeling, 267
  linking CE-QUAL-W2 and WASP/EUTRO models, 266
  temperature modeling, 138
Loss rate, net:
  suspended solids, 59–60
  total phosphorus, 118, 207–208

Maryland coastal bays:
  eutrophication model, 141
  mass transport model, 214–216
  watershed and WASP/EUTRO segmentations, 213
Minnesota River:
  BOD/DO model, 101
  long term CBOD test, 85
  sediment oxygen demand, 101
  tributary to Upper Mississippi River, 30, 73, 192
Mississippi River, *see* Upper Mississippi River
Mixing zone, models:
  Banister River, Virginia, 34
  James Estuary, Virginia, 281
  James River, Virginia, 290

Nitrification, kinetics of, 243
Nitrogen:
  ammonia:
    in nitrification, 88, 127, 243
    in phytoplankton kinetics, 243
  forms in eutrophication model, 243
  half-saturation constant, 228
  nitrate in nitrification, 88, 127
  N/P ratio and, 228
  organic:
    in nitrification, 243
    in phytoplankton kinetics, 243
  periphyton, diurnal DO, 108
  total Kjeldahl, in nitrification, 243
Nitrogenous BOD, NBOD:
  decay rate, 86, 152

relative to BOD, 75
Nonconservative substances:
  estuary, 59, 79, 109, 129
    no net flow, 56–57
  lake, 97, 303
  river, 165–166, 185–187, 207–208, 236, 292–293
Non-point source inputs, nutrients, 257
Norwalk Harbor, Connecticut:
  modeling sediment-water interactions using WASP/EUTRO, 252–253
  two-layer salinity and total suspended solids models, 60
Numerical dispersion:
  finite segment models in, 69
  QUICKEST, 69
Numerical Tagging:
  nitrogen, 251–252
  phosphorus, 243–244
Numerical integration, time variable analysis for, 202
Nutrients:
  phytoplankton growth, effect on, 116
  recycling, 118
  sediment release of, 121

Oxygen:
  diurnal:
    delta method, 105–106
    periphyton nutrient uptake, 107
    relation to phytoplankton DO production, 104
  models:
    anaerobic, river, 127
    DO "sag", 155
    estuary, 139
    lake, 97
    phytoplankton and, 185–187, 232–234
    river, 165, 171
    stream periphyton, 108
  phytoplankton relationship, 104
  saturation, 98–99
  supersaturation, 127
Oxygen deficit:

complete equation, river, 145
unit response, 156
Oxygen depletion, in lake hypolimnion, 97
Oxygen transfer rate, wind effect on, 95

Patuxent Estuary, Maryland:
1-D longitudinal coefficient, 33
CE-QUAL-W2 model segmentation, 64
CE-QUAL-W2 velocity results, 142
phosphorus and nitrogen loads to, 128
post audit of eutrophication model, 129
salinity results from CE-QUAL-W2, 66
SOD and nutrient flux rates, 121
temperature results from CE-QUAL-W2, 65
two-layer eutrophication model, 139
two-layer salinity models, 61, 296
Patuxent River, Maryland:
CBOD deoxygenation rate in, 84
NBOD decay rate in, 86
pH, fluctuation in stream, 108
Phosphorus:
forms in eutrophication models, 243
half-saturation constant, 228
phytoplankton interaction, 243
Photoperiod, 111
Photosynthesis:
average DO production, 103
from chlorophyll measurements, 104
from diurnal DO, 106
light and dark bottle, 104
Phytoplankton:
death rate, 117
extinction coefficient, effect on, 113
growth rate, 112
model kinetics of, 243
settling velocity of, 117
Platte Lake, Michigan, two-layer total phosphorus and DO model, 118

Postaudit:
DO models, 127
finite difference eutrophication model, 129
Potomac Estuary:
eutrophication, model, 95
sediment release fluxes, 121
South Fork South Branch, 22
Pressure, atmospheric, effect on oxygen solubility, 99

QUICKEST, 69
Quinebaug River, Massachusetts, HEC-2 modeling, 27

Rearation, atmospheric, *see also* Oxygen transfer rate
dam reaeration, 93
estuary, 95
river equations, 93
wind effect on, 95
River, models:
CBOD, 145, 155, 157, 165
conservative substances, 29, 31, 181, 204–205
dispersion coefficient, lithium as tracer for, 32
dye dispersion, 290–291
eutrophication, periphyton, 108
eutrophication, phytoplankton, 185–187, 232–234, 236
NBOD, 155
total phosphorus, 207–208
Roanoke River, Virginia:
assimilative capacity, 170
BOD/DO model, 165
BOD wasteload allocations, 167
CBOD deoxygenation rate, 86
DO deficit unit response, 169
time of travel, 162
Rochester Embayment, NY, temperature model, 54
Rock Creek, Pennsylvania:
BOD/DO model, 155
DO deficit response, 156

Rock Creek, Pennsylvania (*cont'd*)
    fecal coliform model, 157
    hydraulic geometry, 151
    total phosphorus model, 157

Sacramento-San Joaquin (Western)
        Delta, California:
    1-D salinity modeling, 46
    dye dispersion modeling, 46
    longitudinal dispersion, 32
    two-layer salinity model, 56–57
    two-layer turbidity maximum model,
        59
Salinity:
    effect on oxygen solubility, 99
    longitudinal dispersion coefficient,
        tracer for, 33, 46
    mass transport model, 56–61, 66,
        216, 219–220, 294
Santa Fe River, New Mexico, benthic
        algae model, 108
Secchi depth:
    empirical relationship to chlorophyll
        a, 113
    extinction coefficient relationship,
        113
Sediment:
    oxygen demand rates, 121
    nutrient release flux rates, 121
Sediment oxygen demand (SOD):
    measured rates, 121
    modeling sediment-oxygen interac-
        tions, 122–124, 252–257
Solar radiation:
    average during photoperiod, 112
    depth variation, 112
    phytoplankton growth, effect on,
        111
St. Croix River, Minnesota, Long term
        CBOD test, 85
Stratification:
    two-layer, lake, 97
    vertical, estuary, 49–61
Streeter-Phelps DO "sag," 170
Suspended solids:

estuary, models, 59–60
    net loss from water column, 58

Temperature:
    stratification
        in lake, 39–42
        in river, 54
Temperature, models, 65, 138
Time of travel:
    Kalamazoo River, Michigan, 28
    Roanoke River, Virginia, 162
    Upper Mississippi River, 198
TMDL, modeling for, example, 15–20
Transquaking River, Maryland, TMDL
        modeling, 15
Turbidity maximum, 59

Unit response, DO deficit, 156, 242
Upper Mississippi River:
    CBOD deoxygenation rate in, 81,
        86
    conductivity model, 31, 205
    DO deficit unit response, 242
    EUTRO model, 232–234, 236
    hydraulic geometry of, 196–197
    light extinction in, 115
    long-term CBOD test, 85
    major tributary flow to, 199
    numerical tagging, 249
    post audit of BOD/DO model of,
        127
    sediment oxygen demand in,
        101–102
    temperature and light extinction for
        EUTRO model of, 226
    time of travel, 198
    total dissolved solids model, 204
    total phosphorus model, 207–208
    tributary flows and nutrient loads to,
        222
    water quality data (1976 and 1988),
        89

Velocity, river flow relationship, 24,
        175, 196–197

Velocity, net:
    loss of suspended solids, 56
    total phosphorus, 118

Waste load allocation (WLA):
    BOD/DO modeling for, 155, 167,
        185–187
    nutrient modeling for, 139–141,
        232–234, 236, 253
    for total phosphorus, 118, 157,
        207–208
Water quality end points, 11–13

Western Delta, Suisun Bay,
        California:
    hydraulic model, 44
    turbidity maximum model, 59
While Lake, Michigan:
    light extinction, 114
    temperature distribution, depth, 39
    vertical diffusion coefficient, 43
    vertical temperature profiles, 41
Wicomico Estuary, Maryland, DYN-
        HYD model, 261–265
Wind, effect on, oxygen transfer, 95

Printed and bound by CPI Group (UK) Ltd, Croydon, CR0 4YY

27/10/2024

14580313-0004